Digital Design and Modeling with VHDL and Synthesis

SELECTED TITLES

Distributed Objects: Methodologies for Customizing Operating Systems
Nayeem Islam

Software Engineering Risk Management
Dale Karolak

Digital Design and Modeling with VHDL and Synthesis
K.C. Chang

Industrial Strength Software: Effective Management Using Measurement
Lawrence Putnam and Ware Myers

Unified Objects: Object-Oriented Programming Using C++
Babak Sadr

Interconnection Networks: An Engineering Approach
Jose Duato, Sudhakar Yalamanchili, and Lionel Ni

Meeting Deadlines in Real-Time Systems: The Rate Monotonic Analysis
Daniel Roy and Loic Briand

Executive Briefings

Controlling Software Development
Lawrence Putnam and Ware Myers

Digital Design and Modeling with VHDL and Synthesis

K. C. Chang

IEEE Computer Society Press
Los Alamitos, California

Washington • Brussels • Tokyo

Library of Congress Cataloging-in-Publication Data

Chang, K.C. (Kou-Chuan), 1957–
Digital design and modeling with VHDL and synthesis / K.C. Chang.
p. cm.
Includes bibliographical references.
ISBN 0-8186-7716-3
1. Very high speed integrated circuits—Design and construction—Data processing. 2. VHDL (Computer hardware description language) 3. Computer-aided design. I. Title
TK7874.7.C47 1997
621.39 ' 5—dc21
96-45231
CIP

IEEE Computer Society Press
10662 Los Vaqueros Circle
P.O. Box 3014
Los Alamitos, CA 90720-1314

IEEE Computer Society Press Order Number BP07716
Library of Congress Number 96-45231
ISBN 0-8186-7716-3

Additional copies may be ordered from:

IEEE Computer Society Press
Customer Service Center
10662 Los Vaqueros Circle
P.O. Box 3014
Los Alamitos, CA 90720-1314
Tel: +1-714-821-8380
Fax: +1-714-821-4641
Email: cs.books@computer.org

IEEE Service Center
445 Hoes Lane
P.O. Box 1331
Piscataway, NJ 08855-1331
Tel: +1-908-981-1393
Fax: +1-908-981-9667
mis.custserv@computer.org

IEEE Computer Society
13, Avenue de l'Aquilon
B-1200 Brussels
BELGIUM
Tel: +32-2-770-2198
Fax: +32-2-770-8505
euro.ofc@computer.org

IEEE Computer Society
Ooshima Building
2-19-1 Minami-Aoyama
Minato-ku, Tokyo 107
JAPAN
Tel: +81-3-3408-3118
Fax: +81-3-3408-3553
tokyo.ofc@computer.org

Editor-in-Chief: Mohamed Fayad
Publisher: Matt Loeb
Acquisitions Editor: Bill Sanders
Developmental Editor: Cheryl Smith
Advertising/Promotions: Tom Fink
Production Editor: Lisa O'Conner
Printed in the United States of America

The Institute of Electrical and Electronics Engineers, Inc

Dedicated to

My parents

Chia-Shia & Jin-Swei

My wife

Tsai-Wei

and My children

Alan, Steven, and Jocelyn

Preface

To keep up with advances in semiconductor technology, design complexity and design methodology have to be improved. The traditional gate-level schematic capture approach cannot meet the shortened design cycle and market window. A higher level of abstraction to capture designs is not an option, but a requirement. For this reason Hardware Description Languages (HDL) such as Verilog and VHDL were developed. Synthesis tools take these HDLs and automatically translate them into schematics, thereby enabling designers to manage more complex and bigger designs in shorter design schedules.

VHDL stands for VHSIC (Very High Speed Integrated Circuits) Hardware Description Language. Designers have interpreted VHDL as meaning Very **Hard** Description Language (VHDL). The major goal of this book is to define VHDL as Very **Handy** Description Language for designing and modeling digital circuits.

This book is the result of the author's practical experience in both design and teaching. Many of the design techniques and design considerations illustrated throughout the chapters are examples of real designs. The author's teaching experience has led to a step-by-step presentation which addresses common mistakes and hard-to-understand concepts in a way that eases learning.

VHDL is introduced with practical design examples, simulation waveforms, and schematics so that readers can better understand their correspondence and relationships. Unique features of the book include the following:

1. There are more than 115 complete examples of 6200 lines of VHDL code in the book. Every line of VHDL code is analyzed, and most of them are simulated and synthesized with simulation waveforms and schematics shown. VHDL codes, simulation waveforms, and schematics are shown together, allowing readers to grasp the concepts easier and faster.
2. Every line of VHDL code has an associated line number for easy reference and discussion. The font used for the VHDL code portion is "Courier" because each character of this font occupies the same width, which allows them to line up vertically. This is close to how an ASCII file appears when readers type in their VHDL code with any text editor.
3. The VHDL code examples are carefully designed to illustrate various VHDL constructs, features, practical design considerations, and techniques. The examples are complete so that readers can assimilate overall ideas more easily.
4. Challenging exercises are provided at the end of each chapter so that readers can put into practice the ideas and information offered in the chapter.
5. A complete design project from concept to final timing verification is provided to demonstrate the entire design process.

6. All VHDL syntax and constructs are discussed with complete examples.
7. Practical design techniques for finite state machines, synthesis processes, and test benches are discussed with complete examples.
8. VHDL'93 important updates are discussed with complete examples.

BOOK ORGANIZATION

This book is divided into 14 chapters and 3 appendixes. All VHDL constructs are discussed in Chapters 1 through 7 and Chapter 10. Chapters 8 through 13 are dedicated to practical design examples. Chapter 14 summarizes VHDL'93 updates. Appendix A is a quick VHDL syntax reference with a complete example. Appendix B summarizes the declarations at each region. Appendix C is a VHDL'93 grammar and syntax reference. The details of each chapter are as follows:

Chapter 1 gives a quick overview of VHDL history and usage. The synthesis terminology is defined and illustrated with a small example. The comparisons of the traditional schematic design approach and the top-down design approach with VHDL and synthesis are discussed.

Chapter 2 summarizes VHDL basic elements such as lexical elements, identifiers, reserved words, operators, packages, entities, architectures, objects, types, and attributes.

Chapter 3 introduces the VHDL simulation concepts of the process concurrency, event driven simulation, delta delay, process sensitivity list, and signal. It also gives an overview where sequential statements and concurrent statements can appear.

Chapter 4 discusses all sequential statements and provides complete examples and applications.

Chapter 5 discusses all VHDL concurrent statements with complete examples and applications.

Chapter 6 presents VHDL subprograms and packages. The concepts of overloading and bus resolution functions are discussed. Practical examples are provided.

Chapter 7 discusses VHDL libraries, design units, and configurations. Complete examples are provided.

Chapter 8 focuses on the important issues of writing VHDL for synthesis. Many examples are presented to infer latched, flipflops, tristate buffers, and combinational circuits. The synthesis process is discussed and illustrated with an actual example from VHDL coding to simulating with the backannotated timing delay.

Chapter 9 illustrates writing VHDL for finite state machines. Complete examples, simulations, and synthesized schematics are provided.

Chapter 10 discusses more behavioral modeling of file text I/O, ROM, RAM, pad models, guarded block, guarded signal, disconnect, and null constructs.

Chapter 11 presents a small design case with techniques for implementing test benches.

Chapter 12 illustrates the design of an ALU (arithmetic logic unit) with VHDL code, simulation, and synthesis.

Chapter 13 describes a design case study, which allows the reader to go through the complete design process from the design concept, VHDL implementation, verification, test bench, layout, and post-layout verification.

Chapter 14 summarizes the important updates of VHDL'93. Many examples and actual simulations are provided.

AUTHOR'S NOTE

Readers who have more software than hardware experience usually have no problem writing another programming language. Their challenges are usually found in the following areas:

1. Process concurrency
2. Sequential statements regions and concurrent statements regions
3. Hardware implication—schematic and simulation waveform
4. Timing concepts
5. Mixing other program languages with VHDL, especially with regard to reserved words and delimiters

Readers who have more hardware than software experience find their challenges in the following areas:

1. VHDL syntax
2. Writing VHDL to imply hardware
3. Sequential statements regions and concurrent statements regions
4. Taking advantages of VHDL constructs
5. Software concepts such as subprograms, packages, libraries, and configurations.

In this book, the VHDL syntax is introduced with BNF (Backus Normal Form). BNF is referred as the production rule and has the format: *left_hand_side* ::= *right_hand_side*. The symbol ::= (two semicolons and one equal sign) means "can be replaced." Thus, *left_hand_side* ::= *right_hand_side* can be read as "the *left_hand_side* can be replaced with the *right_hand_side*." The following are special notations used in the *right_hand_side*:

[...]	Anything inside the square bracket is optional. It can appear as 0 or 1 time.
{ ... }	Anything inside the bracket can be repeated, 0 or any number of times.
... | ...	The vertical bar indicates the item to the left or right of the vertical bar can be used.
if	VHDL reserved words are printed in **bold face**.
___	Underscored portions indicate the VHDL'93 updates.
	Any other special character such as comma (,) and symbol (:=) would be part of the actual text in the syntax.

For example, the if statement can be defined as follows:

if_statement ::=
  [if_label :] **if** condition **then**
    sequence_of_statements
  { **elsif** condition **then**
    sequence_of_statements }
  [**else**
    sequence_of_statements]
  end if [if_label] ;

The preceding if statement BNF shows reserved words **if**, **then**, **else**, **elsif**, **end**. It can have an optional label (with the colon :), the **if** condition **then** can appear only one time, followed by sequence of statements (that can be further replaced), the **elsif** clause can be repeated 0 or any number of times since it is inside the { } bracket, the **else** clause can appear 0 or 1 times since it is inside the [ ] square bracket. It is closed with the reserved words **end if**, followed by the optional label (the same label as in the beginning with no colon :), and ended with the semicolon ;.

The next BNF example shows that an instantiation list can have one or more labels (separated by commas), or the reserved word **others**, or the reserved word **all**, but not combinations of any two items from the three choices separated by the vertical bars.

instantiation_list ::= instantiation_label { , instantiation_label } | **others** | **all**

ACKNOWLEDGMENTS

This book would have been impossible without the following people: Ron Fernandes edited the entire first manuscript. Bill Cormier, Erik Oson, and Carl Wagner of Synopsys supported and provided technical assistance with Synopsys tools. Dear friends Hanchi and Clare Lin, Pinchung and Debra Chen, Johnny and Sunkoo To, Scott and Amy Weaver, Tony and Mayda Eng continuously encouraged me when I did not see the end of this adventure. John Cooley provided test vectors used in Chapter 11.

My wife Tsai-Wei, and children Alan, Steven, and Jocelyn were patient and sacrificed many of our weekends.

The IEEE Computer Society Press editorial team, Bill Sanders, Mohamed Fayad, Cheryl Smith, and Lisa O'Conner contributed many valuable comments. Additionally, Joni Harlan from Bookmark Media did a skillful and thorough editing of the manuscript.

Ross Barta, Gary Nelson, Warren Snapp, and Mike White gave management support.

The author deeply appreciates all of these people.

Contents

Chapter 1

Introduction

1.1 WHAT IS VHDL?

In 1980 the US government developed the Very High Speed Integrated Circuit (VHSIC) project to enhance the electronic design process, technology, and procurement, spawning development of many advanced integrated circuit (IC) process technologies. This was followed by the arrival of VHSIC Hardware Description Language (VHDL), which was endorsed by IEEE in 1986 in its attempt at standardization. By December of 1987 the IEEE 1076.1 standard for VHDL was approved and a VHDL Language Reference Manual (LRM) was published.

This book will be based primarily on the 1987 standard, since it is the foundation for most VHDL simulators and synthesis tools currently on the market. It is also the foundation for the VHDL Initiative Towards ASIC Libraries (VITAL) in detailed function and timing modeling of primitive components. However, the 1987 standard was updated in 1993 and must also be addressed. This is done in Chapter 14. VHDL'87 will refer to the 1987 standard and VHDL'93 will refer to the 1993 standard.

1.2 VHDL ADVANTAGES

VHDL offers the following advantages for digital design:

Standard: VHDL is an IEEE standard. Just like any standard (such as graphics X-window standard, bus communication interface standard, high-level programming languages, and so on), it reduces confusion and makes interfaces between tools, companies, and products easier. Any development to the standard would have better chances of lasting longer and have less chance of becoming obsolete due to incompatibility with others.

Government support: VHDL is a result of the VHSIC program; hence, it is clear that the US government supports the VHDL standard for electronic procurement. The Department of Defense (DOD) requires contractors to supply VHDL for all Application Specific Integrated Circuit (ASIC) designs.

Industry support: With the advent of more powerful and efficient VHDL tools has come the growing support of the electronic industry. Companies use VHDL tools not only with regard to defense contracts, but also for their commercial designs.

Portability: The same VHDL code can be simulated and used in many design tools and at different stages of the design process. This reduces dependency on a set of design tools whose limited capability may not be competitive in later markets. The VHDL standard also transforms design data much easier than a design database of a proprietary design tool.

Modeling capability: VHDL was developed to model all levels of designs, from electronic boxes to transistors. VHDL can accommodate behavioral constructs and mathematical routines that describe complex models, such as queuing networks and analog circuits. It allows use of multiple architectures and associates with the same design during various stages of the design process. As shown in Figure 1.1, VHDL can describe low-level transistors up to very large systems. This book will concentrate on using VHDL for the design of ASICs and digital blocks.

Reusability: Certain common designs can be described, verified, and modified slightly in VHDL for future use. This eliminates reading and marking changes to schematic pages, which is time consuming and subject to error. For example, a parameterized multiplier VHDL code can be reused easily by changing the width parameter so that the same VHDL code can do either 16 by 16 or 12 by 8 multiplication.

Technology and foundry independence: The functionality and behavior of the design can be described with VHDL and verified, making it foundry and technology independent. This frees the designer to proceed without having to wait for the foundry and technology to be selected.

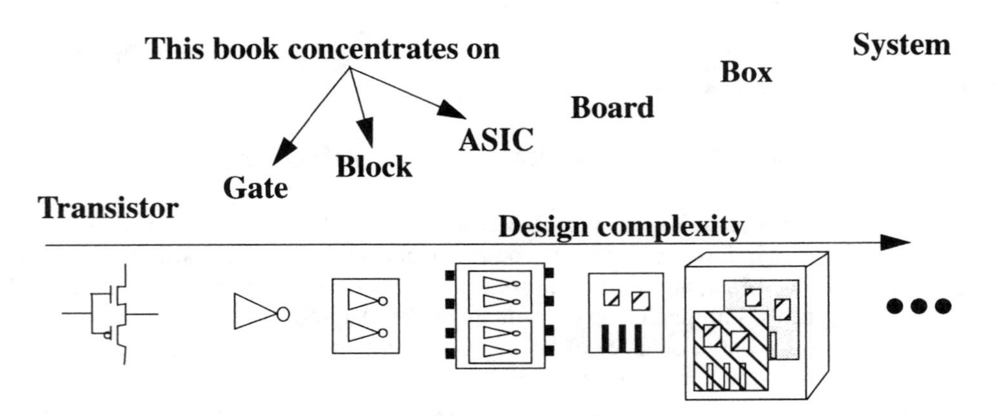

FIGURE 1.1 VHDL modeling capability.

Documentation: VHDL is a design description language which allows documentation to be located in a single place by embedding it in the code. The combining of comments and the code that actually dictates what the design should do reduces the ambiguity between specification and implementation.

New design methodology: Using VHDL and synthesis creates a new methodology that increases the design productivity, shortens the design cycle, and lowers costs. It amounts to a revolution comparable to that introduced by the automatic semi-custom layout synthesis tools of the last few years.

1.3 WHAT IS LOGIC SYNTHESIS?

Synthesis, in the domain of digital design, is a process of translation and optimization. For example, layout synthesis is a process of taking a design netlist and translating it into a form of data that facilitates placement and routing, resulting in optimizing timing and/or chip size. Logic synthesis, on the other hand, is the process of taking a form of input (VHDL), translating it into a form (Boolean equations and synthesis tool specific), and then optimizing in terms of propagation delay and/or area. For example, Figure 1.2 shows the schematic view of the synthesis tool specific internal form translated from a VHDL code. Note that there are some blocks to generate sequential circuits, such as flipflops, and some blocks to generate combinational circuits.

After the VHDL code is translated into an internal form, the optimization process can be performed based on constraints such as speed, area, power, and so on. Figure 1.3 depicts the schematic, after the circuit is optimized.

1.4 NEW DESIGN METHODOLOGY

Introducing VHDL and synthesis enables the design community to explore a new design methodology. The traditional design approach, as shown in Figure 1.4a, starts with drawing schematics and then performs functional and timing simulation based on the same schematic. If there is any design error, the process iterates back to update schematics. After the layout, functions and back annotated timing are verified again with the same schematics.

The VHDL-based design approach is illustrated in Figure 1.4b. The design is functionally described with VHDL. VHDL simulation is used to verify the functionality of the design. In general, modifying VHDL source code is much faster than changing schematics. This allows designers to make faster functionally correct designs, to explore more architecture trade-offs, and to have more impact on the designs. After the functions match the requirements, the VHDL code is synthesized to generate schematics (or equivalent netlists). The netlist can be used to layout the circuit and to verify the timing requirement (both before or after the layout). The design changes can be made by modifying VHDL code or changing the constraints (timing, area, and so on) in the synthesis. This new design approach and methodology has improved the design process by shortening design time, reducing the number of design iterations, and increasing the design complexity that designers can manage.

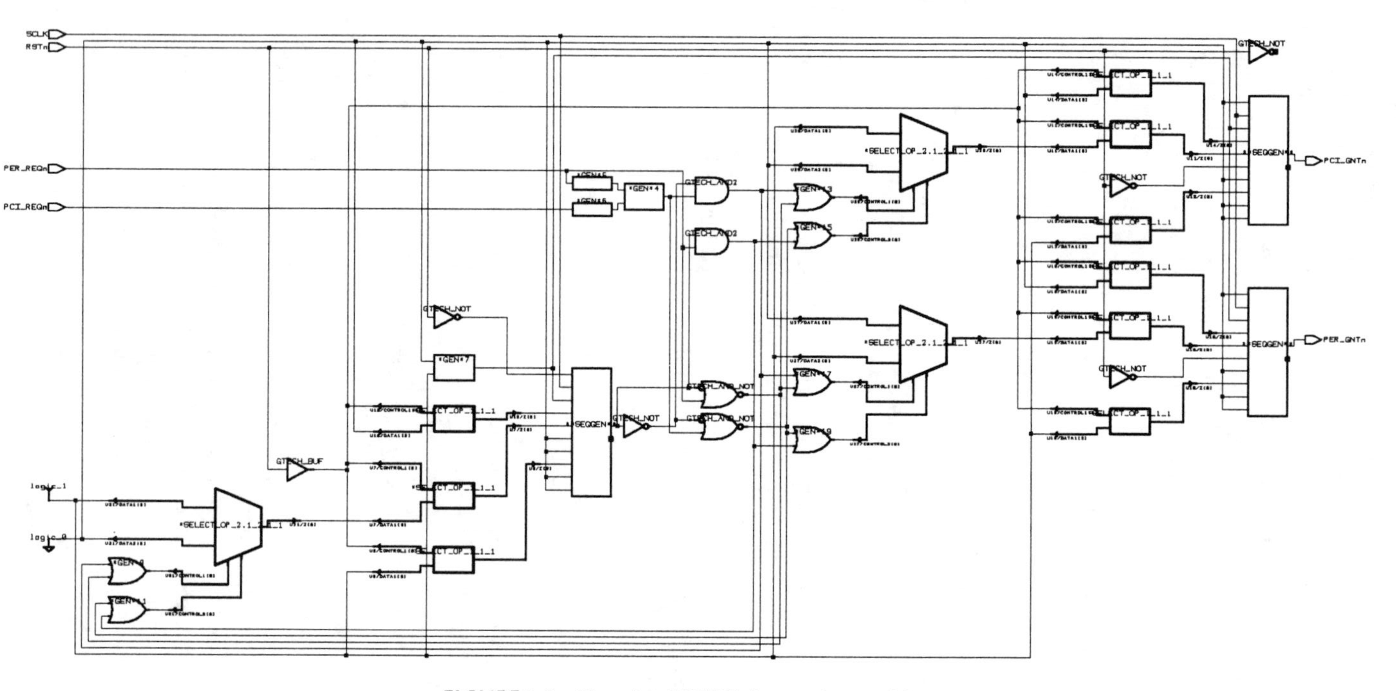

FIGURE 1.2 Translated VHDL into an internal form.

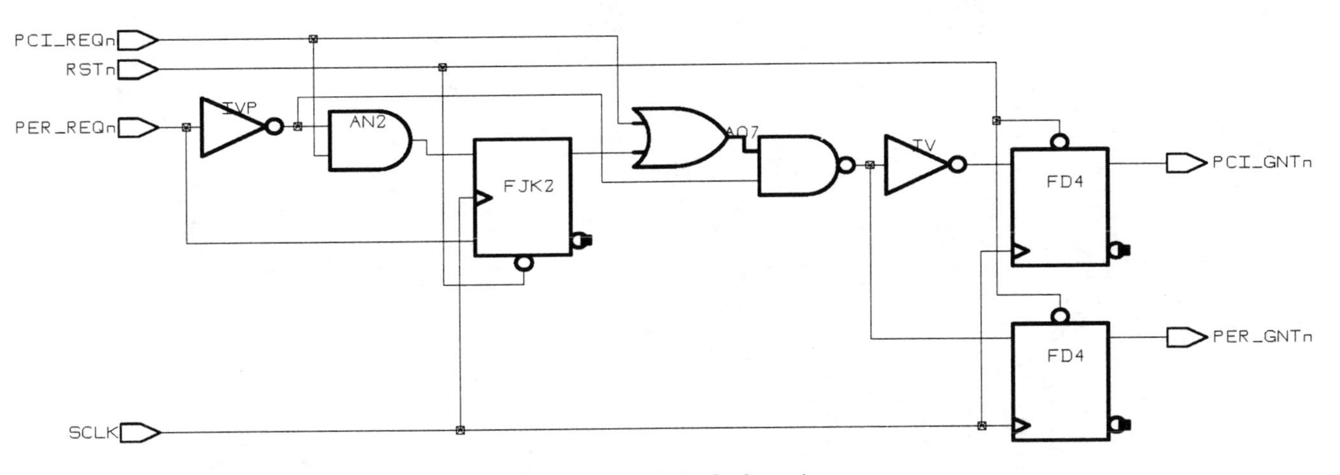

FIGURE 1.3 Optimized schematic.

1.5 BOOK OVERVIEW

The book has 14 chapters. This chapter gives a quick overview of VHDL history and usage. Chapter 2 summarizes VHDL basic language elements. Chapter 3 introduces the VHDL simulation concepts. Chapters 4 and 5 discuss VHDL sequential and concurrent statements respectively. Chapter 6 presents VHDL subprograms and packages. Chapter 7 discusses VHDL library and design unit concepts and configuration. Chapter 8 points out important issues of writing VHDL for synthesis. Chapter 9 illustrates writing VHDL for finite state machines. Chapter 10 discusses more behavioral modeling of file text I/O, ROM, RAM, pad models, guarded block, guarded signal, disconnect, and null constructs. Chapter 11 presents a small design case with test bench design techniques. Chapter 12 discusses an ALU design example. Chapter 13 describes a design case study that goes through the complete design process from the design concept, VHDL implementation, verification, test bench, layout, and post-layout verification. Chapter 14 summarizes VHDL'93 important updates.

In this book, many design examples with VHDL are used for simulation and synthesis illustrations. The original VHDL code is imported into the book with no modification so that readers can see the entire VHDL code (not just part of it). Every line of VHDL code has an associated line number for easy reference and discussion. The font used for the VHDL code portion is "Courier"—each character occupies the same width and characters line up vertically. This would be close to how an ASCII file appears when readers type in their VHDL code with any text editor. The schematics are generated with Synopsys synthesis tools and then imported. Note that the schematic can be different from what you may implement manually from your synthesis output. This is due to different technology component libraries and constraints that are used in the synthesis process. However, the synthesized schematic should represent and correspond functionally with the VHDL code. Three purposes are served by including the entire VHDL code, simulation, and synthesized schematic: (1) Typing

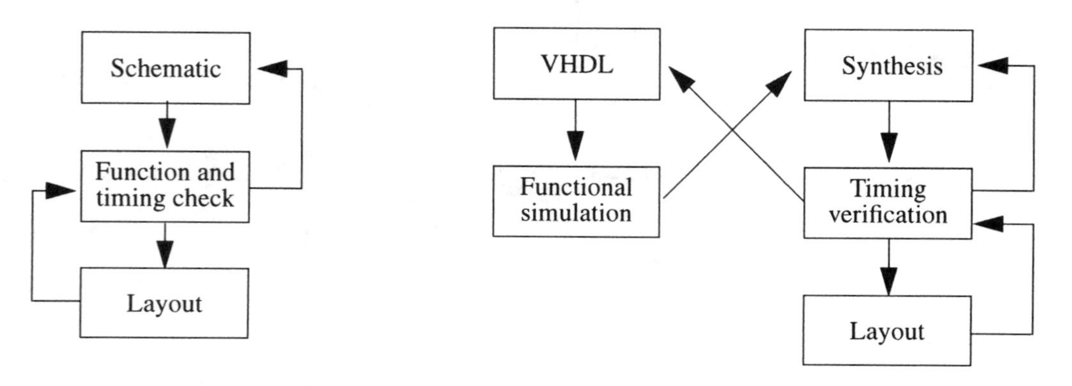

FIGURE 1.4 Design process comparison.

errors in the VHDL code are avoided; (2) readers can use the code as examples for reference, exercises, and to build upon them; (3) readers will see the correspondence among VHDL code, simulation, and synthesis results.

1.6 EXERCISES

1.1 What do VHSIC, VHDL, and VITAL stand for?

1.2 What advantages can we expect when using VHDL in digital designs? Do you agree on the advantages discussed in the text?

1.3 What is synthesis? Have you seen the terminology of "layout" synthesis? Can we say that C language compiler is a synthesis tool for C programming language?

1.4 In software design, the trend has been moving from writing machine dependent assembly languages to writing machine independent standard C high-level programming language. In hardware design, library dependent schematics can be created manually or library and technology independent VHDL code can be written to describe equivalent functions. Are there any similarities between software and hardware design? Do you agree that more and more digital designs will be using VHDL and synthesis just like the trend in software design?

Chapter 2

VHDL Basics

In this chapter, we will introduce some basic VHDL constructs to assist our discussions. As we add more foundations, we will revisit more features of the constructs and their applications.

2.1 LEXICAL ELEMENTS, SEPARATORS, AND DELIMITERS

VHDL text is a sequence of separate **lexical elements** which can be used as any of the following: delimiter, identifier, reserved word, abstract literal, character literal, string literal, bit string literal, comment.

Adjacent lexical elements are separated by **separators** such as space characters, format effectors, or the end of a line. Note that a space character is a separator, except within a comment, a string literal, or a space character literal. At least one separator is required between an identifier and a literal.

A **delimiter** is either one of the following special characters

&‘ ( ) * + , - . / : ; < = > |

or a **compound delimiter** each composed of two adjacent special characters

=> ** := /= >= <= <>

For example, the VHDL statement A <= B and C; has six lexical elements “A”, “<=”, “B”, “and”, “C”, “;”. Two of the six lexical elements are delimiters “<=” (compound delimiter) and “;”. Note, there are spaces as separators between adjacent lexical elements, but it is not necessary to have a separator between a delimiter (such as ;) and a lexical element (such as C).

A **comment** starts with two adjacent hyphens and extends to the end of the line. It can appear on any line of a VHDL description and does not affect the execution of a simulation module. The following are valid examples:

```
----------------------------------------------------------- Start of comment examples
A <= B; -- This comment appears after a valid VHDL statement.
-- This comment is too long and must be broken.
--Each line starts with two hyphens. No space is required after two hyphens.
```

2.2 IDENTIFIERS

Identifiers are used as user names and as reserved words. Each identifier starts with a letter followed by any number of letters or digits. A single underline character can be inserted between a letter or digit and an adjacent letter or digit. An underline character as the last character is not a valid identifier. Two adjacent underline characters are not allowed in identifiers. There is no distinction between an uppercase letter and a lowercase letter within an identifier. For example, AbC7 is the same as aBC7. All characters within an identifier are significant including underline characters. For example, A_3 is not the same as A3. Space character are not allowed within an identifier since they are separators.

Valid identifiers :

carry_OUT
Dim_Sum
Count7SUB_2_goX

Invalid identifiers :

7AB (starts with a digit)
A@B (special character @)
SUM_ (end with an underline)
PI__A (two underline characters)

2.3 RESERVED WORDS

As in many other programming languages, VHDL has words with special meanings called reserved words or keywords. These words can be confused with look-a-like words, so care should be taken. For example: **else**, **if**, **end**, **downto**, and **to** are reserved words, while elseif, endif, and down are not. Reserved words such as **wait** and **select** deserve your caution as they are tempting as user identifiers. The following is a list of the reserved words:

abs	access	after	alias	all
and	architecture	array	assert	attribute
begin	block	body	buffer	bus
case	component	configuration	constant	disconnect
downto	else	elsif	end	entity
exit	file	for	function	generate
generic	guarded	if	in	inout

is	label	library	linkage	loop
map	mod	nand	new	next
nor	not	null	of	on
open	or	others	out	package
port	procedure	process	range	record
register	rem	report	return	select
severity	signal	subtype	then	to
transport	type	units	until	use
variable	wait	when	while	with
xor				

In VHDL'93, the following reserved words are added:

group	impure	inertial	literal	postponed
pure	reject	rol	ror	shared
sla	sll	sra	srl	unaffected
xnor				

To ensure upward compatibility, do not use VHDL'93 reserved words when your VHDL tools are still VHDL'87 compliance as they are not be recognized as such by VHDL'87. The VHDL'93 reserved words will be discussed in Chapter 14.

2.4 LITERALS

VHDL literals are categorized as shown in Figure 2.1.

A decimal or based literal can be an integer, a real, or a real with an exponent.

Integer literals: 21, 0, 1E2, 3e4, 123_000.
Real literals: 11.0, 0.0, 0.468, 3.141_592_6.
Real literals with exponent: 1.23E-11, 1.0E+4, 3.024E+23.

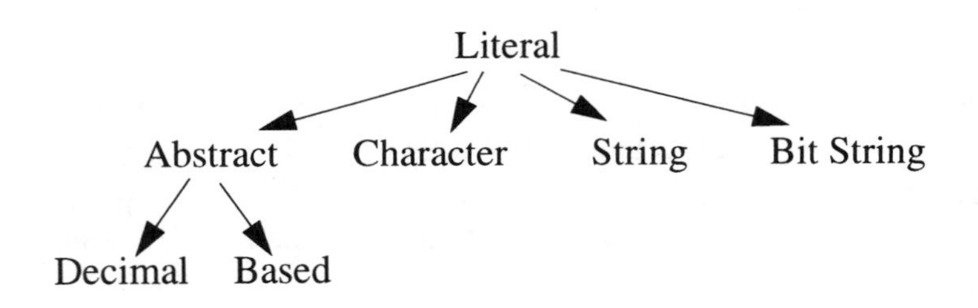

FIGURE 2.1 VHDL Literals.

The exponent letter E, can be written either in lower- or uppercase with the same meaning. The underline character inserted between adjacent digits of a decimal literal does not affect the value of the literal. An exponent for an integer literal must not have a minus sign –. 2E-3 is not an integer. Zero exponents are allowed for integer literals. Note that a space is a separator. It is not allowed in an abstract literal, not even in the exponent portion.

A based literal is an abstract literal expressed in a form that specifies the base explicitly. The base must be at least two and at most sixteen—for example, 2#1111_1100#, 16#fc#, 016#0FC#, 7#510#. They have the same integer value of 252. 2#0110_0000# and 16#6#E1 have the same integer value of 96.

Note that the exponent has the same base (16#6#E1 means 6*16**1). An underline character inserted between adjacent digits does not affect the value. Letters A through F, with the exponent character E, can be written as lowercase, with the same meaning. Real literals 16#F.E#E+1 and 2#1.1111_110#E7 have the same value of 254.0. Each digit used in a base literal must be less than the base.

A character literal is formed by enclosing one of the graphic characters, including the space, between two apostrophe characters. Examples are :'A' 'a' '%' '''(apostrophe) ' ' (space).

Note that lowercase and uppercase character literals are not the same—'A' is different from 'a'.

A string literal is formed by a sequence of graphic characters (possibly none) enclosed between two quotation characters used as string brackets. If a quotation character is to be represented, a pair of adjacent quotation characters must be written at the corresponding place within the string literals. For example: "Setup time violation" ""(empty) " "(space) "A" """""" (quotation).

Note that a string literal must fit in the same line. Concatenation (&) can be used to form a longer string literal.

Bit string literals are formed by sequences of extended digits such as 0 through 9, A through F, or a through f enclosed between two quotations as bit string brackets, preceded by a base specifier of B, O, or X (lowercase is the same) for binary, octal, or hexadecimal. An underline character inserted between adjacent digits of a bit string literal does not affect the value. The length of a bit string literal is the number of values of type BIT in the sequence represented. For example: X"F_FF", O"7777", b"1111_1111_1111". All represent the same value of 12 bits.

2.5 PACKAGE AND TYPE

A type declaration is used in VHDL to allow designers the flexibility to specify user objects. A package is a VHDL unit that can include a number of types, subtypes, constants, procedures, and functions declarations. VHDL has a number of types, subtypes, and functions predefined. For example, the following package STANDARD (refer to line 1, where STANDARD is the name of the package), which is assumed to exist (and is visible, hence, there is no need to reference the package in the VHDL code), with some types such as BOOLEAN, BIT, CHARACTER, INTEGER, NATURAL, REAL, TIME, STRING, BIT_VECTOR, are declared.

```
1     package STANDARD is
2       type BOOLEAN   is (FALSE, TRUE);
3       type BIT       is ('0', '1');
4       type CHARACTER is (
5          NUL, SOH, STX, ETX, EOT, ENQ, ACK, BEL,
6          BS,  HT,  LF,  VT,  FF,  CR,  SO,  SI,
7          DLE, DC1, DC2, DC3, DC4, NAK, SYN, ETB,
8          CAN, EM,  SUB, ESC, FSP, GSP, RSP, USP,
9
10         ' ', '!', '"', '#', '$', '%', '&', ''',
11         '(', ')', '*', '+', ',', '-', '.', '/',
12         '0', '1', '2', '3', '4', '5', '6', '7',
13         '8', '9', ':', ';', '<', '=', '>', '?',
14
15         '@', 'A', 'B', 'C', 'D', 'E', 'F', 'G',
16         'H', 'I', 'J', 'K', 'L', 'M', 'N', 'O',
17         'P', 'Q', 'R', 'S', 'T', 'U', 'V', 'W',
18         'X', 'Y', 'Z', '[', '\', ']', '^', '_',
19
20         '`', 'a', 'b', 'c', 'd', 'e', 'f', 'g',
21         'h', 'i', 'j', 'k', 'l', 'm', 'n', 'o',
22         'p', 'q', 'r', 's', 't', 'u', 'v', 'w',
23         'x', 'y', 'z', '{', '|', '}', '~', DEL);
24
25      type SEVERITY_LEVEL is (NOTE, WARNING, ERROR, FAILURE);
26      type INTEGER is range -2147483648 to +2147483647;
27      type REAL is range -0.1797693134862e+309 to
+0.1797693134862e+309;
28      type TIME is range -9223372036854775807 to +9223372036854775807
29         units
30           fs;                   -- femtosecond
31           ps  = 1000 fs;        -- picosecond
32           ns  = 1000 ps;        -- nanosecond
33           us  = 1000 ns;        -- microsecond
34           ms  = 1000 us;        -- millisecond
35           sec = 1000 ms;        -- second
36           min =   60 sec;       -- minute
37           hr  =   60 min;       -- hour
38         end units;
39      function NOW return TIME;
40      subtype NATURAL  is INTEGER range 0 to INTEGER'HIGH;
41      subtype POSITIVE is INTEGER range 1 to INTEGER'HIGH;
42      type STRING     is array ( POSITIVE range <> ) of CHARACTER;
43      type BIT_VECTOR is array ( NATURAL  range <> ) of BIT;
44    end STANDARD;
```

Note that the range of INTEGER, REAL, and TIME (lines 26, 27, 28) are implementation dependent. A **subtype** is a “constraining subset” of an existing type. For example, POSITIVE type is a subtype of INTEGER type with a constraining (smaller) range. The range of INTEGER in a particular implementation may be determined from the ‘LOW and ‘HIGH attributes as shown in lines 40 and 41.

Types BOOLEAN, BIT, CHARACTER, and SEVERITY_LEVEL are predefined enumerated types with all the possible values listed inside or enclosed by a pair of parenthesis and separated by commas. Type BIT_VECTOR is declared as an unconstrained array with a compound delimiter <> (no space between the two brackets).

VHDL types can be categorized as in Figure 2.2. The scalar type consists of built-in integers and real types and their derived subtypes. Physical types, as shown in the preceding VHDL code, (lines 28 through 38) define TIME physical type. Note that there is a separator space between the abstract literal (1000) and the unit name (ns) as in line 33. Another application would be to define RESISTANCE with units such as ohms, kohms, and so on.

As shown in line 39, a function NOW is declared which returns the current simulation time by the simulator. The implementation of that function is not shown in the package STANDARD. This is a way of "hiding" implementation. The users can use the package without knowing the detailed implementation of the function. The VHDL package body is one of the places that a function body or a procedure body can be defined. The following package EXAMPLE shows that procedure MUX21 is declared in the package and the procedure body is defined in the package body. This will be discussed further in Chapter 6.

```
1     package EXAMPLE is
2        procedure MUX21(
3           signal SEL  : in  bit;
4           signal DIN0 : in  bit;
5           signal DIN1 : in  bit;
6           signal DOUT : out bit);
7     end EXAMPLE;
8
9     package body EXAMPLE is
10       procedure MUX21(
11          signal SEL  : in  bit;
12          signal DIN0 : in  bit;
13          signal DIN1 : in  bit;
14          signal DOUT : out bit) is
15       begin
16          case SEL is
17             when '0'    => DOUT <= DIN0;
18             when others => DOUT <= DIN1;
19          end case;
20       end MUX21;
21    end EXAMPLE;
```

Some valid type declaration examples are shown here. Lines 1 to 3 show enumerated type examples. Line 5 declares a subtype RAM_ADDRESS by constraining an existing type ADDRESS defined in line 4. Lines 7 to 9 depict ways of defining 8-bit types. Line 10 declares a two-dimensional array type that is unconstrained. Line 11 declares a subtype by constraining the bounds on the existing type declared in line 10. Note that line 16 is an incomplete type declaration. This is required for the access type to reference it as shown in line 17. The complete CELL type is defined in lines 18 through 21. File types will be discussed further in Chapter 10.

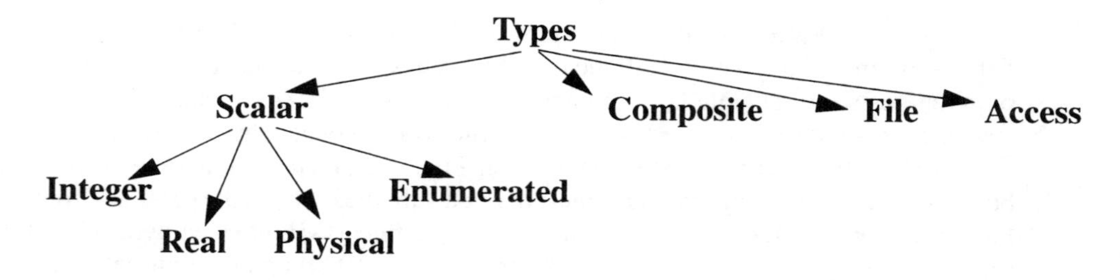

FIGURE 2.2 VHDL Types.

```
1    type LOGIC4 is ('X', '0', '1', 'Z');
2    type STATE is (IDLE, ADDR, DATA);
3    type LIGHT is (GREEN, YELLOW, RED);
4    type ADDRESS is range 0 to 16#FFFF#;
5    subtype RAM_ADDRESS is ADDRESS range 16#F000# to 16#FFFF#;
6    subtype BIT_INDEX is integer range 7 downto 0;
7    subtype BYTE is BIT_VECTOR(BIT_INDEX);
8    type A_BYTE is array (7 downto 0) of bit;
9    type B_BYTE is array (7 downto 0) of bit;
10   type TWO_DIMENSION is array (NATURAL range <>, NATURAL range <>) of
     bit;
11   subtype WINDOW is TWO_DIMENSION (0 to 80, 0 to 100);
12   type RISC is record
13      OPCODE : A_BYTE;
14      ADDR1  : ADDRESS;
15   end record;
16   type CELL;
17   type POINTER is access CELL;
18   type CELL is record
19      ID   : INTEGER;
20      PTR  : POINTER;
21   end record;
```

Note that types A_BYTE and B_BYTE are considered to be distinct types even though they are textually identical. A subtype does not define a new type. It is *based* on an existing type. We shall say that the subtype and its base type have the same base type. VHDL is a strongly typed language so that an object cannot be assigned to another object belonging to different base types without type conversion. For example, an object A is of type LOGIC4 (declared in line 1) and an object B is type BIT, it is not legal in VHDL syntax to assign B to A even though type BIT is a subset of type LOGIC4.

2.6 OBJECT DECLARATION

object_declaration ::=

constant identifier_list : subtype_indication [:= expression] ;

variable identifier_list : subtype_indication [:= expression] ;
signal identifier_list : subtype_indication [signal_kind] [:= expression] ;
signal_kind ::= **register** | **bus**
identifier_list ::= identifier { , identifier }

An object in VHDL contains (has) a value of a given type. An object declaration declares an object of a specified type. There are three classes of objects: constants, signals, and variables. An object declaration is called a single object declaration if its identifier list has a single identifier; it is a multiple object declaration if the identifier list has two or more identifiers (separated by a comma ',').

A VHDL constant is similar to the constant in other high-level programming languages. The value of a constant cannot be updated. The following are constant declaration examples:

```
1    constant PERIOD  : time    := 100 ns;
2    constant PI      : real    := 3.14159;
3    constant WIDTH   : integer := 32;
4    constant DEFAULT : bit_vector(0 to 3) := "0101";
```

A variable declaration declares a variable of the specified type. The following are variable declaration examples. Line 5 is a multiple object (variable) declaration of ROW and COLUMN variables. A VHDL variable is similar to an identifier used in other high-level programming languages. Its value can be initialized and is updated immediately when the variable assignment statement is executed.

```
5    variable ROW, COLUMN : integer range 0 to 31;
6    variable COUNT       : positive := 100;
7    variable MEMORY      : TWO_DIMENSION (0 to 15, 0 to 31);
```

The concept of a VHDL signal is somewhat special. It is not normally seen in other high-level programming languages. For now, let's consider that a signal is similar to a physical wire that connects various pieces together. We will discuss variables and signals in more detail in Chapter 4. The following are examples of signal declarations. Line 12 shows an example of an alias declaration which declares an alternative name for an existing object.

```
8    signal CLK, RESETn  : bit;
9    signal COUNTER      : integer range 0 to 31;
10   signal RAM          : TWO_DIMENSION (0 to 10, 0 to 15);
11   signal INSTRUCTION  : bit_vector(15 downto 0);
12   alias  OPCODE       : bit_vector(3 downto 0) is INSTRUCTION(15
     downto 12);
```

If a signal kind appears in a signal declaration, the signal(s) is(are) guarded signals of the kind indicated. Guarded signals will be discussed in Chapter 10.

If ":=" followed by an expression is specified, the expression is the default value of the object (s). For a variable or a signal of a subtype T, without the default value specified, the default value is given as T'left (the left-most value of its type). For example, value '0' is the default value for an object of the BIT type. Value FALSE is the default value for an object of BOOLEAN. If ":=" followed by an expression is not present for a constant declaration, the constant is called a deferred constant. Such a constant declaration can only appear in a package declaration. The corresponding full constant declaration which defines the value of the constant must appear in the package body.

2.7 ENTITY AND ARCHITECTURE

An entity declaration defines the interface between a design entity and the environment in which it is used. An architecture body defines the body of a design entity. It specifies the relationships (behavior, functionality) between the inputs and outputs of a design entity. As shown in Figure 2.3, design entity NANDXOR (line 1) has three inputs (A, B, C) and one output (D) as its interface, which is defined in the port list as shown in lines 2 through 5. Line 6 closes the entity declaration.

An architecture body defines the body of a design entity. It specifies the relationships between the inputs and outputs of a design entity and may be expressed in terms of structure, data flow, or behavior. More than one architecture body may correspond to a given entity declaration. For example, the VHDL code below shows three architectures (named RTL1, RTL2, and RTL3) for the same entity NANDXOR.

```
1     entity NANDXOR is
2        port (
3           A, B   : in  bit;
4           C      : in  bit;
5           D      : out bit);
6     end NANDXOR;
7
8     architecture RTL1 of NANDXOR is
9     begin
10       D <= (A nand B) xor C;
11    end RTL1;
12
13    architecture RTL2 of NANDXOR is
```

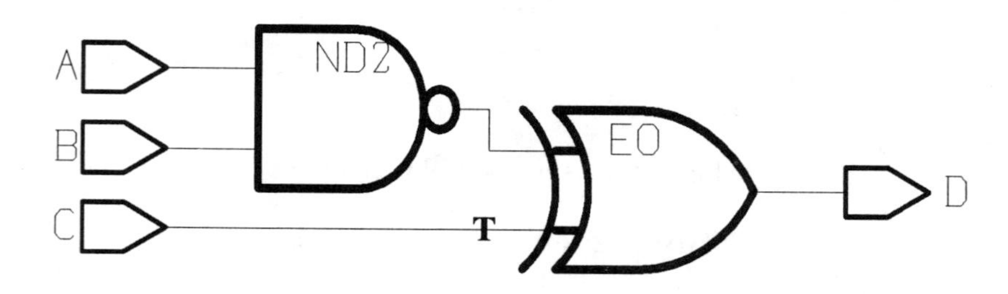

FIGURE 2.3 NANDXOR.

```
14    begin
15       process (A, B, C)
16       begin
17          if (C = '0') then
18             D <= A nand B;
19          else
20             D <= A and B;
21          end if;
22       end process;
23    end RTL2;
24
25    architecture RTL3 of NANDXOR is
26       signal T : bit;
27    begin
28       T <= A nand B;
29       p1 : process (T, C)
30       begin
31          D <= T xor C;
32       end process p1;
33    end RTL3;
```

2.8 PREDEFINED ATTRIBUTES

VHDL attributes are used to attach information to ease the coding and to pass on information to other design tools. There are both predefined attributes and user-defined attributes. For example, the predefined **event** attribute associated with a signal CLK (written as CLK'event) is a BOOLEAN value TRUE when the value of CLK is changed. To check whether it is at the rising edge of the clock CLK, the BOOLEAN expression, (CLK'event and CLK = '1'), can be used. An example of a user-defined attribute can be the critical weighting factor for a given signal so that the layout tool can use this information to determine the critical nets for placement and routing. The user-defined attributes will be discussed in Chapter 10. Here, we would first introduce the predefined attributes for types, subtypes, and arrays. The attributes related to signals will be discussed in Chapter 3.

The following attributes are associated with types and subtypes:

TS'left, TS'right, TS'low, TS'high

They return the left bound, right bound, low bound, and high bound, respectively. For example, type A_BYTE is array (7 downto 0) of bit;, A_BYTE'left = A_BYTE'high = 7. A_BYTE'low = A_BYTE'right = 0.

TS'base

It returns the base type of TS. It should be used with another attribute such as TS'base'low.

TS'pos(X)

It returns the position within the discrete or physical type of the value of the parameter X.

TS'val(X),

It returns the value at the position within the discrete or physical type of the value of the parameter X.

TS'succ(X), TS'pred(X)

They return the value at the position within the discrete or physical type whose position is one greater and one less of the value of the parameter X, respectively.

TS'leftof(X), TS'rightof(X)

They return the value at the position within the discrete or physical type whose position is to the left and to the right of the value of the parameter X, respectively.

The following attributes are associated with array objects or constrained array subtypes:

A'left(N), A'right(N), A'high(N), A'low(N), A'length(N)

The parameter is optional with default as 1 to indicate one-dimensional array. They return the left bound, right bound, lower bound, right bound, the number of values of the Nth index of the array object or subtypes, respectively. For example,

signal RAM : TWO_DIMENSION (0 to 10, 0 to 15);

RAM'left(2)= RAM'low(2) = RAM'low = RAM'low(1) = RAM'left = RAM'left(1) = 0

A'range(N), A'reverse_range(N)

The parameter is optional with default as 1 to indicate one-dimensional array. They return the range and the reverse range of the Nth index of the array object or subtypes respectively. For example, RAM'range(2) is 0 to 15, RAM'reverse_range is 10 downto 0. These ranges can be used in a loop statement to be discussed in Chapter 5.

2.9 NAMES AND OPERATORS

Names are used to denote declared (either implicitly or explicitly) objects. The following examples are commonly used names:

Simple names are identifiers such as COUNTER, CLK.

Indexed names are used to denote array elements such as RAM(1, 3) and INSTRUCTION(5).

Slice names are portions of objects such as INSTRUCTION(7 downto 0).

Selected names are used to select objects of a record, an access object, or a library. DATE1.MONTH denotes the MONTH element of a record or an access type. PACK.VDD can refer a global signal VDD in package PACK. WORK.DESIGN can denote a design entity DESIGN in library WORK.

Attribute names can be used to denote predefined or user defined attributes such as CLK'stable(3 ns), INSTRUCTION'range, and GLUE'PLACEMENT_GROUP.

VHDL uses the following operators, which are listed from the lowest precedence to the highest precedence. Parentheses can be used to change the order of precedence. Note that xnor is added in VHDL'93.

```
1    Logical                              and   or    nand   nor   xor
2    Relational                           =     /=    <      <=    >     >=
3    Adding and Concatenation             +     -     &
4    Sign (unary)                         +     -
5    Multiplying                          *     /     mod    rem
6    Exponent, absolute, complement       **    abs   not
```

Note that an expression "A nand B nand C" is not valid. It does not represent a 3-input NAND gate. To imply a 3-input NAND gate, "not (A and B and C)" can be used. "A nand (B nand C)" is not the same as "(A nand B) nand C". "A nand B nand C" is considered a syntax error to avoid ambiguity.

2.10 EXERCISES

2.1 Which of the following identifiers is not legal and state why? (A_A, A_, 9U, A%B, P8__3, e4, wait, in, OU_T)

2.2 Which of the following is not a legal integer literal and why? (12_000, 1E-0, 2#1101#, 5#44#, 3#3# 16#F# E3).

2.3 Which of the following is not a legal floating point literal and why? (1__0.00, 100_.0, 1.1E-2, 2#10.11#, 16#F#E2, 1_0.0).

2.4 Which of the following is not a legal bit string literal and why? (2"110", B"1101", O"047", H"ABcd", 10"99", "0101").

2.5 In Section 2.6, VHDL code line 10, can we declare signal RAM : TWO_DIMENSION; without constraining the ranges?

2.6 For the following VHDL code, correct syntax errors and explain the reasons. Use a VHDL analyzer to verify your answers.

```
1    entity EXP is
2    end EXP;
3
4    architecture RTL of EXP is
5       signal A, B, C, D, E, F : bit;
6       signal X, Y, Z, S, T, R : bit;
7    begin
8       R <= A and B and C;
9       S <= B or C or D;
10      T <= A and B or D;
11      Y <= C nor E nor F;
12      X <= A and not B and not C;
13      Z <= F nand E nand B;
14   end RTL;
```

2.7 What is the default value of an object of type integer if the default value is not specified in the object declaration?

Chapter 3

VHDL Modeling Concepts

3.1 THE CONCEPT OF THE SIGNAL

In Chapter 2, VHDL objects of constants, variables, and signals were introduced. The concept of a constant object implies that its value is not changed. The concept of a variable object is similar to an identifier of any other high-level programming languages. We are only interested in its value after some computations. Its value is updated immediately after the computations. Its past value is ignored. However, in the hardware design, we are also interested in the history of the circuit. How does the circuit get to the current state? We want to know what the circuit state is at what instance of time. For example, a D-type flipflop output Q will not change its value until the rising edge of the clock input even though the flipflop D input value changes before the rising edge of the clock input. The Q output value has its **present value** and a **future value** scheduled to be updated after the rising edge of the clock. Another important concept is the timing delay caused by the hardware. For example, the output value of an actual NAND gate will not change at the exact same time when the input values of the NAND gate change. The output value will change after some delay time even though the delay may be very small (such as 0.3 ns). In other words, the NAND gate output holds its present value for 0.3 ns, then schedules a change to its future value after 0.3 ns is elapsed. We say, "a signal has an **event**" when the signal value is changed. The update of a signal value and a variable value will be discussed in more detailed in the next chapter.

Another hardware feature that causes some concern is the connection of wires to several outputs of the circuits. For example, the outputs of an NAND gate and an OR gate may be connected together as shown in Figure 3.1. Signal Y is driven by two sources. Each source is a **driver**. A signal has multiple drivers if the signal is driven by multiple sources. The value of Y in the actual hardware depends on the technology (CMOS, and so on) that implements the circuit. In VHDL simulation, the value is determined by a bus resolution function, which will be discussed in Chapter 6.

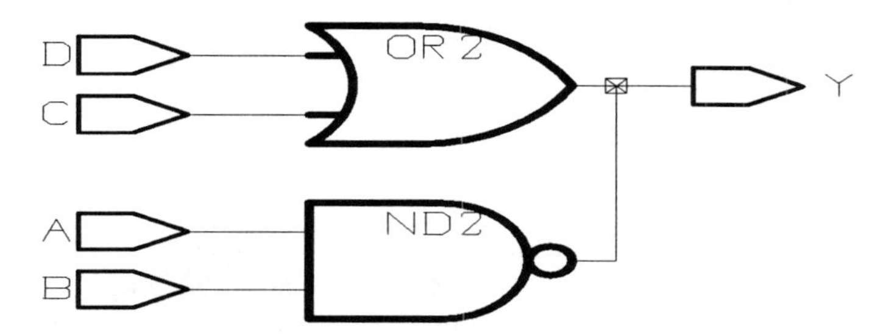

FIGURE 3.1 A signal is driven by multiple sources.

3.2 PROCESS CONCURRENCY

Many things in the real world can occur concurrently. For example, we can read a book and listen to music at the same time. Digital electronic circuits can also occur concurrently. For example, two different NAND and NOR gates can operate at the same time. VHDL is designed to model the concept of concurrency. Basically, we can partition the system into several concurrent processes that would conceptually be running concurrently. Figure 3.2 shows a circuit schematic with three inputs and one output.

To describe the circuit, the following VHDL can be used:

```
1     entity NANDXOR is
2        port (
3           A, B   : in  bit;
4           C      : in  bit;
5           D      : out bit);
6     end NANDXOR;
7
8     architecture RTL of NANDXOR is
9        signal T : bit;
10    begin
11       p0 : T <= A nand B after 2 ns;
```

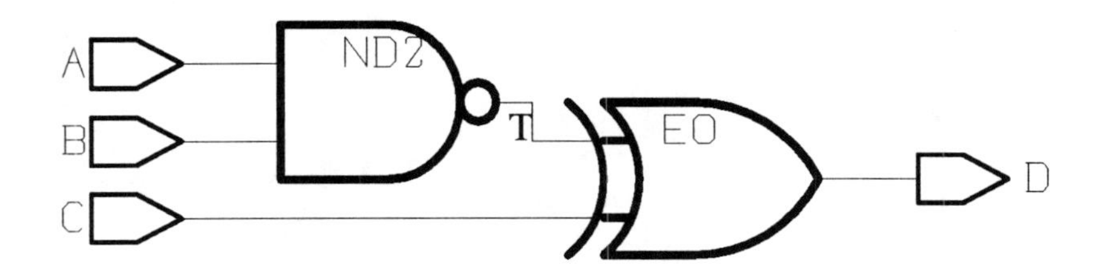

FIGURE 3.2 NANDXOR schematic.

```
12    p1 : process (T, C)
13       begin
14          D <= T xor C after 3 ns;
15       end process p1;
16    end RTL;
```

We know that the circuit input and output are specified in the entity (lines 1 to 6), while the circuit function is described in the architecture (lines 8 to 16). To describe the NAND gate, line 11 is used and is labeled p0. The exclusive-or gate is described in lines 12 to 15 and is labeled p1. We can assume that this circuit has two gates and is described with two concurrent processes p0 and p1. Each process can be in one of three states: running, active, and suspended. A process is "**running**" when the simulator is executing the process. A process is "**active**" when the process is waiting for the simulator to execute it. A process is "**suspended**" when it is not "running" or "active." Figure 3.3 shows the basic process state transition diagram.

Most simulators are running on a single processor; therefore, only one process can be "running" at any instance of time. Processes that need to be executed are put into a queue. The processor selects a process from the queue of "active" processes to be the "running" process and executes the VHDL statements. A process continues to run until the process is suspended. Another "active" process is then selected from the queue to be the "running" process. Note that after a "running" process is "suspended," it may trigger another "suspended" process to be placed in the "active" processes queue where it waits to be executed. When the "active" process queue is empty, signal values are scheduled to be updated. A "suspended" process can become an "active" process when any one of its driving signals is changed. For example, as long as signals A and B are not changed, process p0 ("suspended") is not activated and evaluated. When signal A or B is changed, then process p0 is moved from the "suspended" state to the "active" state and eventually to the "running" state. Signal T value may be changed due to the change on signal A or B. When signal T value is changed, then process p1 will be activated and moved from the "suspended" state to the "active" state and to the "running" state. The VHDL simulator changes the state of a process based on the completion of an execution of a running process, scheduler, and signal event.

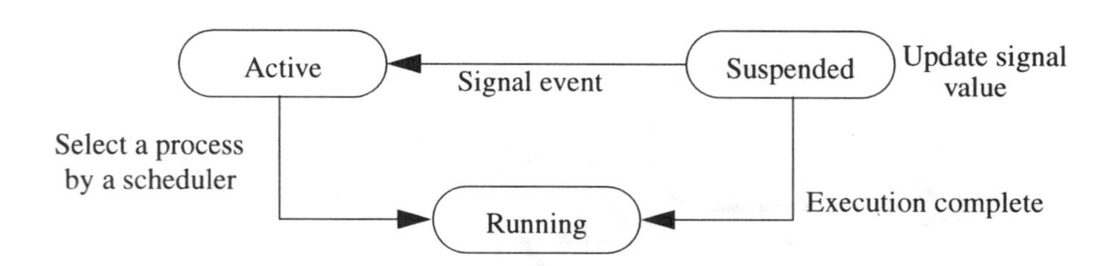

FIGURE 3.3 Process state transition.

The execution of a model consists of an initialization phase followed by the simulation cycle, which is the repetitive execution of concurrent processes in the description of the model. Figure 3.4 shows the relationship of the initialization phase, simulation cycles, and the simulation time. During the initialization phase, each signal (A, B, C) is set to a default value as its left-most value ('0') of its type (bit). Each concurrent process is evaluated once until it is suspended (when the process reaches either an explicit or implicit wait condition). For example, process p0 causes T to be evaluated to value '1', and then suspended. Process p1 will evaluate D to value '1' and then it is suspended. D value will be scheduled to update its value at time 5 ns (2 ns delay for the process p0 (NAND gate) and 3 ns delay for the process p1 (XOR gate). After the initialization phase is done, both processes p0 and p1 are suspended. Signals values T and D are updated at time 2 ns and 5 ns, respectively, when the simulation time is beyond 5 ns.

The simulation cycle starts when either signal changes its value. For example, at time 10 ns, input signal C is changed from '0' to '1'. Process p1 is activated, run, and suspended. Output signal D is changed at time 13 ns (10 ns plus 3 ns delay) when the simulation time is beyond 13 ns. The next simulation cycle happens at time 20 ns when input signal A is changed from '0' to '1'. Process p0 is activated, run, and suspended. However, signal T value is not changed. Process p1 will not be evaluated. At time 30 ns, input signal B is changed from '0' to '1'. Another simulation cycle starts. Process p0 is activated, run, and suspended. Signal T is updated at time 32 ns (30 ns plus 2 ns delay). This causes process p1 to be activated, run, and suspended. Output signal D is updated at time 35 ns (32 ns plus 3 ns) when the simulation time is beyond 35 ns. At time 40 ns, input signal C is changed from '1' to '0'. A new simulation cycle starts, which causes process p1 to be activated, run, and suspended. Output signal D is updated at time 43 ns (40 ns plus 3 ns) when the simulation cycle is beyond 43 ns. Figure 3.5 shows the simulation waveform after the simulation has run for 50 ns.

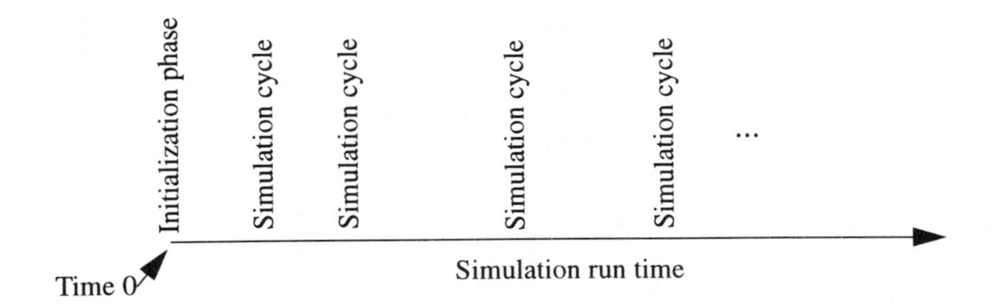

FIGURE 3.4 VHDL simulation of initialization phase and simulation cycle.

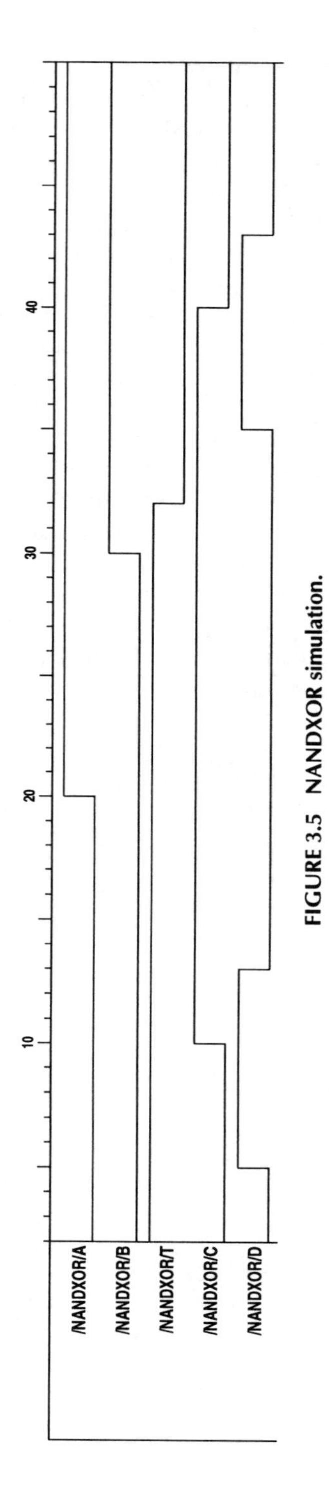

FIGURE 3.5 NANDXOR simulation.

3.3 DELTA TIME

The following VHDL is another architecture (called DELTA) of the entity NANDXOR. It is identical to architecture RTL of NANDXOR, except the assignment statements do not have the after 2 (3) ns clause. Compare Figure 3.5 and Figure 3.6 simulation waveforms. They have the same input patterns causing the input signals to be changed exactly the same way. Note that signals T and D are changed exactly at the same time as input signals changed. For example, at time 30 ns, input signal B changed, which causes signal T to change, and then causes output signal D to change. They all happen exactly at the same simulation running time.

```
1    architecture DELTA of NANDXOR is
2       signal T : bit;
3    begin
4       p0 : T <= A nand B;
5       p1 : process (T, C)
6       begin
7          D <= T xor C;
8       end process p1;
9    end DELTA;
```

Let's look at the waveform in Figure 3.6 at time 30 ns closer at Figure 3.7. Just before 30 ns, processes p0 and p1 are suspended. At time 30 ns, input signal B has an event. A simulation cycle starts. It causes process p0 to be active, running, and then suspended. Signal T is changed to '1'. It is reasonable to say that signal T is changed after signal B is changed. We shall refer to this as a delta time. The event on signal T causes process p1 to be active, running, and then suspended. Signal D is changed. Again, signal D is changed after signal T was changed. Another delta time is assumed. Delta time delay is a simulation concept. From the simulation running time point of view, it is 0 ns. This is the reason that Figure 3.6 shows signals B, T, and D are changed at the same time at 30 ns.

When process p1 is suspended at 2 deltas after 30 ns, no more signals are changed. The VHDL simulator then advances the simulation running time to the next signal event at time 40 ns. Figure 3.8 shows the simulation run time versus delta delay. During the simulation cycle, the VHDL simulator evaluates processes with delta delays that do not advance the simulation run time. We can assume that the simulator goes up in "delta delay" axis. After all processes are in the suspended state, the simulator advances simulation run time. Note that the numbers of delta delays at each instant of time are not necessarily the same. At some point in time, more signals may be changed, and more processes may be active. If the simulator continues to increase the delta delay without advancing simulation run time (delta delay goes up infinitely), the simulator may go into an infinite loop.

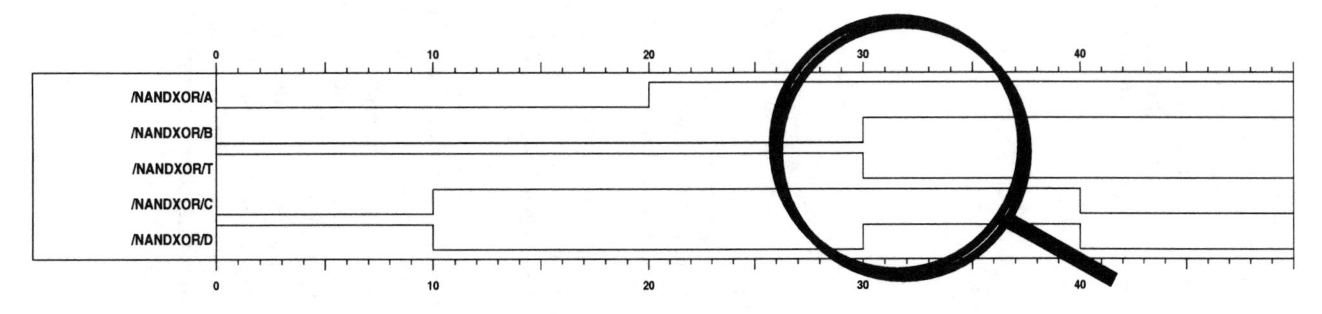

FIGURE 3.6 NANDXOR simulation waveform without after clause.

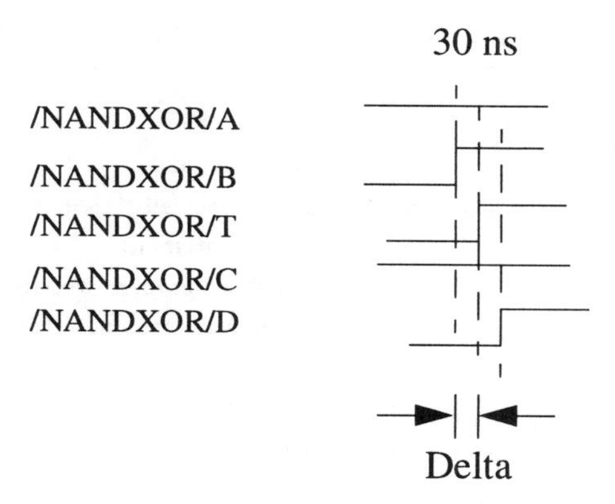

FIGURE 3.7 Delta delay.

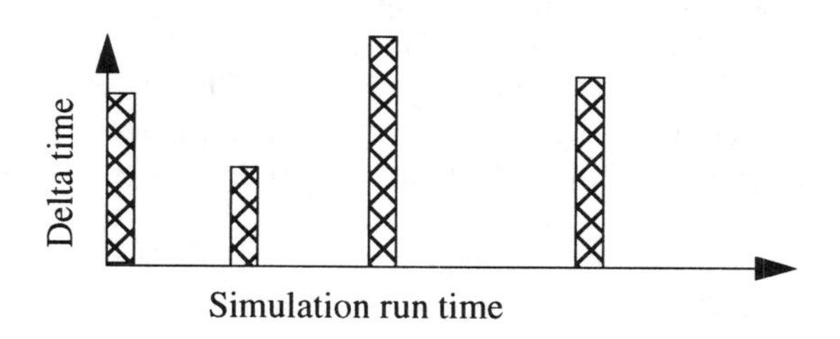

FIGURE 3.8 Simulation run time versus delta delay.

3.4 CONCURRENT AND SEQUENTIAL STATEMENTS

Besides having the capability of modeling concurrent processes, VHDL can also model sequential behavior by means of a set of constructs as most high-level programming languages can. In VHDL, executable statements are categorized as concurrent statements and sequential statements. Concurrent statements are used to model concurrent behaviors. VHDL concurrent statements are block, process, generate, component instantiation, signal assignment, assert, and procedure call statements. Most sequential statements are inherent from other high-level languages such as if, case, loop, and assignment statements. VHDL sequential statements are if, case, loop, assert, signal assignment, variable assignment, next, null, return, exit, procedure call, and wait statements. Figure 3.9 shows all types of VHDL concurrent and sequential statements.

Sequential statements are used to define algorithms for the execution of a subprogram (function and procedure) or a process. They execute in the order in which

VHDL Concurrent Statements	VHDL Sequential Statements
block statement	if statement
process statement	case statement
generate statement	loop statement
procedure call statement	**procedure call statement**
assert statement	**assert statement**
signal assignment statement	**signal assignment statement**
component instantiation statement	variable assignment statement
	null statement
	exit statement
	wait statement
	return statement
	next statement

FIGURE 3.9 VHDL concurrent and sequential statement table.

they appear. The details of each sequential and concurrent statement will be discussed in Chapters 4 and 5. To combine the sequential and concurrent statement in the same VHDL model, VHDL coding rules (for lack of a better name) need to be followed. The following VHDL code shows the general VHDL model. Note that the entity does not have any port signals. The architecture statement part (between **begin** [line 7] and **end** [line 9]) can also be empty. Actually, the VHDL code is valid without any syntax error.

```
1    entity OVERALL is
2    end OVERALL;
3
4    architecture RTL of OVERALL is
5       -- architecture declarative part
6    begin
7       -- architecture statement part
8    end RTL;
```

The important VHDL coding rules are as follows:

1. Only concurrent statements can be inside the architecture statement part.
2. Sequential statements can only appear inside the procedure and function body and inside the process statement.
3. Signals are used to communicate among concurrent processes.
4. Local variables can only be declared inside the procedure and function body and the process statement. They are not visible outside of the procedure, function, and process statement.

Figure 3.10 summarizes the relationship of the architecture statement part, concurrent statements, sequential statements, subprograms, and the declaration of signals and variables.

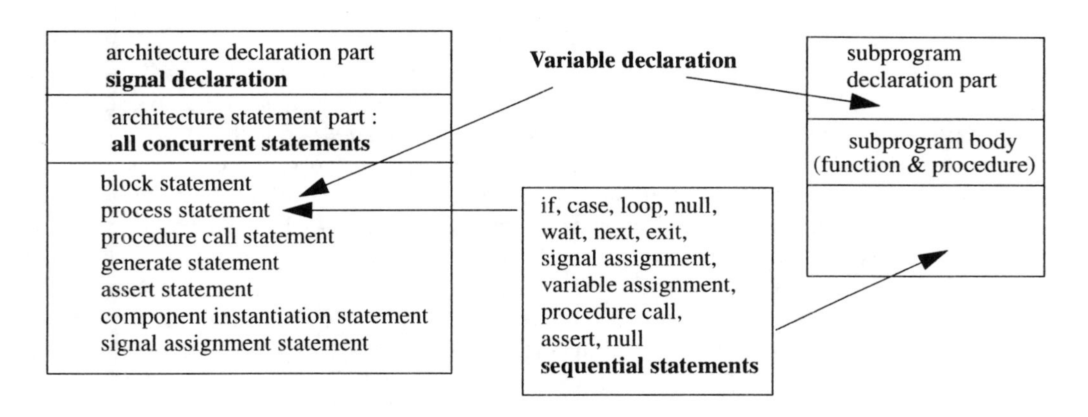

FIGURE 3.10 Concurrent and sequential statements relationship.

Note that VHDL concurrent process statements refer to a VHDL code. The process as we discussed in the last section of process concurrency refers to the process in terms of the VHDL simulator, which is not the same as a VHDL process statement. At least one concurrent process can be generated by a VHDL simulator from any one of the VHDL concurrent statements. The VHDL process statement is just one of the concurrent statements. For example, in the NANDXOR architecture, there are two VHDL concurrent statements. One is a signal assignment statement in line 11. This concurrent signal assignment statement will generate a concurrent process by a VHDL simulator. The other is a concurrent process statement from line 12 to line 15. This will generate another concurrent process by a VHDL simulator. Line 14 is a sequential signal assignment statement which is inside a process statement. Note also that signal assignment, procedure call, and assert statements can be sequential or concurrent, depending on where they appear in the concurrent statement region or sequential statement region.

As an example of the VHDL coding rule 3, Figure 3.11 shows the overall NANDXOR VHDL entity, architecture, and processes communication with signals (not variables) relationship. Note that T is a signal. A, B, C, D declared as ports of the entity are considered as signals too.

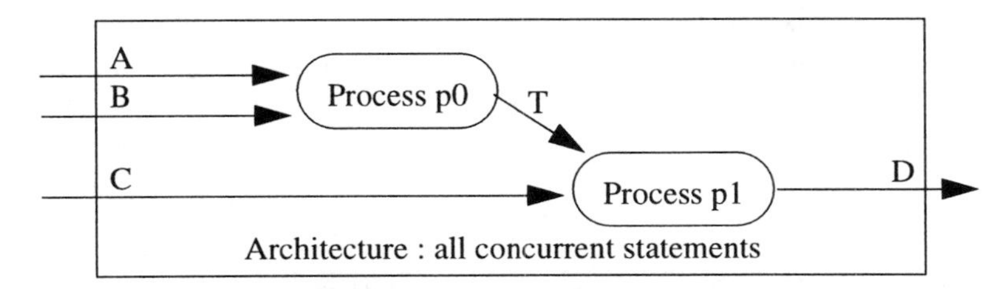

FIGURE 3.11 Signals used to communicate among processes.

3.5 PROCESS ACTIVATION BY A SIGNAL EVENT

The preceding NANDXOR VHDL code example illustrates how a signal event activates a concurrent process. The NANDXOR VHDL code has two concurrent statements. Process p0 is a concurrent signal assignment statement. It is activated whenever there is an event on any one of the signals appearing in the right hand-side of the assignment "<=" delimiter. For example, in line 4 of DELTA architecture, the sensitivity list of the process p0 is A and B. The second process is a concurrent process statement. The concurrent process statement has the **sensitivity list**, which are the signals appearing inside the parenthesis after the keyword "**process**". For example, in line 5 of DELTA architecture, the sensitivity list of the process p1 is T and C.

If none of the signal in the sensitivity list has an event, the process statement remains at the suspended state. If we modify the DELTA architecture to the following code by deleting signal C from the sensitivity list of the process p1, then line 5 of SLIST architecture has only T inside the parenthesis. At the initialization phase, process p0 and p1 are evaluated and then suspended. Process p1 will be suspended at line 8 and waiting for the event of T (sensitivity list). We can then say that the process statement is implicitly waiting for an event on any signal in the process sensitivity list.

```
1    architecture SLIST of NANDXOR is
2       signal T : bit;
3    begin
4       p0 : T <= A nand B;
5       p1 : process (T)
6       begin
7          D <= T xor C;
8       end process p1;
9    end SLIST;
```

Figure 3.12 depicts the simulation waveform of the SLIST architecture of the NANDXOR entity. Note that process p1 has only signal T as its sensitivity list. Therefore, process p1 will only be activated by the event of signal T. The event on signal C would not activate process p1 to be evaluated. This explains the reason that the value of signal D is not changed at time 10 ns, 40 ns, 60 ns. The SLIST architecture does not model the NANDXOR design correctly.

Process p1 will run forever if signal T is again taken out from the process sensitivity list. In other words, the process statement is not waiting for anything to execute the statement inside. Another way to explicitly specify the suspension of the process statement is by a wait statement, which will be discussed in Chapter 4. A process statement requires an explicit wait statement or a process sensitivity list but not both.

3.6 SIGNAL-VALUED AND SIGNAL-RELATED ATTRIBUTES

In Chapter 2, we discussed the attributes for types, subtypes, and arrays. The attributes attached to a signal are described in this section. There are two categories of attributes

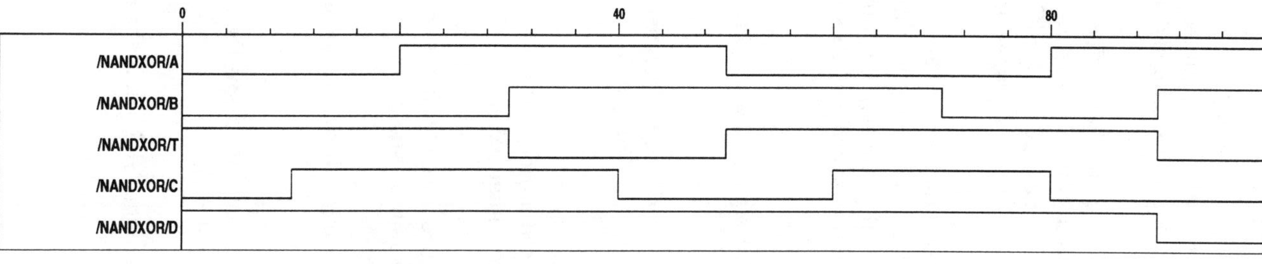

FIGURE 3.12 Wrong simulation waveform due to sensitivity list.

for signals. One category of attributes results in another signal. The other category of attributes results in a value. Each of the following attributes attached to a signal results in another signal.

SIG'delayed(T)
: It defines a signal which is the signal SIG delayed by time T. The parameter is optional. T = 0 ns is the default if parameter T is not specified. SIG'delayed would have the same value as SIG after a delta delay.

SIG'stable(T)
: It defines a BOOLEAN signal whose value is TRUE if signal SIG has not had an event for the length of time T. The parameter is optional. T = 0 ns is the default if parameter T is not specified. SIG'stable would be FALSE during the simulation cycle when SIG is *changed* and then returns to TRUE.

SIG'quiet(T)
: It defines a BOOLEAN signal whose value is TRUE if signal SIG has not had a transaction (not active) for the length of time T. The parameter is optional. T = 0 ns is the default if parameter T is not specified. SIG'stable would be FALSE during the simulation cycle when SIG is *assigned to* and then returns to TRUE.

SIG'transaction
: It defines a BIT typed signal whose value toggles each time a transaction occurs on signal SIG.

The following attributes are defined for signal objects, but they are not signals themselves, they are values with a particular type.

SIG'event
: It is a BOOLEAN typed attribute. It is true if an event occurs on signal SIG during the current simulation cycle.

SIG'active
: It is a BOOLEAN typed attribute. It is true if a transaction occurs on signal SIG during the current simulation cycle.

SIG'last_event
: It is a TIME typed attribute. It returns the amount of time elapsed since the last event on signal SIG.

SIG'last_active
: It is a TIME typed attribute. It returns the amount of time elapsed since the last transaction on signal SIG.

SIG'last_value
: It returns the value of signal SIG before the last event on signal SIG.

3.7 EXERCISES

3.1 Is it possible for a VHDL simulator to run forever without advancing simulation time? For example, Figure 3.13 shows a NAND gate with its output connecting to one of its inputs.

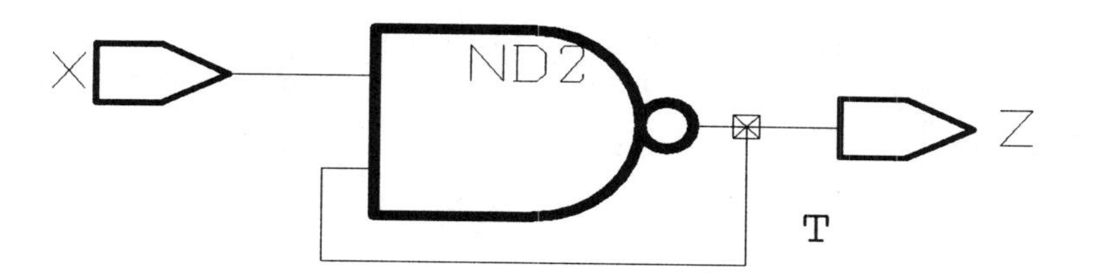

FIGURE 3.13 A feedback loop.

The following is the corresponding VHDL model. What would happen in the initialization phase? If we change the input signal X to '1' and run a VHDL simulator, what would happen? Try it on your VHDL simulator.

```
1    entity RUNAWAY is
2       port (
3          X      : in  bit;
4          Z      : out bit);
5    end RUNAWAY;
6    architecture RTL of RUNAWAY is
7       signal T : bit;
8    begin
9       T <= X nand T;
10      Z <= T;
11   end RTL;
12
```

3.2 IEEE 1076 standard does not specify how the "active" processes are placed in queue. This is left for specific implementation. If you were to design a VHDL simulator, how would you select a process from the "active" processes to run? Is there a better scheduling algorithm? Why?

3.3 What is the difference between a transaction and an event for a signal? What is the difference between a signal-valued attributes SIG'quiet and a signal SIG'stable for a signal SIG?

3.4 If the execution order of the "active" processes is not fixed (implementation dependent), is it possible that the same VHDL model will get different results from a different run with the same simulator and same test vectors? Will the results be the same for different VHDL simulators?

3.5 What VHDL statements can be either sequential or concurrent?

3.6 Explain the simulation waveform result of SLIST architecture for every signal event.

3.7 What would happen if we delete signal T from the sensitivity list of process p1 of SLIST architecture so that the sensitivity list of process p1 is empty? Run a VHDL simulator to verify your answer.

Chapter 4

Sequential Statements

VHDL sequential statements are similar to statements of other high-level programming languages which are executed sequentially. These statements can only appear inside a process statement or inside a subprogram body (function or procedure body). VHDL sequential statements are:

variable assignment statement
signal assignment statement
if statement
case statement
loop statement
next statement
exit statement
null statement
procedure call statement
return statement
assertion statement
wait statement

4.1 VARIABLE ASSIGNMENT STATEMENT

variable_assignment_statement ::= target := expression ;

The *variable assignment statement* replaces the current value of a variable with the new value specified by the expression. The variable and the expression must be of the same base type. Variables are similar to objects in other high-level programming languages in which the variable gets its new value when the variable assignment statement is executed. This is not the same as signals, which are discussed in the next section.

In VHDL, local variables can only be declared in the declaration region of a process statement and a subprogram (function or procedure). The following are some examples of variable assignment statements:

```
1     entity VASSIGN is
2     end VASSIGN;
3
4     architecture RTL of VASSIGN is
5        signal A, B, J   : bit_vector(1 downto 0);
6        signal E, F, G   : bit;
7     begin
8        p0 : process (A, B, E, F, G, J)
9           variable C, D, H, Y : bit_vector(1 downto 0);
10          variable W, Q       : bit_vector(3 downto 0);
11          variable Z          : bit_vector(0 to 7);
12          variable X          : bit;
13          variable DATA       : bit_vector(31 downto 0);
14       begin
15          C          := "01";
16          X          := E nand F;
17          Y          := H or J;
18          Z(0 to 3)  := C & D;
19          Z(4 to 7)  := (not A) & (A nor B);
20          D          := ('1', '0');
21          W          := (2 downto 1 => G, 3 => '1', others => '0');
22          DATA       := (others => '0');
23       end process;
24    end RTL;
```

In line 15, variable C gets a constant value. The expressions in lines 16 and 17 use BOOLEAN operators that can take both single bit value (E, F) and array (H, J). In line 18, the target is a slice (continuous index portion) of an array and the expression uses a concatenation operator (&) to have 4 bits. Line 19 shows a target slice with combinations of concatenation and BOOLEAN operators. Line 20 shows an aggregate expression which uses positional mapping. Line 21 uses a name mapping aggregate and keyword **others** is used. It must appear as the last choice if it exists inside the aggregate. Line 22 sets all bits of DATA to '0' without using a long string which easily loses the right count of '0's such as "00000000000000000000000000000000".

Note that local variables are only visible inside the process or subprogram which is declared. VHDL'93 defines another class of variables called **shared** variables which can be shared (visible) with processes and subprograms. Shared variables will be discussed in Chapter 14.

4.2 SIGNAL ASSIGNMENT STATEMENT

signal_assignment_statement ::=
    target <= [**transport**] waveform_element { , waveform_element };

waveform_element ::=
  value_expression [**after** time_expression] | **null** [after time_expression]

The value expression in the right-hand side of the *signal assignment statement* is similar to the variable assignment statement. The signal assignment delimiter is "<=", while ":=" is used for variable assignment statements. The optional keyword **transport** specifies a *transport delay* rather than an *inertial delay*. Inertial delays are characteristic of switching circuits. A pulse with a duration shorter than the switching time of the circuit will not be transmitted. The following VHDL code shows the difference of a transport delay and an inertial delay.

```
1     entity DELAY is
2     end DELAY;
3
4     architecture RTL of DELAY is
5        signal A, B, X, Y : bit;
6     begin
7        p0 : process (A, B)
8        begin
9           Y <= A nand B after 10 ns;
10          X <= transport A nand B after 10 ns;
11       end process;
12       p1 : process
13       begin
14          A <= '0', '1' after 20 ns, '0' after 40 ns,
15               '1' after 60 ns;
16          B <= '0', '1' after 30 ns, '0' after 35 ns,
17               '1' after 50 ns;
18          wait for 80 ns;
19       end process;
20    end RTL;
```

Process p0 has two signal assignment statements. Signal X is assigned with the transport delay, while signal Y has the default of inertial delay. Process p1 generates signals A and B waveforms. Figure 4.1 shows the simulation waveform. Note that signal B has a pulse of 5 ns width and the inertial delay of Y is 10 ns. The 5 ns pulse of signal B does not transmit to signal Y. Signal X shows the effect of the 5 ns pulse of signal B due to a transport delay that transmits waveforms without the consideration of the switching time. Both signals X and Y have delay of 10 ns with respect to the event of signal A or B.

The execution of a signal assignment statement affects the projected output waveforms representing the current and future values of signal drivers. A driver of a signal is a container for a projected output waveform. Each process that assigns to a given signal contains a driver for that signal implicitly. A signal's value is a function of the current values of its drivers. Each waveform element produces a transaction for a driver. A transaction is determined by the current time and the value of the time expression. The following VHDL code illustrates how a signal driver is handled:

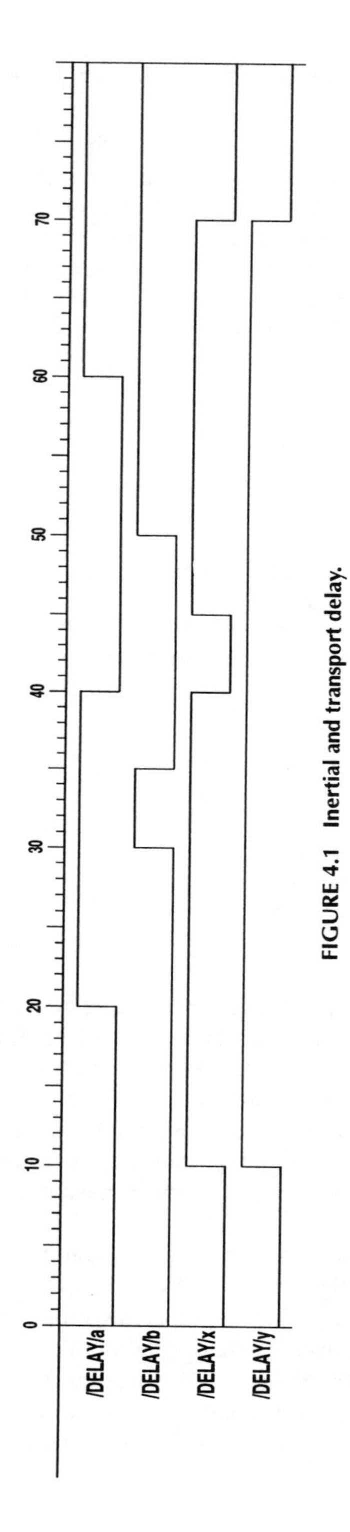

FIGURE 4.1 Inertial and transport delay.

```
1    entity DRIVER is
2    end DRIVER;
3
4    architecture RTL of DRIVER is
5       signal A : integer;
6    begin
7       pa : process
8       begin
9          A <= 3, 5 after 20 ns, 7 after 40 ns, 9 after 60 ns;
10         wait for 30 ns;
11         A <= 2, 4 after 20 ns, 6 after 40 ns, 8 after 60 ns;
12         wait for 50 ns;
13      end process;
14   end RTL;
```

Lines 9 and 11 assign signal A with various values and time expressions. When line 9 is executed, the driver of signal A contains four transaction pairs (value and time) as (3, 0 ns), (5, 20 ns), (7, 40 ns), (9, 60 ns). Line 10 waits for 30 ns. When the current simulation time is after the transaction, the signal is updated as the value in the transaction (values 3 and 5 are shown before time at 30 ns). After line 10 is executed and done, line 11 is executed at time 30 ns. The previous old transactions will be discarded and new transactions will be added as ((2, 0 ns), (4, 20 ns), (6, 40 ns), (8, 60 ns)). Line 12 waits for 50 ns. The first 50 ns of the driver values (2, 4, 6) are shown. The end of the process is reached and the execution continues at line 9. Figure 4.2 shows the value of signal A in a waveform.

The null waveform element produces a *null transaction*. This will be discussed with the topic of the **guarded** signal in Chapter 10.

It is important to know the difference between a local variable and a signal in writing VHDL. They are summarized as follows:

1. Local variables are declared and only visible inside a process or a subprogram. Signals cannot be declared inside a process or a subprogram.
2. The new value of a local variable is immediately updated when the variable assignment statement is executed. A signal assignment statement updates the signal driver. The new value of the signal is updated as a function of the signal drivers when the process is suspended. This is the reason that a signal requires a two-list implementation by the simulator. One list remembers its current value, type, and so on. The other list contains the drivers that determine the future new value. From the VHDL simulator point of view, variables are cheaper to implement. They require less memory. It is more efficient because the evaluation of drivers is not needed.
3. Only signals can be used to communicate among concurrent statements. Ports declared in the entity are signals. Subprogram arguments can be signals or variables (the detailed of function and procedure will be discussed in Chapter 6).
4. In describing a digital circuit, a signal is usually used to indicate an interconnect (net in a schematic). Local variables are usually used as a temporary value in an algorithm description of a function. For example, a local variable

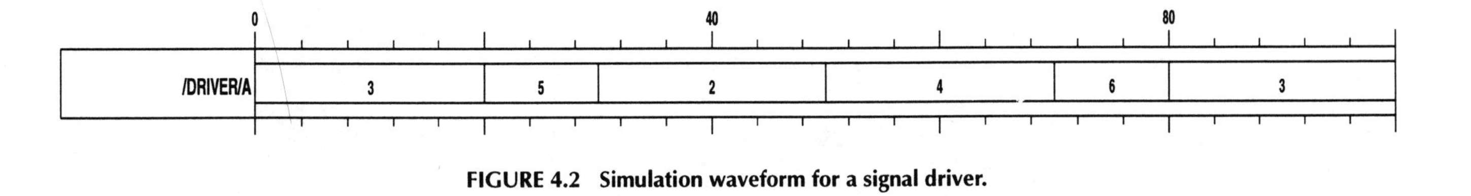

FIGURE 4.2 Simulation waveform for a signal driver.

can be used to calculate a parity for a 32-bit bus with a loop statement, and then assign the local variable to a signal. A local variable is also very useful to factor out common parts of complex equations to reduce the mathematical calculation.

5. The right-hand side of a variable assignment statement is an expression. There is no associated time expression. The right-hand side of a signal assignment statement is a sequence of waveform elements with associated time expressions.

The following VHDL code illustrates some differences of local variables and signals.

```
1     entity SIGVAL is
2        port (
3           CLK, D   : in  bit;
4           FF2, FF3 : out bit;
5           Y        : out bit_vector(7 downto 0));
6     end SIGVAL;
7
8     architecture RTL of SIGVAL is
9        signal FF1, SIG0, SIG1    : bit;
10    begin
11       p0 : process (D, SIG1, SIG0)
12          variable VAR0, VAR1 : bit;
13       begin
14          VAR0 := D;
15          VAR1 := D;
16          SIG0 <= VAR0;
17          SIG1 <= VAR1;
18          Y(1 downto 0) <= VAR1 & VAR0;
19          Y(3 downto 2) <= SIG1 & SIG0;
20          VAR0 := not VAR0;
21          VAR1 := not VAR1;
22          SIG0 <= not VAR0;
23          SIG1 <= not D;
24          Y(5 downto 4) <= VAR1 & VAR0;
25          Y(7 downto 6) <= SIG1 & SIG0;
26       end process;
27       p1 : process
28       begin
29          wait until CLK'event and CLK = '1';
30          FF1 <= D;
31          FF2 <= FF1;
32       end process;
33       p2 : process
34          variable V3 : bit;
35       begin
36          wait until CLK'event and CLK = '1';
37          V3  := D;
38          FF3 <= V3;
```

```
39        end process;
40     end RTL;
```

Lines 14 to 17 assign local variables VAR0, VAR1 and signals SIG0, SIG1. They are then assigned to Y(1 downto 0) and Y(3 downto 2), respectively, in lines 18 and 19. Lines 20 to 23 again assign local variables VAR0, VAR1 and signals SIG0, SIG1. They are then assigned to Y(5 downto 4) and Y(7 downto 6), respectively, in lines 24 and 25. Note that lines 16 and 17 have no effect (the transactions created by lines 16 and 17 are discarded) because SIG0 and SIG1 are again assigned in lines 22 and 23. The assignment statements in lines 19 (and 25), SIG0 (and SIG1) would have the same current values since the new value of SIG0 (and SIG1) has not been updated. Since the value of a local variable is updated immediately when the variable assignment statement is executed, the value of VAR0 (and VAR1) in line 18 is not the same as the value of VAR0 (and VAR1) in line 24. Figure 4.3 shows the simulation waveform. Note that Y(7 downto 6) is always the same as Y(3 downto 2). Y(5 downto 4) is opposite to the value of Y(1 downto 0).

The preceding VHDL code also shows another two process statements labeled p1 (lines 27 to 32) and p2 (lines 33 to 39). The only difference between these two processes is in lines 30 and 37: process p1 uses a signal FF1, while process p2 uses a local variable V3. Both processes have a wait statement that is waiting for the rising edge of the clock. In lines 37 and 38, V3 gets the value of D immediately and then assigns to FF3. FF3 will clock in the value of D. In line 30, FF1's driver (the new value has not been updated) gets the current value of D. In line 31, FF2's driver gets the current value of FF1. In other words, the value of D is delayed two clocks to go to FF2, while D is delayed one clock to go to FF3. Note the difference in the above waveform. From the hardware point of view, D is going through two flipflops to go to FF2 while there is only one flipflop between D and FF3. Figure 4.4 shows the synthesized schematic.

A local variable is often used as a temporary variable to be shared by more than one expression. The following VHDL code shows an example of this application. Signals Y and Z have a common term (B*C + D*E*F + G) as shown in lines 9 and 10 of architecture RTL. This common term can be extracted to a local variable and shared by signals Y and Z in architecture RTL1 (lines 20 to 22). Architecture RTL2 shows the common term is assigned to a signal (line 32). Signals Y and Z in lines 33 and 34 would not get the current value of (B*C + D*E*F + G), but the previous value. Architectures RTL and RTL1 would have the same behavior, which is not the same as architecture RTL2.

```
1     entity TEMP is
2     end TEMP;
3
4     architecture RTL of TEMP is
5        signal A, B, C, D, E, F, G, Y, Z : integer;
6     begin
7        p0 : process (A, B, C, D, E, F, G)
8        begin
9           Y <= A + (B*C + D*E*F + G);
10          Z <= A - (B*C + D*E*F + G);
```

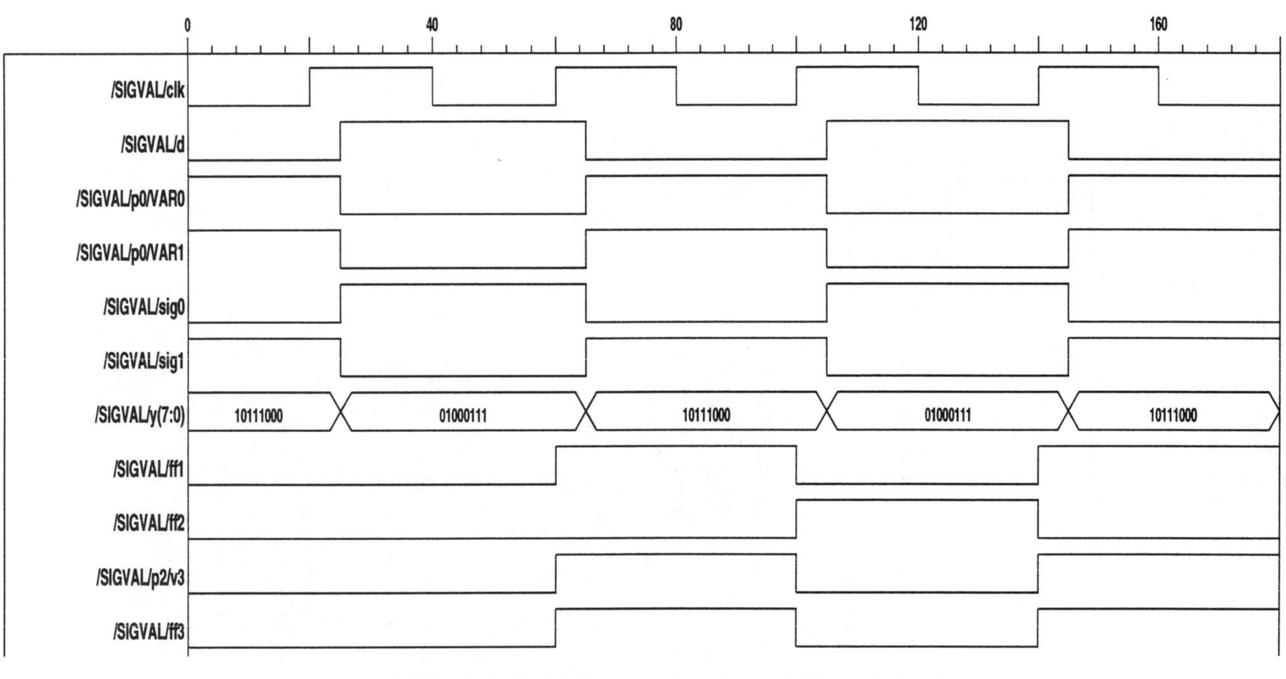

FIGURE 4.3 Simulation waveform for variables and signals.

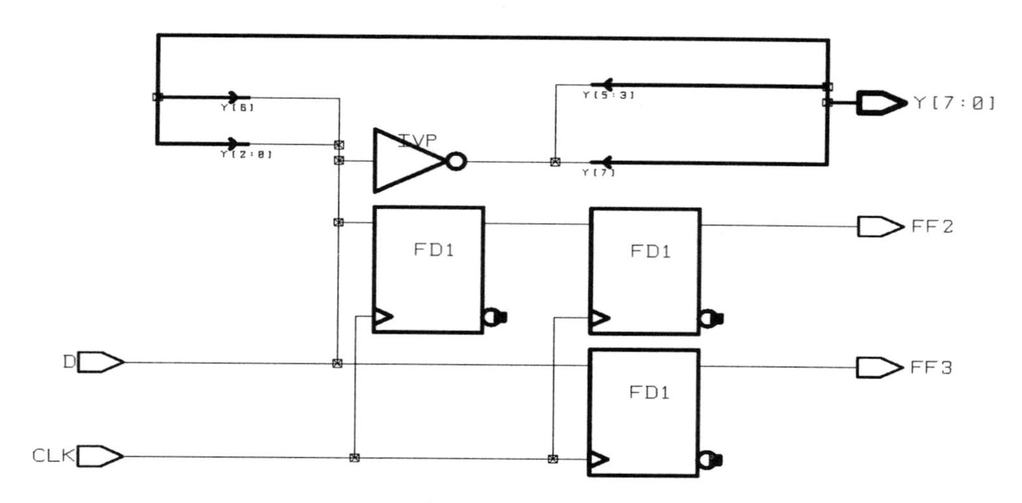

FIGURE 4.1 Synthesized schematic for signals and variables.

```
11  end process;
12  end RTL;
13
14  architecture RTL1 of TEMP is
15     signal A, B, C, D, E, F, G, Y, Z : integer;
16  begin
17     p0 : process (A, B, C, D, E, F, G)
18        variable V : integer;
19     begin
20        V := (B*C + D*E*F + G);
21        Y <= A + V;
22        Z <= A - V;
23     end process;
24  end RTL1;
25
26  architecture RTL2 of TEMP is
27     signal A, B, C, D, E, F, G, Y, Z : integer;
28     signal V : integer;
29  begin
30     p0 : process (A, B, C, D, E, F, G)
31     begin
32        V <= (B*C + D*E*F + G);
33        Y <= A + V;
34        Z <= A - V;
35     end process;
36  end RTL2;
```

4.3 IF STATEMENT

VHDL *if statement* is similar to other high-level programming languages' *if statements*. It selects execution of one or none of the enclosed sequences of sequential statements based on the evaluated condition.

if_statement ::= **if** condition **then**
sequence of sequential statements
{ **elsif** condition **then**
sequence of sequential statements }
[**else**
sequence of sequential statements]
end if;

The condition expression must be of type BOOLEAN. Note that there can be none, one, or more than one **elsif** portion in one if statement. The **else** portion can appear only none or one time. **Elsif** (not *elseif*) is a keyword. There is a space between keywords **end** and **if** (not *endif* as one word) to close the if statement. The following VHDL code models a 5-bit counter using if statements.

```
1     entity IFSTMT is
2        port (
3           RSTn, CLK, EN, PL : in  bit;
4           DATA              : in  integer range 0 to 31;
5           COUNT             : out integer range 0 to 31);
6     end IFSTMT;
7
8     architecture RTL of IFSTMT is
9        signal COUNT_VALUE : integer range 0 to 31;
10    begin
11       p0 : process (RSTn, CLK)
12       begin
13          if (RSTn = '0') then
14             COUNT_VALUE <= 0;
15          elsif (CLK'event and CLK = '1') then
16             if (PL = '1') then
17                COUNT_VALUE <= DATA;
18             elsif (EN = '1') then
19                if (COUNT_VALUE = 31) then
20                   COUNT_VALUE <= 0;
21                else
22                   COUNT_VALUE <= COUNT_VALUE + 1;
23                end if;
24             end if;
25          end if;
26       end process;
27       COUNT <= COUNT_VALUE;
28    end RTL;
```

As shown in the preceding entity declaration, the 5-bit counter has input ports RSTn, CLK, EN, PL, and DATA. The output port COUNT gets the value of the counter. RSTn is the asynchronous reset, and CLK is the clock input signal. The asynchronous reset behavior is done by checking whether RSTn = '0' (line 13) and initializing the counter to 0 (line 14) before checking the rising edge of the clock (line 15). The if statement implies the priority sequence of checking BOOLEAN expressions.

Lines 16 to 24 are another if statement that models the synchronous behavior of parallel load and count enable of the counter. Note that parallel load (PL) has higher priority than the count enable (EN) since PL = '1' is checked first in line 16. EN = '1' is checked only when PL = '1' is not TRUE (line 18). Figure 4.5 shows the simulation waveform. At time 110, both PL and EN are '1', and the counter is loaded with the DATA value (30). At time 190, the counter does not count up and hold its value since EN is '0'.

COUNT is declared as mode **OUT** (line 5). VHDL does not allow the port signal of mode **OUT** to be read. That means the signal name COUNT cannot appear in the VHDL expression to be evaluated. A temporary signal COUNT_VALUE is declared in line 22. Line 27 is then assign COUNT_VALUE to the output port COUNT outside the process. Figure 4.6 depicts a synthesized schematic of the 5-bit counter using the preceding VHDL code.

4.4 CASE STATEMENT

A *case statement* selects execution one of a number of alternative sequences of sequential statements. The chosen alternative is defined by the value of the expression.

```
case_statement ::=    case expression is
                          when choice(s) =>
                              sequence of sequential statements
                          [ when choice(s) =>
                              sequence of sequential statements ]
                      end case;
```

The expression in the case statement must be of a discrete type or of a one-dimensional character array type (whose values can be represented as string or bit string literals). The choice must be of the same type as the expression. All possible choices of the expression must be covered exactly once. No two separate alternative can overlap one value of the expression type. The choice **others** is to cover the value(s) not covered in the previous alternatives; it should be the last alternative, if it exists, and must cover at least one value. The following VHDL code shows an example of a case statement.

```
1     package PACK is
2        type month_type is (JAN, FEB, MAR, APR, MAY, JUN,
3                            JUL, AUG, SEP, OCT, NOV, DEC);
4     end PACK;
5
6     use work.PACK.all;
7     entity CASESTMT is
8        port (
9           MONTH    : in  month_type;
10          LEAP     : in  boolean;
11          DAYS     : out integer);
```

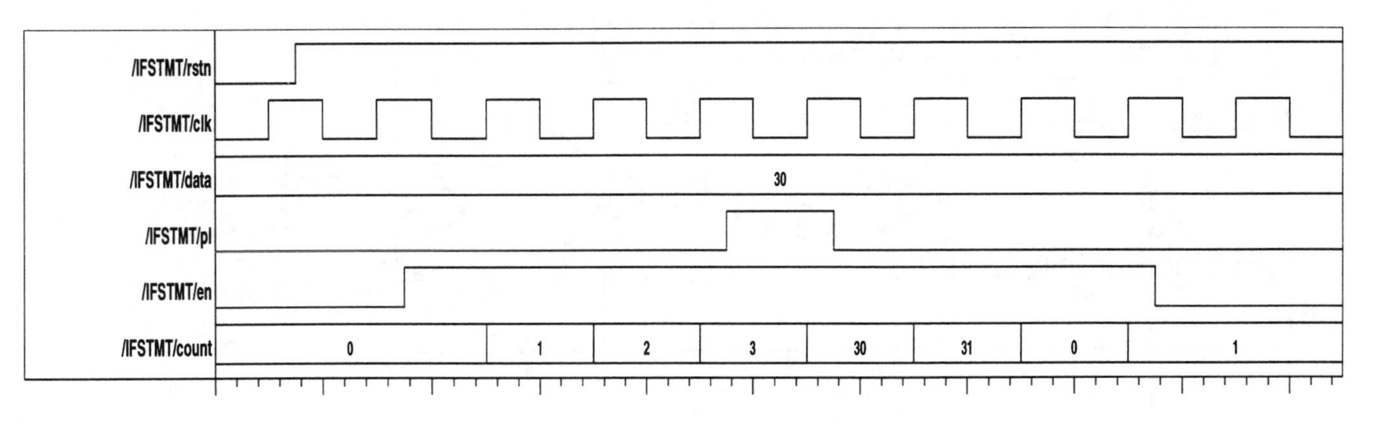

FIGURE 4.5 5-bit counter simulation waveform.

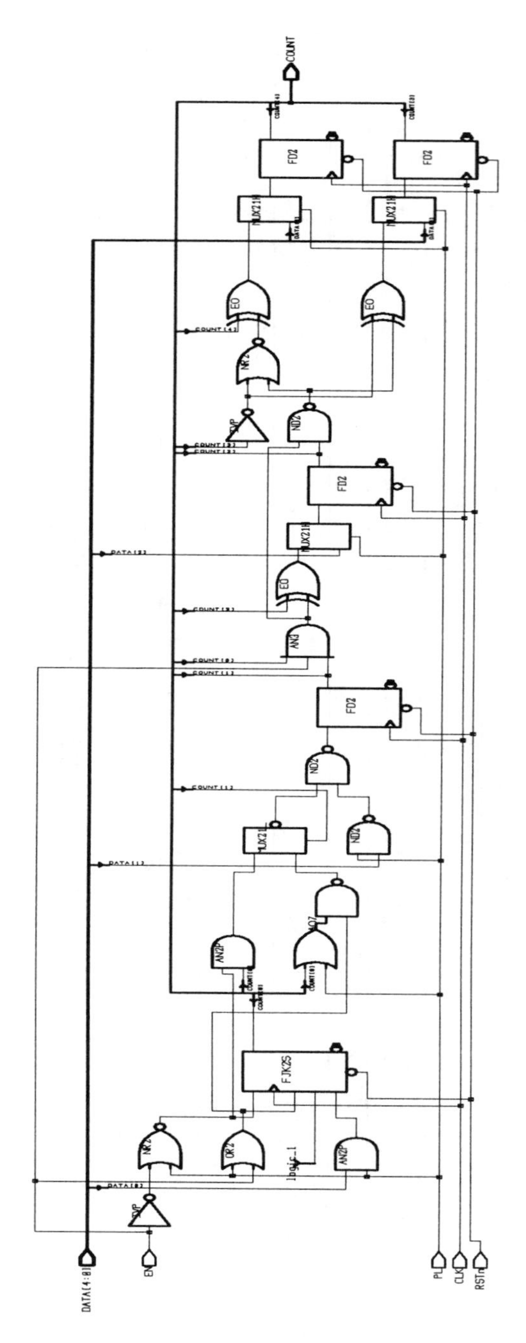

FIGURE 4.6 5-bit counter synthesized schematic.

```
12   end CASESTMT;
13
14   architecture RTL of CASESTMT is
15   begin
16      p0 : process (LEAP, MONTH)
17      begin
18         case MONTH is
19            when FEB =>
20               if LEAP then
21                  DAYS <= 29;
22               else
23                  DAYS <= 28;
24               end if;
25            when APR | JUN | SEP | NOV =>
26               DAYS <= 30;
27            when JUL to AUG =>
28               DAYS <= 31;
29            when others =>
30               DAYS <= 31;
31         end case;
32      end process;
33   end RTL;
```

The function of this VHDL code is to input the month, represented as three-letter codes, and a BOOLEAN input to indicate whether it is a leap year or not. The output shows how many days in that month. Lines 1 to 4 define a package with an enumerated month_type. This package is referenced in line 6 so that the port MONTH in the entity can use the month_type.

The preceding case statement example shows various ways of using the choices. Line 19 has a unique value as a choice. Line 25 shows a choice that covers four different values (vertical bar '|' indicates or). The purpose of line 27 is to illustrate a choice covering a continuous range of values (line 27 and 28 can be deleted and its choice can be covered by the **others** choice in line 29. The **others** choice appears as the last choice in line 29). Case statements are suitable for modeling finite state machines and microprocessor instructions. Figure 4.7 shows simulation waveform for the entity CASESTMT.

4.5 LOOP STATEMENT

loop_statement ::=
    [loop_label :] [**while** condition | **for** identifier **in** discrete_range] **loop**
        sequence of sequential statements
end loop [loop_label] ;

VHDL *loop statements* are similar to looping constructs in other high-level programming languages (DO statements in FORTRAN). The loop label is optional. When the loop label is specified after the keywords **end loop**, the loop label must be

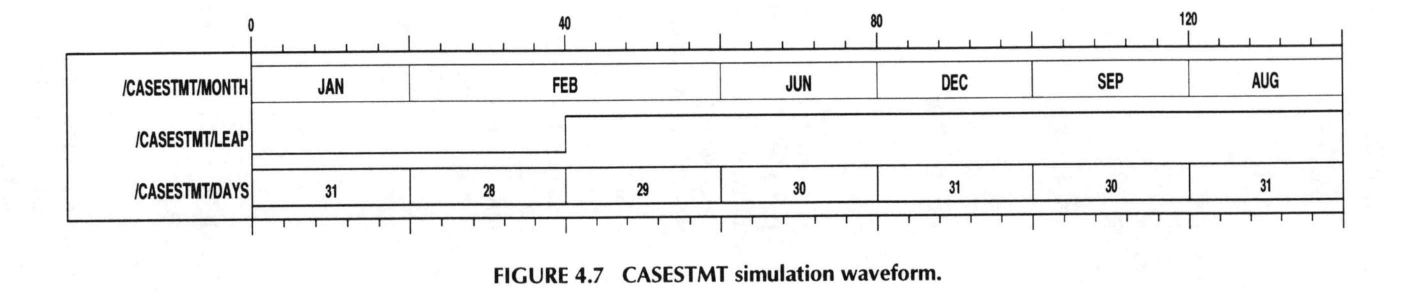

FIGURE 4.7 CASESTMT simulation waveform.

repeated in the beginning of the loop statement. The iteration scheme can be (1) **while** iteration scheme, (2) **for** iteration scheme, (3) no iteration scheme.

When the ***while iteration*** is used, the condition is first evaluated. If the condition is TRUE, the sequence of sequential statements is executed, otherwise, the loop statement is completed.

When the ***for*** *iteration scheme* is used, the identifier specifies the *looping parameter* with the base type of the discrete range. The looping parameter is used as a constant inside the scope of the loop statement, and it cannot be the target of an assignment statement. It is not visible outside the scope of the loop (not the same as in FORTRAN and C). The discrete range is first evaluated. If the range is a null range, the execution of the loop statement is complete. Prior to each iteration, the corresponding value of the discrete range is assigned to the looping parameter.

Besides the iteration scheme, the completion of the loop statement can also be a consequence of a exit, next, or return (in a function or a procedure) statement. They are often used when a loop statement without an iteration scheme specifies repeated execution of the sequence of the sequential statements.

The looping identifier need not be declared when the *for looping scheme* is used. The looping identifier is not visible outside of the loop statement. A common mistake is trying to use the looping identifier outside the loop statement. The VHDL analyzer would flag the looping identifier as not declared. The user then declares the looping identifier as a local variable and tries to use it outside of the loop statement. The following VHDL code shows such an example:

```
1     entity LOOPSTMT is
2     end LOOPSTMT;
3
4     architecture RTL of LOOPSTMT is
5        type arytype is array (0 to 9) of integer;
6        signal A : arytype := (1, 2, 3, 4, 11, 6, 7, 23, 9, 10);
7        signal TOTAL : integer := 0;
8     begin
9        p0 : process (A)
10          variable sum : integer := 0;
11          variable i   : integer := 20;
12       begin
13          sum := 0;
14          loop1 : for i in 0 to 9 loop
15             exit loop1 when A(i) > 20;
16             next when A(i) > 10;
17             sum := sum + A(i);
18          end loop loop1;
19          if i = 20 then
20             TOTAL <= -33;
21          else
22             TOTAL <= sum;
23          end if;
24       end process;
25    end RTL;
```

The above VHDL code declared local variable i (by mistake), which is used as the looping identifier. The example is used to demonstrate that the looping identifier i is not visible outside the loop statement, and the local variable i is not the same as the looping identifier i. The local variable is initialized to be 20. Lines 19 to 23 check the value of local variable i and assign the signal TOTAL. Figure 4.8 shows the simulation waveform with various values of array A. Note that local variable i is always 20, and the TOTAL is always –33. The index range "0 to 9" in line 14 can also be replaced by A'range attribute.

4.6 NEXT STATEMENT

next_statement ::= **next** [loop_label] [**when** condition] ;

A VHDL *next statement* is used to complete the execution of one iteration of an enclosing loop statement. If a loop label exists in the next statement, it must be enclosed by a loop statement with the same loop label, and the next statement applies to that loop statement. If the loop label is not specified, it always applies to the innermost level of the loop statements. For the execution of the next statement, the condition (if present) is first evaluated. The current iteration of the loop is terminated when the condition is TRUE, or if there is no condition. A next statement example is shown in line 16 of the preceding VHDL code.

4.7 EXIT STATEMENT

exit_statement ::= **exit** [loop_label] [**when** condition] ;

A VHDL *exit statement* is used to complete the execution of an enclosing loop statement. If the loop label exists in the exit statement, it must be enclosed by the loop statement with the same loop label, and the exit applies to that loop statement. If the loop label is not specified, the exit always applies to the innermost level of the loop statements. For the execution of the exit statement, the condition (if present) is first evaluated. The exit would take place when the condition is TRUE, or if there is no condition.

The following VHDL code shows examples of exit statements inside two nested loop statements. Line 4 declares a two-dimensional array type with five rows and four columns of integers. Lines 5 through 9 define a constant TABLE with the two-dimensional array type just declared. The VHDL code determines whether there are at least two rows of the TABLE such that the sum of each element of the row is greater than 10.

```
1    entity EXITSTMT is
2    end EXITSTMT;
3    architecture BEH of EXITSTMT is
4       type matrix is array (1 to 5, 1 to 4) of integer;
```

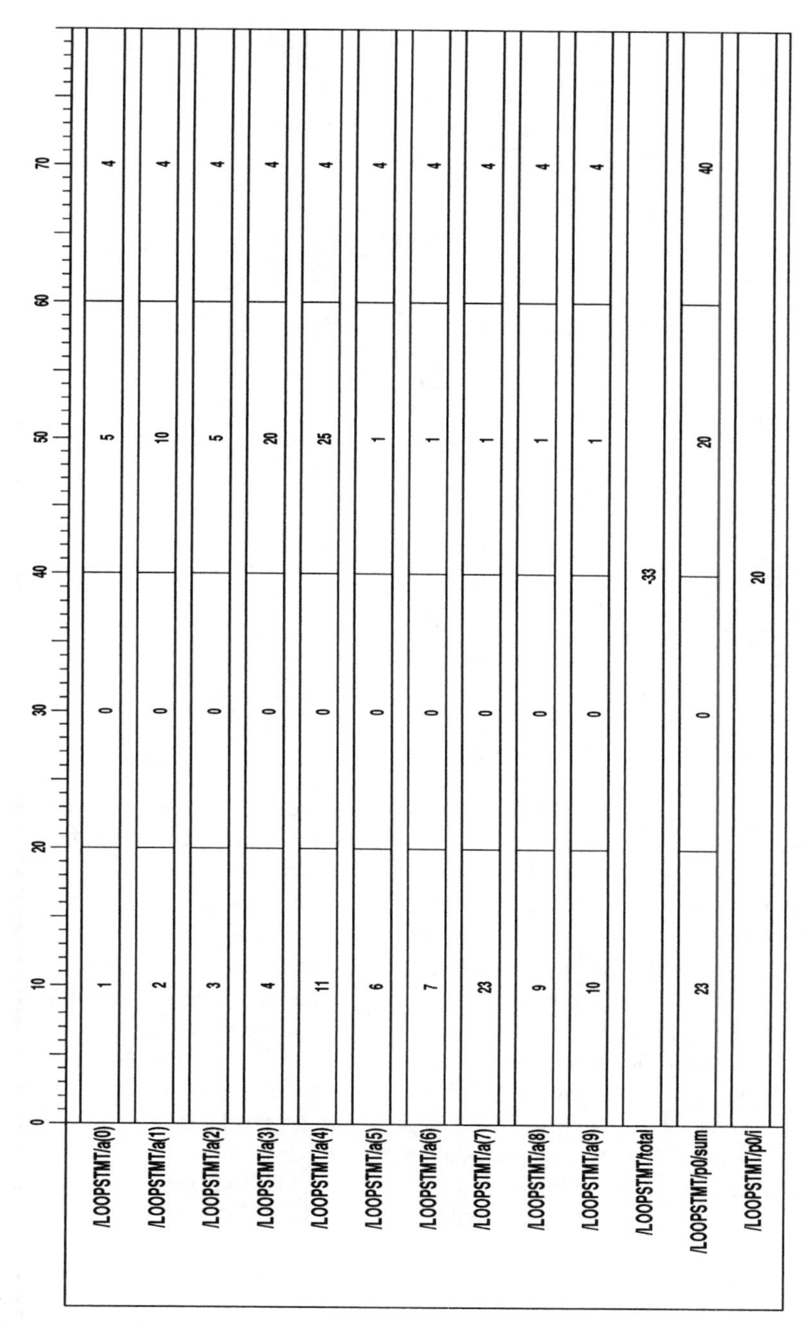

FIGURE 4.8 Simulation waveform for the LOOPSTMT.

```
5      constant TABLE : matrix := ( (1, 2, 3, 4),
6                                   (2, 8, 1, 0),
7                                   (8, 5, 3, 7),
8                                   (3, 0, 2, 1),
9                                   (1, 1, 0, 2) );
10   begin
11      p0 : process
12         variable NUMROW, ROWSUM   : integer := 0;
13         variable ROWDONE, ALLDONE : bit;
14      begin
15         ALLDONE := '0';
16         outloop : for i in matrix'range(1) loop
17            ROWSUM  := 0;
18            ROWDONE := '0';
19            inloop : for j in matrix'range(2) loop
20               ROWSUM := ROWSUM + TABLE (i, j);
21               if (ROWSUM > 10) then
22                  NUMROW := NUMROW + 1;
23                  exit outloop when NUMROW = 2;
24                  exit; -- get out of inloop
25               end if;
26               wait for 20 ns;
27            end loop inloop;
28            ROWDONE := '1';
29            wait for 20 ns;
30         end loop outloop;
31         ALLDONE := '1';
32         wait for 60 ns;
33      end process;
34   end BEH;
```

This is done with a *process statement* in lines 10 through 33. Note that there is no process sensitivity list after the reserved word **process** in line 10. Lines 11 and 12 declare variables. ROWNUM is used to indicate how many rows we have found which have sums greater than 10. ROWSUM is used to add each element of a row. It is set to 0 in line 17 before each row is calculated. ROWDONE is used to indicate that a row is done. ALLDONE is used to indicate that the result is obtained. Basically, there are two loop statements. The outer loop statement labeled outloop starts at line 16 and finishes at line 30. It goes through each row with the attribute matrix'range(1), which is 1 to 5. The inner loop statement starts at line 19 and finishes at line 27. It goes through each element of the row with attribute matrix'range(2), which is 1 to 4. Line 20 finds the sum. Lines 21 through 25 are an if statement to check whether the sum is greater than 10. If the number of rows having sums greater than 10 is two, the result is obtained. There is no need to continue. The exit statement in line 23 would jump out of the outer loop statement when NUMROW is equal to 2. Line 24 exits the row if the sum is already greater than 10. There is no need to add the rest of the element in the same row. The exit statement will exit the inner loop statement and continue to the next row.

To show the effect of the exit statements, *wait statements* (to be discussed in Section 4.12) are used so that 20 ns is used to examine each element. Figure 4.9 is a waveform from the simulation of the preceding VHDL code. From time 0 to 100 ns, the sum of the first row is 10. All five elements are examined. Variable ROWDONE is high from 80 ns to 100 ns. For row 2, only three elements (2, 8, 1) are examined, and the sum is already greater than 10. There is no need to continue on that row. Variable NUMROW is increased by 1. ROWDONE is high from time 140 ns to 160. For row 3, only two elements are necessary to be added to know if it is greater than 10. The number of rows that have sums greater than 10 is two. It jumps out of both loop statements. Variable ALLDONE is set to '1'. Note that variable ROWDONE is still '0' because line 28 is not executed. At time 240 ns, line 32 is completed. The execution continues to line 15. It resets variable ALLDONE to '0'. Variable NUMROW retains its value of 2 because it is only initialized when it is declared in line 11. It is not assigned by a variable assignment statement as is the variable ALLDONE in line 15.

4.8 NULL STATEMENT

null_statement ::= **null** ;

A VHDL *null statement* performs no action. It is used to explicitly specify that no action is to be performed. This is especially useful in the case statement when no action is required for a choice (all possible values of the expression must be covered by choices in the case statement). An example of a null statement is shown in line 15 of entity IFWAIT VHDL code in Section 4.12.

4.9 PROCEDURE CALL STATEMENT

A VHDL *procedure call statement* consists of the procedure name with arguments mapped (if exist) in the parenthesis. It is a statement while a *function call* appears in an expression. The following VHDL code shows a function declared in lines 4 to 11 and a procedure declared in lines 12 to 23. Line 31 shows a function call inside an expression, and line 32 shows a procedure call statement. Lines 34 to 40 show a process statement which assigns values to signals for testing the function and the procedure. Figure 4.10 shows the simulation waveform. We will discuss function and procedure in more detailed in a later chapter.

```
1     entity RETURNSTMT is
2     end RETURNSTMT;
3     architecture RTL of RETURNSTMT is
4        function BOOL2BIT (BOOL : in boolean) return bit is
5        begin
6           if BOOL then
7              return '1';
8           else
```

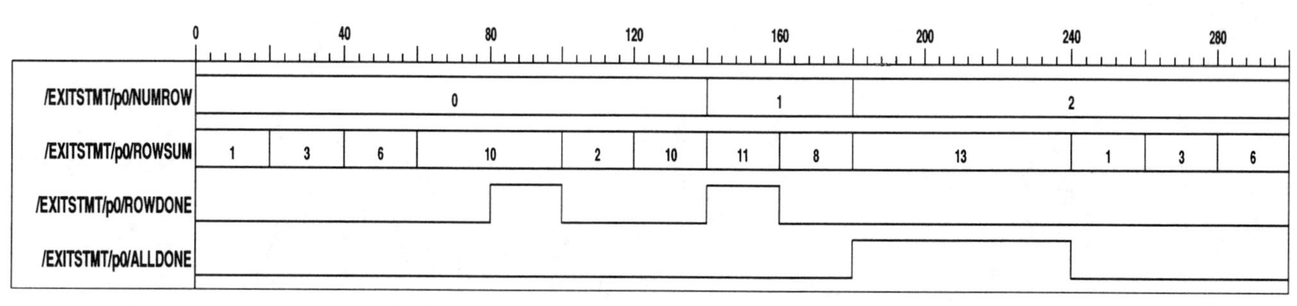

FIGURE 4.9 Simulation waveform for the EXITSTMT VHDL code.

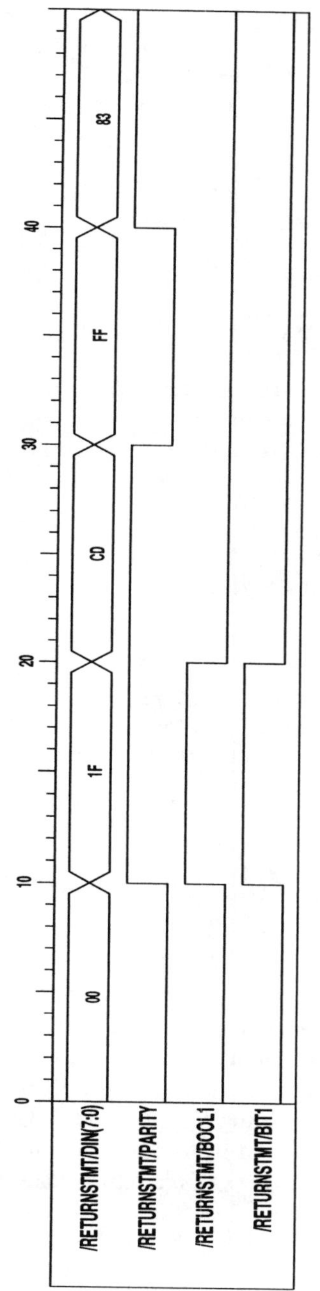

FIGURE 4.10 Simulation waveform for procedure and function.

```
9              return '0';
10     end if;
11        end BOOL2BIT;
12        procedure EVEN_PARITY (
13           signal DATA   : in  bit_vector(7 downto 0);
14           signal PARITY : out bit) is
15           variable temp : bit;
16        begin
17           temp := DATA(0);
18           for i in 1 to 7 loop
19              temp := temp xor DATA(i);
20           end loop;
21           PARITY <= temp;
22           return;
23        end EVEN_PARITY;
24
25        signal DIN    : bit_vector(7 downto 0);
26        signal BOOL1  : boolean;
27        signal BIT1, PARITY : bit;
28     begin
29        doit : process (BOOL1, DIN)
30        begin
31           BIT1 <= BOOL2BIT(BOOL1);
32           EVEN_PARITY(DIN, PARITY);
33        end process;
34        vector : process
35        begin
36           BOOL1 <= TRUE after 10 ns, FALSE after 20 ns;
37           DIN   <= "00011111" after 10 ns, "11001101" after 20 ns,
38                    "11111111" after 30 ns, "10000011" after 40 ns;
39           wait for 50 ns;
40        end process;
41     end RTL;
```

4.10 RETURN STATEMENT

A VHDL *return statement* is used to complete the execution of the innermost enclosing function or procedure body. It is allowed only within the body of a function or a procedure. A return statement in a function must have an expression that evaluates to the same base type as the type declared in the function after the reserved word **return** as shown in line 4 of the preceding VHDL code. Examples of return statements for a function are shown in lines 7 and 9. A return statement in a procedure must not have an expression (line 22). The syntax of the return statement is as follows:

return_statement ::= **return** [expression];

Note that a return statement is not required in a procedure body. Line 22 can be removed and the execution of the procedure body would be completed. However, more return statements can be used with other sequential statements (such as if, case statements) to control the return of the function or procedure.

4.11 ASSERTION STATEMENT

A VHDL *assertion statement* checks that a specified condition is TRUE and reports an error if the condition is FALSE.

assertion_statement ::= **assert** condition [**report** expression] [**severity** expression] ;

The report expression is an expression of a STRING type to be reported (usually on the screen by the simulator). A default value for the report expression is "Assertion violation" if the report clause is not specified. Severity expressions have the predefined type SEVERITY_LEVEL that specifies the severity level of the assertion. Recall that type SEVERITY_LEVEL is defined in the package STANDARD in Chapter 2. It is an enumerated type with four values NOTE, WARNING, ERROR, and FAILURE. The default severity level is ERROR if the severity clause is not specified. A VHDL simulator can terminate simulation, based on the user-specified SEVERITY_LEVEL (for example, FAILURE), when an assertion statement asserts a higher SEVERITY_LEVEL. The following VHDL code shows examples of assertion statements used in a simple D-type flipflop for checking setup and hold time violation:

```
1    entity ASRTSTMT is
2    end ASRTSTMT;
3    architecture BEH of ASRTSTMT is
4       constant SETUP : time := 3 ns;
5       constant HOLD  : time := 3 ns;
6       signal   D, CLOCK, Q  : bit;
7    begin
8       setup_check : process (CLOCK)
9       begin
10         assert FALSE report "Event on CLOCK" severity NOTE;
11         if (CLOCK'event and CLOCK = '1') then
12            assert D'STABLE(SETUP)
13               report "D setup error" severity WARNING;
14            if (D'STABLE(SETUP)) then
15               Q <= D;
16            end if;
17         end if;
18      end process;
19      hold_check : process
20      begin
21         wait on D;
22         assert FALSE report "Event on D" severity NOTE;
23         if (CLOCK = '1') then
```

```
24            assert (CLOCK'LAST_EVENT > HOLD)
25               report "D hold error" severity WARNING;
26         end if;
27      end process;
28      vector : process
29      begin
30         CLOCK <= '1' after 10 ns, '0' after 20 ns,
31                  '1' after 30 ns, '0' after 40 ns;
32         D     <= '1' after  8 ns, '0' after 12 ns,
33                  '1' after 15 ns;
34         wait;
35      end process;
36   end BEH;
```

Process setup_check (lines 8 to 18) checks the setup time violation. At the rising edge of the CLOCK (line 11), a signal ATTRIBUTE STABLE (D'STABLE) is used to check whether signal D is stable for the SETUP time. D'STABLE (SETUP) returns a BOOLEAN value. If D'STABLE(SETUP) returns FALSE, the assert statement in lines 12 and 13 will output the string "D setup error" by the VHDL simulator. When there is no setup timing violation, the value of D is clocked into Q as in lines 14 and 15. Line 10 will always cause the report string to be generated because FALSE is used as the assert expression.

Process hold_check (lines 19 to 27) checks the hold time violation. This is checked when CLOCK is '1' (line 23) and D has an event. Another signal ATTRIBUTE LAST_EVENT (CLOCK'LAST_EVENT in line 24) returns the time span, since the signal value was changed. At time 12 ns, CLOCK'LAST_EVENT would return 2 ns. The assert expression evaluates to FALSE. The report expression is generated.

Process vector (lines 28 to 35) assigns values to signal CLOCK and D.

Figure 4.11 shows the simulation waveform. Note that Q is not changed at time 10 ns due to setup timing violation. Q is not changed until at 30 ns. Some of the messages generated by the VHDL simulator, due to assert statements, are shown as follows. Note that each assert message contains the severity level, simulation time, design unit (with architecture name), process, and report expression from the assert statement.

Assertion NOTE at 0 NS in design unit ASRTSTMT(BEH) from process /ASRTSTMT/SETUP_CHECK: "Event on CLOCK"

Assertion NOTE at 8 NS in design unit ASRTSTMT(BEH) from process /ASRTSTMT/HOLD_CHECK: "Event on D"

Assertion NOTE at 10 NS in design unit ASRTSTMT(BEH) from process /ASRTSTMT/SETUP_CHECK: "Event on CLOCK"

Assertion WARNING at 10 NS in design unit ASRTSTMT(BEH) from process /ASRTSTMT/SETUP_CHECK: "D setup error"

Assertion NOTE at 12 NS in design unit ASRTSTMT(BEH) from process /ASRTSTMT/HOLD_CHECK: "Event on D"

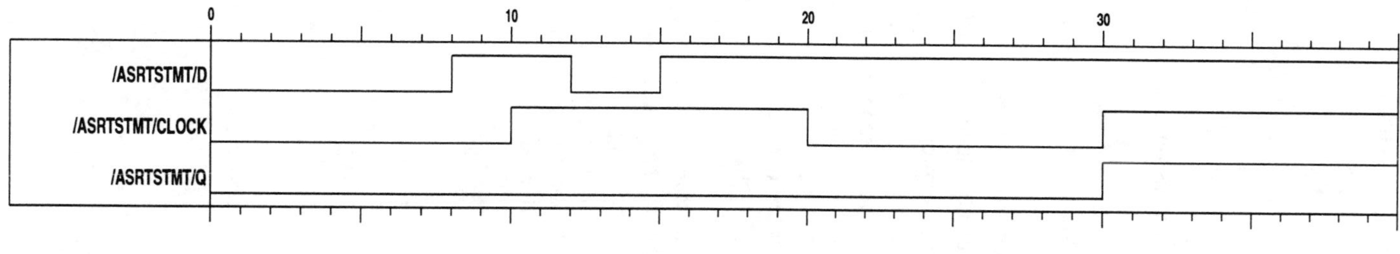

FIGURE 4.11 D flipflop simulation with setup and hold check.

Assertion WARNING at 12 NS in design unit ASRTSTMT(BEH) from process /ASRTSTMT/HOLD_CHECK: "D hold error"

Assertion NOTE at 15 NS in design unit ASRTSTMT(BEH) from process /ASRTSTMT/HOLD_CHECK: "Event on D"

4.12 WAIT STATEMENT

A *wait statement* causes the suspension of a process statement or a procedure. It is defined as follows:

> wait_statement ::= **wait** [sensitivity_clause] [condition_clause] [timeout_clause] ;
>
> sensitivity_clause ::= **on** sensitivity_list
>
> conditional_clause ::= **until** boolean_expression
>
> timeout_clause ::= **for** time_expression

The condition clause specifies a condition that must be met for the process to continue execution. If no condition clause appears, the condition clause **until FALSE** is assumed.

The timeout clause specifies the maximum amount of time the process will remain suspended at this wait statement. If no timeout clause appears, the timeout clause (STD.STANDARD.TIME'HIGH - STD.STANDARD.NOW) is used.

The execution of a wait statement causes the timeout expression to be evaluated to determine the timeout interval. The execution of the process statement is suspended. The suspended process will resume, *at the latest,* immediately after the timeout interval has expired. If an event occurs on any signal in the sensitivity list (before the timeout interval has expired), the condition clause is evaluated. If the value of the condition is TRUE, the process will resume. The following are examples of wait statements.

```
1     entity WAITSTMT is
2        port (
3           CLOCK         : in  bit;
4           A, B, C, D    : in  bit;
5           Q, W, X, Y, Z : out bit);
6     end WAITSTMT;
7
8     architecture BEH of WAITSTMT is
9        constant PERIOD : time := 30 ns;
10    begin
11       dff : process
12       begin
13          wait until CLOCK'event and CLOCK = '1';
14          Q <= D;
15       end process;
```

```
16        nand0 : process (A, B)
17        begin
18           W <= A nand B;
19        end process;
20        nand1 : process
21        begin
22           X <= A nand B;
23           wait on A, B;
24        end process;
25        more : process
26        begin
27           wait on A, B until (C or D) = '0' for 100 ns;
28           Y <= A or B;
29           wait until A = '0' for PERIOD;
30           wait on C for 50 * PERIOD;
31           wait on A until B = '1';
32           Z <= A or (C and D);
33           wait for 2 * PERIOD;
34        end process;
35        forever : process
36        begin
37           -- do something
38           if (A and B and C and D) = '1' then
39              wait;
40           else
41              wait for PERIOD;
42           end if;
43        end process;
44     end BEH;
```

Process dff (lines 11 to 15) can be used to model a D-type flipflop. The process is waiting for the event of the CLOCK and the value of the CLOCK is '1' (rising edge of the clock). We can say that Q gets the value of D when the rising edge of the CLOCK occurs.

Process nand0 (lines 16 to 19) and process nand1 (lines 20 to 24) are exactly the same if we make W and X the same. Process nand0 has a sensitivity list of A and B inside the parenthesis after the keyword **process** in line 16. Process nand1 has a sensitivity list in the wait statement in line 23. Due to the fact that every process would be evaluated until it is suspended in the initialization phase by the VHDL simulator, they are identical. If we exchange lines 22 and 23, they would not be the same during the initialization phase.

Process more (lines 25 to 34) illustrates more examples of wait statements. Note that more than one wait statement can be inside the same process statement. It is also important to remember that there must be a way to suspend a process to avoid an infinite simulation loop. For a process statement, we can have either a sensitivity list after the keyword **process** (inside the parenthesis) or a wait statement, *but not both.*

The wait statement in line 39 has no sensitivity clause, condition clause, or timeout clause. The process would be suspended forever when line 39 is executed. It can be used when the execution of a process is no longer needed.

It is also important to be careful when wait statements are used in different conditions (alternatives) of if (or case) statements. As shown in the following VHDL code, line 15 is a null statement. There is no wait when SEL(1) = '1'. The process will run forever when this condition is TRUE in the if statement.

```
1    entity IFWAIT is
2    end IFWAIT;
3    ----------------------------------------------
4    architecture RTL of IFWAIT is
5       signal CLK : bit;
6       signal SEL : bit_vector(1 downto 0);
7    begin
8       p0 : process
9       begin
10         if SEL = "00" then
11            wait until CLK'event and CLK = '1';
12         elsif SEL = "01" then
13            wait for 60 ns;
14         else
15            null;
16         end if;
17      end process;
18   end RTL;
```

4.13 EXERCISES

4.1 There is a tendency to overuse loop statements. Can you write a VHDL code for a 32-bit shift-left register without a loop statement that has input ports RSTn, CLK, PL, EN, DATA (32 bits), SI, and an output port SO which is the output of the left-most bit of the register? When asynchronous reset RSTn = '0', the register is initialized to all '0'. When the rising edge of the clock, if PL = '1', DATA is parallel loaded to the register. If EN = '1', the shift-left operation is performed (the left-most bit is dropped and SI goes to the right-most bit of the register). PL has higher priority than EN. Simulate your VHDL code.

4.2 Write VHDL code to have input DIN and output DOUT with 8-bits each. DOUT is the 2's complement of DIN. Hint: scan through from LSB (Least Significant Bit), DOUT(i) is the same as DIN(i) until '1' is encountered, after that DOUT(i) <= not DIN(i). Simulate your VHDL code.

4.3 In the IFSTMT VHDL code, COUNT_VALUE is declared as a signal. What changes are required to declare COUNT_VALUE as a local variable and still retain the same behavior? Simulate your updated VHDL code and synthesize.

4.4 In PCI (Peripheral Components Interconnect), even parity is used across the address data bus AD (32 bits) and command byte enable CBEn (4 bits). Write VHDL code to have AD and CBEn as inputs and PARITY as output. Simulate your VHDL code.

4.5 In the WAITSTMT VHDL code, what would happen to the process labeled *forever* if lines 38, 40, 41, and 42 are deleted?

4.6 Write a process statement to generate a clock signal with a period of 30 ns and 50 percent duty cycle.

4.7 In VHDL, where can a local variable be declared?

4.8 What is the reason that the right-hand side of the variable assignment statement does not take a waveform that has associated time expression?

4.9 What are the differences and similarities between a local variable and a signal in VHDL?

4.10 Can you name all the VHDL sequential statements? Where do they appear?

4.11 Which sequential statement may be labeled? Is this optional or required?

4.12 Can a process have a sensitivity list and a wait statement inside the process statement body? Why?

4.13 Can a process have no sensitivity list and no wait statement inside the process statement body? Why and what would happen? Can you give an example?

4.14 Write a process statement with more than one wait statement in the process statement body.

4.15 In other high-level programming languages (FORTRAN, C, PASCAL), is an object such as an array or an integer similar to a local variable or a signal? It seems that we can translate a C function to a VHDL process statement with objects replaced by variables. Discuss the possibility, restrictions, and challenges.

4.16 The following VHDL code models a JK flipflop. Note that Q is an output port which cannot be read (evaluated). Which architecture(s) implement(s) the correct function of a JK flipflop? Verify your answers and reasons by simulating the code.

```
1    entity JKFF is
2       port (
3          CLK, RSTn, J, K : in  bit;
4          Q               : out bit);
5    end JKFF;
6    ----------------------------------------------
7    architecture RTL of JKFF is
8       signal FF : bit;
9    begin
10      process (CLK, RSTn)
11         variable JK : bit_vector(1 downto 0);
12      begin
13         if (RSTn = '0') then
14            FF <= '0';
15         elsif (CLK'event and CLK = '1') then
16            JK := J & K;
17            case JK is
18               when "01" => FF <= '0';
19               when "10" => FF <= '1';
20               when "11" => FF <= not FF;
21               when "00" => FF <= FF;
22            end case;
23         end if;
24      end process;
25      Q <= FF;
26   end RTL;
```

```
27   ------------------------------------------------
28   architecture RTL1 of JKFF is
29      signal FF : bit;
30   begin
31      process (CLK, RSTn)
32      begin
33         if (RSTn = '0') then
34            FF <= '0';
35            Q  <= '0';
36         elsif (CLK'event and CLK = '1') then
37            if (J = '1') and (K = '1') then
38               FF <= not FF;
39            elsif (J = '1') and (K = '0') then
40               FF <= '1';
41            elsif (J = '0') and (K = '1') then
42               FF <= '0';
43            end if;
44         end if;
45         Q <= FF;
46      end process;
47   end RTL1;
48   ------------------------------------------------
49   architecture RTL2 of JKFF is
50      signal FF : bit;
51   begin
52      process (CLK, RSTn)
53      begin
54         if (RSTn = '0') then
55            FF <= '0';
56            Q  <= '0';
57         elsif (CLK'event and CLK = '1') then
58            if (J = '1') and (K = '1') then
59               FF <= not FF;
60            elsif (J = '1') and (K = '0') then
61               FF <= '1';
62            elsif (J = '0') and (K = '1') then
63               FF <= '0';
64            end if;
65            Q <= FF;
66         end if;
67      end process;
68   end RTL2;
69   ------------------------------------------------
70   architecture RTL3 of JKFF is
71   begin
72      process (CLK, RSTn)
73         variable FF : bit;
74      begin
75         if (RSTn = '0') then
76            FF := '0';
```

```
77              Q  <= '0';
78           elsif (CLK'event and CLK = '1') then
79              if (J = '1') and (K = '1') then
80                 FF := not FF;
81              elsif (J = '1') and (K = '0') then
82                 FF := '1';
83              elsif (J = '0') and (K = '1') then
84                 FF := '0';
85              end if;
86              Q <= FF;
87           end if;
88        end process;
89     end RTL3;
```

4.17 The architecture RTL of JKFF declares a variable JK in line 11 of 2 bits. Local variable JK is assigned as J & K (concatenation of J and K) in line 16. The case statement in line 17 uses JK. Can we just have "case J & K is" in line 17 so that the local variable can be avoided? We can define a subtype of BIT2 as "subtype BIT2 is bit_vector(1 downto 0);". Can we then replace line 17 as "case BIT2'(J & K) is" or "case BIT2(J & K) is"? Which one is correct? Explain.

4.18 Write an entity and architecture VHDL code to have a 64-bit input bus DIN and an output 64-bit bus DOUT such that DOUT is in reversing order of DIN (the left-most bit of DOUT is the right-most bit of DIN).

4.19 To improve execution efficiency, the computation inside a loop statement is usually optimized. For example, computation which is independent of the looping variable can be taken out of the loop. How can we take advantage of this concept and modify the following VHDL code to be more efficient? Compare the number of computations (such as the number of additions and multiplications).

```
1      package LOOPPACK is
2         type arytype is array (0 to 9) of integer;
3      end LOOPPACK;
4
5      use work.LOOPPACK.all;
6      entity LOOPCONS is
7         port (
8            A, B  : in  arytype;
9            C, D  : in  integer;
10           SUM   : out integer);
11     end LOOPCONS;
12
13     architecture RTL of LOOPCONS is
14     begin
15        p0 : process (A, B, C, D)
16           variable total : integer;
17        begin
18           total := 0;
19           for i in A'range loop
20              total := total + A(i) * B(i) + C * D;
```

```
21          end loop;
22          SUM <= total;
23       end process;
24    end RTL;
```

Chapter 5

Concurrent Statements

VHDL *concurrent statements* define a circuit's concurrent behavior. The order of their execution is not affected by their order of appearance inside the *architecture body.* They are activated by the event on signals that are used to communicate among concurrent statements. VHDL concurrent statements are :

process statement
assertion statement
procedure call statement
conditional signal assignment statement
selected signal assignment statement
component instantiation statement
generate statement
block statement

5.1 PROCESS STATEMENT

process_statement ::=
[process_label :] **process** [(sensitivity_list)]
process_declaration_part
begin
sequence of sequential statements
end process [process_label] ;

A process statement is a concurrent statement that defines an independent sequential behavior of some portion of the design described by a sequence of sequential statements. A process label is optional. If the optional process label exists after the end process, the process label should be repeated in the beginning of the process statement.

Note that the subprogram body, type declaration, subtype declaration, constant declaration, variable declaration, file declaration, alias declaration, attribute declaration, attribute specification, and use clause can be in the process declaration part. Signals cannot be declared inside a process. Refer to Appendix B, declaration part table, and compare with other declaration parts.

As with the sensitivity list and the wait statement in the last chapter, the process statement can have *either* a sensitivity list (after the keyword **process**) or wait statement(s) inside the process statement body *but not both.* When the sensitivity list exists, it is assumed to contain an implicit wait statement with a sensitivity clause as the last sequential statement in the process. In the beginning of the simulation, the process is executed until it is suspended (when either an implicit or an explicit wait statement is reached). The execution of a process statement consists of the repetitive execution of a sequence of sequential statements. After the last sequential statement in the process is executed, execution will immediately continue with the first sequential statement in the sequence.

Signal assignment statements in a process statement define a set of drivers for signals. They affect the transactions (value and time) of the signal drivers. The values of the signals are then determined by its drivers and are updated when all concurrent processes are suspended.

5.2 ASSERTION STATEMENT

concurrent_assertion_statement ::= [label :] assertion_statement

The syntax of a concurrent assertion statement is the same as a sequential assertion statement with an optional label. A concurrent assertion statement is equivalent to a passive process statement containing a specified assertion statement (which is a sequential statement inside a process). It would never cause an event to occur when the assertion statement is executed. If the condition is defined as a static expression, it is equivalent to having a process statement that ends with a wait statement with no sensitivity clause, no conditional clause, and no timeout clause; such a process will be executed once in the beginning of the simulation and wait forever. The following VHDL code shows examples of concurrent assertion statements.

```
1     entity ASRTSTMT is
2     end ASRTSTMT;
3     architecture BEH of ASRTSTMT is
4        signal S, T, Y, Z  : bit;
5     begin
6        p0 : process (S, T)
7        begin
8           Y <= S nand T;
9        end process;
10       p1 : process (S, T)
11       begin
12          if (S = '1' and T = '1') then
13             Z <= '0';
14          else
15             Z <= '1';
16          end if;
17       end process;
```

```
18      asrt0: assert Y = Z report "Y and Z not equal"
19               severity NOTE;
20      asrt1 : process (Y, Z)
21      begin
22         assert FALSE report "compare Y and Z" severity NOTE;
23         assert Y = Z report "Y and Z not equal"
24            severity NOTE;
25      end process;
26      asrt2 : assert FALSE
27              report "start to compare Y and Z (do only once)"
28                severity NOTE;
29   end BEH;
```

Both processes p0 and p1 perform BOOLEAN NAND operation on signals S and T to get signals Y and Z. Values of Y and Z should always be identical. Lines 18 and 19 show a concurrent assertion statement to verify Y and Z. It is equivalent to a process statement (lines 20 to 25) with Y and Z as the process sensitivity list (except that line 22 is used to indicate that the process is executed). Lines 26 to 28 illustrate a concurrent assertion statement with a static condition which is executed exactly only once in the beginning of the simulation. Figure 5.1 shows the simulation waveform. Note that waveforms of Y and Z are identical.

The following messages were generated from the assertion statements:

Assertion NOTE at 0 NS in design unit ASRTSTMT(BEH) from process /ASRTSTMT/ASRT2: "start to compare Y and Z (do only once)"
Assertion NOTE at 0 NS in design unit ASRTSTMT(BEH) from process /ASRTSTMT/ASRT1: "compare Y and Z"
Assertion NOTE at 0 NS in design unit ASRTSTMT(BEH) from process /ASRTSTMT/ASRT1: "compare Y and Z"
Assertion NOTE at 40 NS in design unit ASRTSTMT(BEH) from process /ASRTSTMT/ASRT1: "compare Y and Z"
Assertion NOTE at 60 NS in design unit ASRTSTMT(BEH) from process /ASRTSTMT/ASRT1: "compare Y and Z"

Note that the concurrent assertion statement in lines 26 to 28 has a static BOOLEAN expression FALSE. It will be executed only once, generating a message only once as shown.

Note that the idea of using assertion statements to verify (to compare) two signals that are generated from different sources can be expanded. For example, in the design of a multiplier with A, B as 16-bit inputs, and C as a 32-bit output, we can implement a Booth and Wallace tree multiplier architecture. To verify the correctness of its implementation, we can write a simple signal assignment statement as D <= E * F, where E and F are integers converted from A and B, respectively. The assertion can then be done by comparing C with the converted 32-bit value from integer D. Values of A and B can be set up by a process statement with loop statements to generate various combinations (or randomly with special cases). The verification can then be reduced to pass and fail by looking at the output of the assertion statement, without

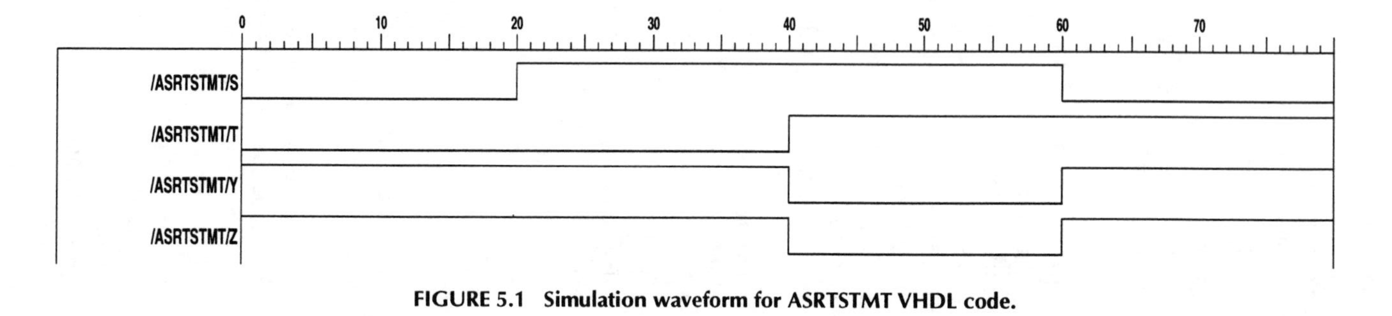

FIGURE 5.1 Simulation waveform for ASRTSTMT VHDL code.

looking at the waveform and using a calculator. The test bench example will be further discussed in Chapter 12.

5.3 CONCURRENT PROCEDURE CALL STATEMENT

concurrent_procedure_call ::= [label :] procedure_call_statement

A *concurrent procedure call statement* represents a process containing the corresponding sequential procedure call statement. Its execution is equivalent to a process statement. If the procedure has formal parameter(s) of mode **in** or **inout** signal(s), it is equivalent to a process statement (no sensitivity list after the keyword **process**), and ends with a wait statement with a sensitivity clause with those signals in the actual parameter (must be static signals). If there is no formal parameter of mode **in** or **inout**, the equivalent process statement has a wait statement with no sensitivity clause, condition clause, or timeout clause as the last sequential statement. In other words, it will be executed exactly only once in the beginning of the simulation and suspended forever.

Each formal parameter of a procedure that is invoked by a concurrent procedure call must be of type constant or signal. Local variables cannot be used since a concurrent procedure call statement appears in the concurrent statements region where local variables are declared inside a process statement and not inside the architecture declaration part.

Concurrent procedure call statements make it possible to declare and create commonly used processes. This reduces the amount of coding and may increase readability. The following VHDL code shows concurrent procedure call examples:

```
1     entity PROCALL is
2        port (
3           DIN  : in  bit_vector(7 downto 0);
4           DOUT : out bit_vector(1 downto 0));
5     end PROCALL;
6     architecture RTL of PROCALL is
7        procedure ANDOR (
8           signal A, B, C, D : in  bit;
9           signal Y          : out bit) is
10       begin
11          Y <= (A and B) or (C and D);
12       end ANDOR;
13    begin
14       ANDOR (A => DIN(7), B => DIN(6), C => DIN(5), D => DIN(4),
15              Y => DOUT(1));
16       label0 : ANDOR (DIN(3), DIN(2), DIN(1), DIN(0), DOUT(0));
17    end RTL;
```

The procedure body ANDOR is declared in lines 7 to 12. Lines 14 and 15 are a concurrent procedure call with name mapping. Line 16 shows an optional label with positional mapping. Figure 5.2 shows the synthesized schematic.

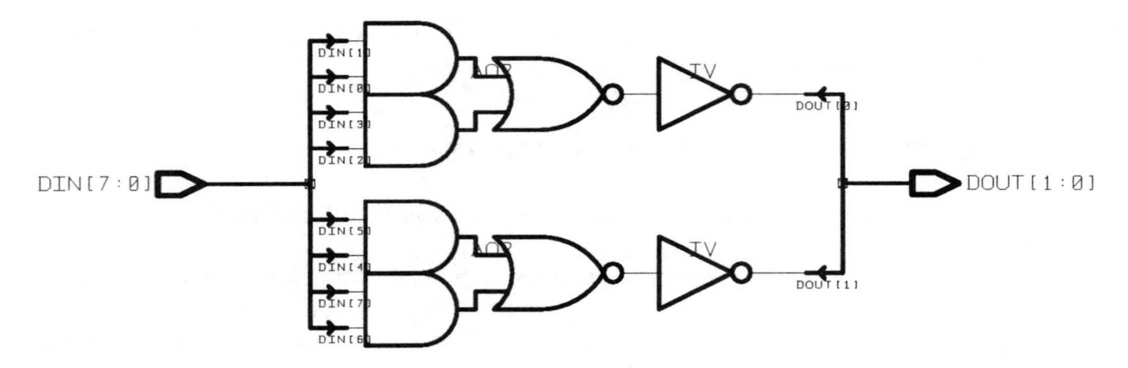

FIGURE 5.2 Synthesized schematic of procedure calls.

The preceding VHDL code shows only the procedure body declared inside the architecture declaration part. There is no procedure declaration specified. Subprogram declarations can be specified inside a package declaration and subprogram bodies can also be specified inside a package body. Procedure declaration and procedure body will be discussed in more detail in Chapter 6.

5.4 CONDITIONAL SIGNAL ASSIGNMENT STATEMENT

In Chapter 4, we discussed the sequential signal assignment statement. Here, we will present VHDL concurrent signal assignment statements that are equivalent to process statements assigning values to signals.

concurrent_signal_assignment_statement ::=
[label :] conditional_signal_assignment | [label :] selected_signal_assignment

There are two forms of the concurrent signal assignment statements. For each form, one or both of the options **guarded** and **transport** can be used. The guarded option will be discussed in Chapter 10. The concept of the transport delay and the inertial delay was discussed in Chapter 4. In this section, we will concentrate on the conditional signal assignment statement. The selected signal assignment statement will be discussed in the next section.

conditional_signal_assignment ::=
[**guarded**] [**transport**] { waveform **when** condition **else** } waveform ;

The conditional signal assignment statement represents a process statement in which the signal transform is an if statement inside a process, and the process has the sensitivity list with all signals appearing on the right-hand side of the assignment statement. The following VHDL code shows architecture RTL1 using a conditional signal assignment statement which is a concurrent statement. Architecture RTL2 uses a sequential if statement inside a process statement. Note that the sensitivity list (line

15) should include all signals on the right-hand side of the conditional signal assignment statement in lines 8 to 11. In general, the concurrent conditional signal assignment statement requires significantly fewer lines of code compared to the equivalent process statement. Another advantage of the concurrent statements is that there is no need to specify the sensitivity list. For a process statement, if any of the signal is missing in the sensitivity list, the behavior would not be the same as expected. It may be hard to catch the error (especially when there are many signals in the sensitivity list) without thorough testing with various combinations of signal values.

```
1     entity CONDSTMT is
2        port (
3           A, B, C, D, SEL : in  bit_vector(1 downto 0);
4           Y               : out bit_vector(1 downto 0));
5     end CONDSTMT;
6     architecture RTL1 of CONDSTMT is
7     begin
8        Y <= A when SEL = "00" else
9             B when SEL = "01" else
10            C when SEL = "10" else
11            D;
12    end RTL1;
13    architecture RTL2 of CONDSTMT is
14    begin
15       process (A, B, C, D, SEL)
16       begin
17          if (SEL = "00") then
18             Y <= A;
19          elsif (SEL = "01") then
20             Y <= B;
21          elsif (SEL = "10") then
22             Y <= C;
23          else
24             Y <= B;
25          end if;
26       end process;
27    end RTL2;
```

5.5 SELECTED SIGNAL ASSIGNMENT STATEMENT

selected_signal_assignment ::=
  with expression **select**
    target <= [**guarded**] [**transport**] { waveform when choices, } waveform choices

The selected signal assignment statement represents an equivalent process statement with a sequential case statement inside and all the signals on the right-hand side of the selected signal assignment statement as the process sensitivity list. The rule for the choice and use of the keyword **others** in a selected signal assignment statement is the same as the sequential case statement. The following VHDL code shows a selected signal assignment statement (lines 8 to 12) and its equivalent process statement (lines 16 to 28). Again, using a concurrent statement reduces the amount of coding and avoids having to check if the right signals are in the sensitivity list.

```
1    entity SELSTMT is
2       port (
3          A, B, C, D, SEL : in  bit_vector(1 downto 0);
4          Y               : out bit_vector(1 downto 0));
5    end SELSTMT;
6    architecture RTL1 of SELSTMT is
7    begin
8       with SEL select
9       Y <= A when "00",
10           B when "01",
11           C when "10",
12           D when others;
13   end RTL1;
14   architecture RTL2 of SELSTMT is
15   begin
16      process (A, B, C, D, SEL)
17      begin
18         case SEL is
19            when "00"   =>
20               Y <= A;
21            when "01"   =>
22               Y <= B;
23            when "10"   =>
24               Y <= C;
25            when others =>
26               Y <= D;
27         end case;
28      end process;
29   end RTL2;
```

5.6 COMPONENT INSTANTIATION STATEMENT

component_instantiation_statement ::=
instantiation_label : component name [generic_map_aspect]
[port_map_aspect] ;

The component instantiation statement can be used to imply the structural organization of a hardware design. This is similar to the hierarchy of a schematic. The subcomponents can be instantiated (with the component instantiation statement) and interconnected by signals. Note that each component instantiation statement requires a label which is not optional. We will discuss optional generic mapping in Chapter 7. The following VHDL code shows examples of component instantiations:

```
1    entity DFF is
2       port (
3          RSTn, CLK, D : in  bit;
4          Q            : out bit);
5    end DFF;
6    architecture RTL of DFF is
7    begin
8       process (RSTn, CLK)
9       begin
10         if (RSTn = '0') then
11            Q <= '0';
12         elsif (CLK'event and CLK = '1') then
13            Q <= D;
14         end if;
15      end process;
16   end RTL;
17   ---------------------------------------------
18   entity SHIFT is
19      port (
20         RSTn, CLK, SI : in  bit;
21         SO            : out bit);
22   end SHIFT;
23   architecture RTL1 of SHIFT is
24      component DFF
25      port (
26         RSTn, CLK, D : in  bit;
27         Q            : out bit);
28      end component;
29      signal T : bit_vector(6 downto 0);
30   begin
31      bit7 : DFF
32         port map (RSTn => RSTn, CLK => CLK, D => SI, Q => T(6));
33      bit6 : DFF
34         port map (RSTn, CLK, T(6), T(5));
35      bit5 : DFF
36         port map (RSTn, CLK, T(5), T(4));
37      bit4 : DFF
38         port map (CLK => CLK, RSTn => RSTn, D => T(4), Q => T(3));
39      bit3 : DFF
40         port map (RSTn, CLK, T(3), T(2));
41      bit2 : DFF
42         port map (RSTn, CLK, T(2), T(1));
43      bit1 : DFF
44         port map (RSTn, CLK, T(1), T(0));
45      bit0 : DFF
46         port map (RSTn, CLK, T(0), SO);
```

```
47    end RTL1;
48    ---------------------------------------------
```

This VHDL code shows an entity (DFF) and architecture of a D-type flipflop from lines 1 to 16. An 8-bit shift register entity is specified in lines 18 to 22. Lines 24 to 28 declare the component DFF. Line 29 declares a 7-bit signal T to be used to connect between adjacent flipflops. The DFF is instantiated eight times to get an 8-bit shift register. Note how the signals are connected (mapped). Each component instantiation statement has a label (such as bit7 in line 31). These labels share the similar concept as instance names in a schematic. The same components are used (instantiated) several times in a schematic. Note that the port mapping can be either name mapping (bit7 and bit4) or positional mapping (bit0, bit1, bit2, bit3, bit5, bit6). In the name mapping method, the position of the parameters are ignored. The schematic in Figure 5.3 shows the synthesized 8-bit shift register.

Suppose we would like to build a 24-bit shift register. We can modify the preceding VHDL code to instantiate DFF 24 times. Another alternative is to instantiate the 8-bit shift register three times so that another level of hierarchy is created, as shown in the following VHDL code:

```
1     entity SHIFT24 is
2        port (
3           RSTn, CLK, SI : in  bit;
4           SO            : out bit);
5     end SHIFT24;
6     architecture RTL5 of SHIFT24 is
7        component SHIFT
8        port (
9           RSTn, CLK, SI : in  bit;
10          SO            : out bit);
11       end component;
12       signal T1, T2 : bit;
13    begin
14       stage2 : SHIFT
15          port map (RSTn => RSTn, CLK => CLK, SI => SI, SO => T1);
16       stage1 : SHIFT
17          port map (RSTn => RSTn, CLK => CLK, SI => T1, SO => T2);
18       stage0 : SHIFT
19          port map (RSTn => RSTn, CLK => CLK, SI => T2, SO => SO);
20    end RTL5;
```

The component SHIFT is declared in lines 7 to 11 and instantiated three times in lines 14 to 19. Figure 5.4 shows the synthesized schematic. Note that the component can also be declared in a package which can be shared and referenced by a use clause in different places.

In component instantiation port mapping, all ports are required to have a mapping. However, keyword **open** can be used to map to a port of mode **out** to indicate

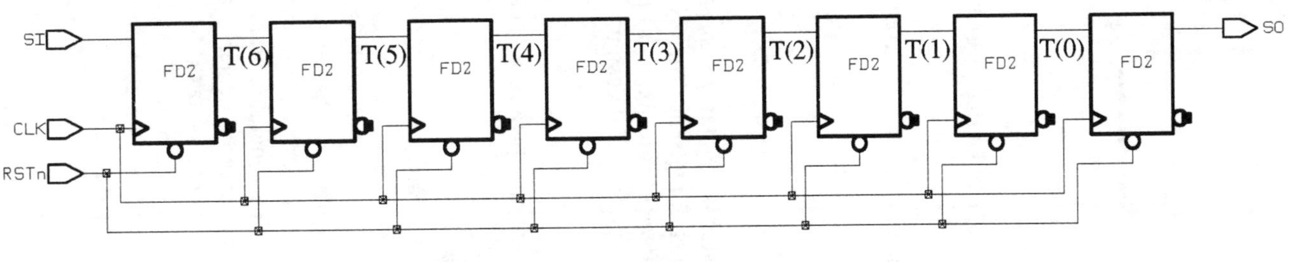

FIGURE 5.3 Synthesized schematic of an 8-bit shift register.

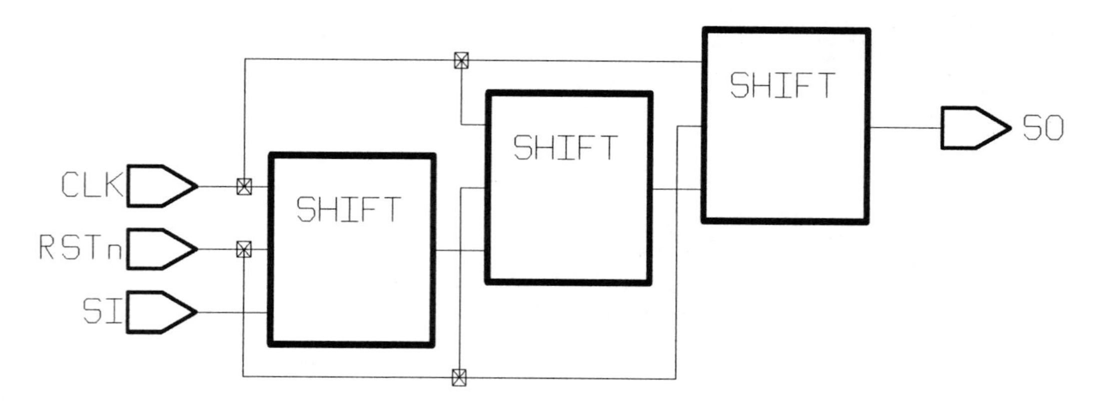

FIGURE 5.4 Synthesized schematic of a 24-bit shift register with three 8-bit shift registers.

that the output port of the subcomponent is not connected to this level and left unconnected.

5.7 GENERATE STATEMENT

The structure of a shift register is regular. As already discussed, we can instantiate DFF N times to make a N-bit shift register. When N is large, it may require more coding. The VHDL *generate statement* is designed to provide a mechanism for iterative or conditional elaboration of a portion of a design.

> generate_statement ::=
> generate_label : **for** generate_parameter_specification **generate** | **if** condition **generate**
> concurrent statements
> **end generate** [generate_label] ;

A generate statement requires a label while it is optional in the end of the statement after the keywords **end generate**. There are two schemes of using a generate statement. One is to use a **for** iteration scheme, and its syntax is the same as in the sequential **for loop** statement. The generate parameter is treated as a constant of the discrete range's base type and cannot be declared as a local variable or signal. The other scheme is the **if** condition whose syntax is the same as in the sequential **if** statement. Even the **for** iteration and **if** condition syntax in the generate statement are the same as the for iteration in the loop statement and if condition in the if statement, users need to notice the following differences:

1. Generate statement is a concurrent statement.
2. *If statements* and *for loop statements* are sequential statements.
3. The *if condition generate statement* does not have **else, elsif** clause.
4. A label is required for a generate statement.

5. Only concurrent statement(s) can appear inside a generate statement. Only sequential statement(s) can appear inside a sequential for loop statement and a sequential if statement.

The 8-bit shift register SHIFT discussed in the last section consists of regular structure. The following VHDL code uses generate statements with different architecture names to describe the same 8-bit shift register.

```
1     architecture RTL2 of SHIFT is
2        component DFF
3        port (
4           RSTn, CLK, D : in  bit;
5           Q            : out bit);
6        end component;
7        signal T : bit_vector(6 downto 0);
8     begin
9        g0 : for i in 7 downto 0 generate
10          g1 : if (i = 7) generate
11             bit7 : DFF
12               port map (RSTn => RSTn, CLK => CLK, D => SI, Q => T(6));
13          end generate;
14          g2 : if (i > 0) and (i < 7) generate
15             bitm : DFF
16                port map (RSTn, CLK, T(i), T(i-1));
17          end generate;
18          g3 : if (i = 0) generate
19             bit0 : DFF
20                port map (RSTn, CLK, T(0), SO);
21          end generate;
22       end generate;
23    end RTL2;
24    ---------------------------------------------
25    architecture RTL3 of SHIFT is
26       component DFF
27       port (
28          RSTn, CLK, D : in  bit;
29          Q            : out bit);
30       end component;
31       signal T : bit_vector(8 downto 0);
32    begin
33       T(8) <= SI;
34       SO   <= T(0);
35       g0 : for i in 7 downto 0 generate
36          allbit : DFF
37             port map (RSTn => RSTn, CLK => CLK, D => T(i+1), Q =>
                T(i));
38       end generate;
39    end RTL3;
```

Both architectures RTL2 (lines 2 to 6) and RTL3 (lines 26 to 30) declare component DFF. These component declarations can be declared in a package so that it can be referenced and shared among architecture RTL1, RTL2, and RTL3. Then these component declarations in architecture RTL1, RTL2, and RTL3 can be deleted. In architecture RTL2, the if generate statement is used inside a for generate statement. A component instantiation statement is inside the if generate statement. They (for generate, if generate, component instantiation) are all concurrent statements. In architecture RTL3, only a for generate statement is used. Note that signal T is 9 bits (7 bits in architecture RTL2). Lines 33 and 34 signal assignment statements make the SI (SO) the same as T(8) (T(0)). The generate parameter i does not need to be declared. All three architectures RTL1, RTL2, and RTL3 of entity SHIFT describe the same behavior and will synthesize to the same schematic.

The preceding examples show a component instantiation statement inside a generate statement and an (if) generate statement inside a (for) generate statement. Note that only concurrent statements are allowed inside a generate statement such as concurrent signal assignment statements, process statements, concurrent procedure calls, and so on.

5.8 BLOCK STATEMENT

block_statement ::= block_label : **block** [(guard_expression)]
    block_header
    block_declarative_part
begin
    concurrent statements
end block [block_label] ;

A block statement defines an internal block representing a portion of a design. The blocks can be hierarchical nested to support design decomposition. A block label is required in a block statement. If the optional label appears after the keywords **end block**, the label must be the same as the block label appearing in the beginning of the block statement before the keyword **block**. Subprogram declarations, subprogram bodies, types, subtypes, constants, signals, aliases, files, declarations of attribute and component, specifications of attribute and configuration, and use clauses are allowed in the block declaration part. Refer to Appendix B for comparisons with other declaration parts. Note that concurrent statements are required in the block statement body. The following VHDL code shows a simple example of a block statement:

```
1    entity BLOCKSTMT is
2       port (
3          RSTn, CLK, SI : in  bit;
4          SO            : out bit);
5    end BLOCKSTMT;
6    architecture RTL5 of BLOCKSTMT is
7       component SHIFT
8       port (
9          RSTn, CLK, SI : in  bit;
```

```
10          SO              : out bit);
11       end component;
12       signal T1, T2 : bit;
13    begin
14       stage2 : SHIFT
15          port map (RSTn => RSTn, CLK => CLK, SI => SI, SO => T1);
16       block0 : block
17       begin
18          stage1 : SHIFT
19             port map (RSTn => RSTn, CLK => CLK, SI => T1, SO => T2);
20          stage0 : SHIFT
21             port map (RSTn => RSTn, CLK => CLK, SI => T2, SO => SO);
22       end block;
23    end RTL5;
```

The preceding VHDL code is exactly the same as a 24-bit shift register SHIFT24, except that we have a block statement (lines 16 to 22) enclosing two component instantiation statements. A hierarchy can then be created, as shown in Figure 5.5. Block0 contains two stages of SHIFT as shown in Figure 5.6.

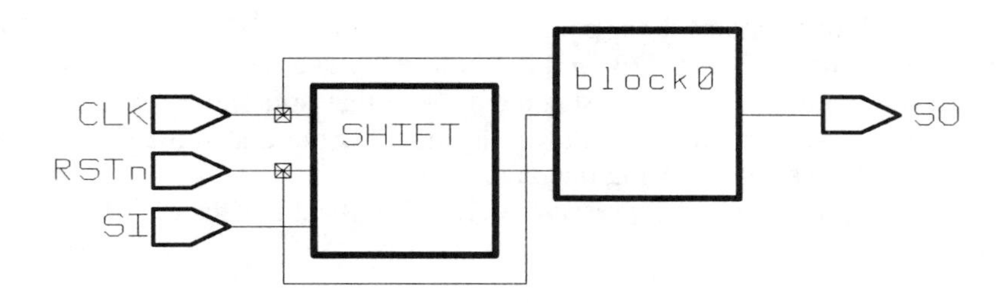

FIGURE 5.5 Synthesized schematic with a block statement.

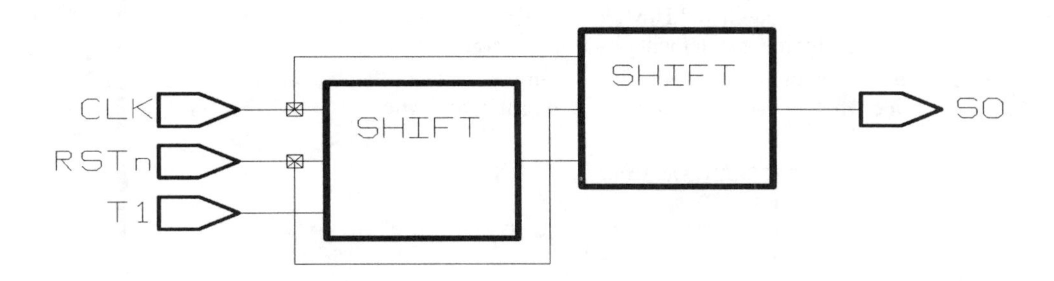

FIGURE 5.6 Schematic for the subcomponent block0.

The block header consists of a port clause (similar to the entity port clause) and a port map clause (same as in the component instantiation statement). They are used to allow local signals inside the block statement and then mapped to the signals connecting to the block statement. The following VHDL code shows such an example:

```
1    entity BKHEAD is
2       port (
3          DA, DB   : in  bit_vector(3 downto 0);
4          DZ       : out bit_vector(3 downto 0));
5    end BKHEAD;
6    architecture BEH of BKHEAD is
7    begin
8       gen4 : for i in 3 downto 0 generate
9          block0 : block
10            port (A, B : in bit; Z : out bit);
11            port map (A => DA(i), B => DB(i), Z => DZ(i));
12         begin
13            Z <= A nand B;
14         end block;
15      end generate;
16   end BEH;
```

The block statement has port signals A, B, and Z as declared in line 10. They are mapped to one bit of DA, DB, and DZ in line 11. The (concurrent) block statement is inside a (concurrent) generate statement. Note that port signals A, B, and Z are used inside the block statement, as shown in line 13. Note that some synthesis tools may have restrictions against using the block header.

The optional guard expression will be discussed in Chapter 10.

5.9 EXERCISES

5.1 What are VHDL concurrent statements?

5.2 Which concurrent statement requires concurrent statements inside?

5.3 Which concurrent statement requires sequential statements inside?

5.4 "S <= A;" is a signal assignment statement. Could it be a sequential statement? Could it be a concurrent statement? How do we tell? If it is a concurrent statement, is it a conditional signal assignment statement or a selected signal assignment statement?

5.5 The following VHDL code shows four examples of the concurrent procedure call. Determine whether each procedure is valid or not, and why.

```
1    entity PROCALL_EX is
2    end PROCALL_EX;
3    architecture RTL of PROCALL_EX is
4       procedure ANDOR (
5          signal A, B, C, D : in  bit_vector(1 downto 0);
6          signal Y          : out bit_vector(1 downto 0)) is
```

```
7        begin
8           Y <= (A and B) or (C and D);
9        end ANDOR;
10       signal DIN, DOUT : bit_vector(7 downto 0);
11       signal X, Y, Z   : bit_vector(1 downto 0);
12    begin
13       call0 : ANDOR (A => DIN(7) & DIN(6), B => DIN(5 downto 4),
14                      C => DIN(3 downto 2), D => DIN(1 downto 0),
15                      Y => DOUT(1 downto 0));
16       call1 : ANDOR (A => DIN(7 downto 6), B => DIN(5 downto 4),
17                      C => DIN(3 downto 2), D => DIN(1 downto 0),
18                      Y => DOUT(3 downto 2));
19       call2 : ANDOR (A => DIN(7 downto 6) and DIN(5 downto 4),
20                      B => DIN(5 downto 4),
21                      C => DIN(3 downto 2), D => DIN(1 downto 0),
22                      Y => DOUT(5 downto 4));
23       call3 : ANDOR (A => X nand Y, B => Z,
24                      C => DIN(3 downto 2), D => DIN(1 downto 0),
25                      Y => DOUT(7 downto 6));
26    end RTL;
```

5.6 Can a local variable be used as an actual parameter for a concurrent procedure call statement?

5.7 Which concurrent statement requires a label? Which concurrent statement can have an optional label? What is the benefit of having a label? What is the reason that some statements require labels while some are optional?

5.8 Can a component be declared inside the process declaration part? Explain why?

5.9 Rewrite the following VHDL code (which is a BCD to 7-segment encoder) to use concurrent signal assignment statements.

```
1     entity IFCASE is
2        port (
3           HEX   : in  bit_vector(3 downto 0);
4           LED   : out bit_vector(6 downto 0));
5     end IFCASE;
6     ---------------------------------------------
7     architecture RTL of IFCASE is
8     begin
9        p0 : process (HEX)
10       begin
11          case HEX is
12             when "0000" => LED <= "1111110";
13             when "0001" => LED <= "1100000";
14             when "0010" => LED <= "1011011";
15             when "0011" => LED <= "1110011";
16             when "0100" => LED <= "1100101";
17             when "0101" => LED <= "0110111";
18             when "0110" => LED <= "0111111";
19             when "0111" => LED <= "1100010";
```

```
20            when "1000" => LED <= "1111111";
21            when "1001" => LED <= "1110111";
22            when "1010" => LED <= "0111001";
23            when "1011" => LED <= "0111101";
24            when "1100" => LED <= "0011001";
25            when "1101" => LED <= "1111001";
26            when "1110" => LED <= "1011111";
27            when others => LED <= "0001111";
28         end case;
29      end process;
30   end RTL;
31   ----------------------------------------------
32   architecture RTL1 of IFCASE is
33   begin
34      p0 : process (HEX)
35      begin
36         if    HEX = "0000" then LED <= "1111110";
37         elsif HEX = "0001" then LED <= "1100000";
38         elsif HEX = "0010" then LED <= "1011011";
39         elsif HEX = "0011" then LED <= "1110011";
40         elsif HEX = "0100" then LED <= "1100101";
41         elsif HEX = "0101" then LED <= "0110111";
42         elsif HEX = "0110" then LED <= "0111111";
43         elsif HEX = "0111" then LED <= "1100010";
44         elsif HEX = "1000" then LED <= "1111111";
45         elsif HEX = "1001" then LED <= "1110111";
46         elsif HEX = "1010" then LED <= "0111001";
47         elsif HEX = "1011" then LED <= "0111101";
48         elsif HEX = "1100" then LED <= "0011001";
49         elsif HEX = "1101" then LED <= "1111001";
50         elsif HEX = "1110" then LED <= "1011111";
51         else                    LED <= "0001111";
52         end if;
53      end process;
54   end RTL1;
```

5.10 The preceding VHDL code shows two architectures for entity IFCASE. Do they describe the same function? Is one of them better than the other? Explain why.

Chapter 6

Subprograms and Packages

Examples for subprograms and packages were discussed in previous chapters. This chapter focuses on subprograms and packages in more detail in resolution functions, subprogram parameter types, object classes (**constant**, **variable**, **signal**), and modes (**in**, **out**, **inout**, **buffer**, **linkage**).

6.1 SUBPROGRAM DECLARATION

A subprogram has two forms: functions and procedures. A procedure call is a statement while a function call returns a value in an expression.

subprogram_declaration ::=
  procedure identifier [(formal_parameter_list)] |
  function designator [(formal_parameter_list)] return type_mark ;

A function designator can be an identifier or a string literal (operator symbol). Note that a string literal cannot be a procedure name. Extra spaces are not allowed in an operator symbol, and the letters can be lowercase, uppercase, or both. For example, "AND", "AnD", "and" are treated the same when they represent function operators.

The formal parameter list is in a form of an interface list which consists of one or more interface declarations. Semicolon ";" is used to separate between interface declarations. There is no ";" after the last interface declaration.

interface_list ::= interface_declaration { ; interface_declaration }
interface_declaration ::=
  [**constant**] identifier_list : [**in**] subtype_indication [:= static_expression] |
  [**signal**] identifier_list : [mode] subtype_indication [**bus**]
    [:= static_expression] |
  { **variable**] identifier_list : [mode] subtype_indication [:= static_expression]
mode ::= **in** | **out** | **inout** | **buffer** | **linkage**

An interface declaration declares an interface object of a specified type. The constant interface objects appear in the generics of a design entity (to be discussed in

Chapter 7), component, block, or subprograms. Signal interface objects appear as ports of a design entity, component, block, or subprograms. Variable interface objects appear only in subprograms.

The keyword **bus** is used for a signal interface object to indicate a guarded signal of kind **bus**. The guarded signal will be discussed in Chapter 10. If the ":=" symbol followed by an expression (must be static) exists, the expression is the default (initial) value of the interface object. When the mode is not specified, mode **in** is assumed.

For the interface object of mode **in**, the value of the interface object can only be read. Any attribute of the interface object can be read, but STABLE, QUIET, DELAYED, and TRANSACTION of a signal interface object cannot be read in a subprogram.

For the interface object of mode **out**, the value of the interface object may be updated. Any of the attributes of the interface object can be read except STABLE, QUIET, DELAYED, TRANSACTION, EVENT, ACTIVE, LAST_EVENT, LAST_ACTIVE, and LAST_VALUE.

For the interface object of mode **inout** or **buffer**, the value of the interface object may be both read and updated. The attributes of the interface object can be read. The interface object of mode **buffer** can only be updated by at most one source.

For the interface object of mode **linkage**, the value of the interface object may be both read and updated.

Formal parameters of subprograms can be constants, variables, and signals object classes. All formal parameters can be the mode of **in**, **inout**, and **out**. If the mode is **in** and no object class is explicitly specified, constant object is assumed. If the mode is **inout** or **out** and no object class is explicitly specified, variable is assumed.

The only mode of function formal parameters is mode **in** (whether the mode is explicit or implicit). The object class must either be constant or signal. Constant is assumed if the object class is not explicitly specified.

In a subprogram call, the actual designator (actual argument) associated with a formal parameter of object class of signal must be a signal. The actual designator associated with a formal parameter of object class of variable must be a variable. The actual designator associated with a formal parameter of object class of constant must be an expression. Note that attributes of an actual parameter are never passed to a subprogram. The attributes used in a subprogram must associate with the formal parameter and both must be valid. Table 6.1 summarizes the interface parameters mode and object class with the required mapping actual parameter.

	Procedure parameter mode **IN**	Procedure parameter mode **INOUT**	Procedure parameter mode **OUT**	Function parameter mode **IN**(only)	Entity port of all modes	Entity generic mode **IN** only
Constant object	expression	**not allowed**	**not allowed**	expression	**not allowed**	expression
Variable object	variable	variable	variable	**not allowed**	**not allowed**	**not allowed**
Signal object	signal	signal	signal	signal	signal	**not allowed**
Default (if object class not specified)	constant	variable	variable	constant	signal	constant

TABLE 6.1 Interface parameter mode, object class, and mapping table.

For parameters of the object class of constant and variable, only the values of actual or formal parameters are copied into and out of the subprogram call. For a parameter of a scalar type (INTEGER, REAL, PHYSICAL, enumerated as discussed in Chapter 2) or an access type, the parameter is passed by copy. The values of formal parameters with mode **in** or **inout** are copied from the associated actual parameters when the subprogram is called. During the completion of the subprogram, the values of actual parameters with mode **out** or **inout** are copied from the associated formal parameters. The parameter of an array or a record type can be implemented the same way of a scalar using copy. Another alternative is to use the "call by reference" that uses the actual parameters. For a (variable) parameter of the file type, a reference to the formal parameter is equivalent to the reference of the actual parameter.

For a formal parameter of the signal object class, references to the signal, the driver of the signal, or both, are passed into the subprogram call. For a signal parameter of mode **in** or **inout**, the actual signal is associated with the formal signal at the start of each subprogram call. During the execution of the subprogram, a reference to the formal signal parameter within an expression (not as a target of an assignment statement) is the same as a reference to the actual signal. For a signal parameter of mode **out** or **inout**, the driver of the actual parameter is associated with the driver of the formal signal parameter. During the execution of the subprogram, the assignment to the driver of the formal signal is equivalent to the assignment to the actual signal driver. The type conversion function cannot be associated with the formal or actual signal parameter. It is an error if attributes STABLE, QUIET, and DELAYED are read within a subprogram.

6.2 SUBPROGRAM BODY

```
subprogram_body ::=    subprogram_declaration is
                               subprogram_declarative_part
                       begin
                               sequential statements
                       end [ procedure identifier or function designator ] ;
```

A subprogram body specifies the execution of the subprogram. The subprogram declaration before the keyword **is** is exactly the same as discussed in the last section. The subprogram declarative part can have a subprogram body, declaration of type (subtype, constant, variable, file, alias, attribute), attribute specification, and use clause. Refer to Appendix B for comparisons with other declaration parts.

A separate subprogram declaration before the subprogram body is optional. The subprogram declaration of the subprogram body acts as the declaration. All subprograms can be called recursively. The following VHDL code shows an example of a recursive function to calculate the factorial number. In line 11, the same function FACTORIAL is called within its own function body. Note also that it has the subprogram body, but does not have the separate subprogram declaration. Figure 6.1 shows the simulation result.

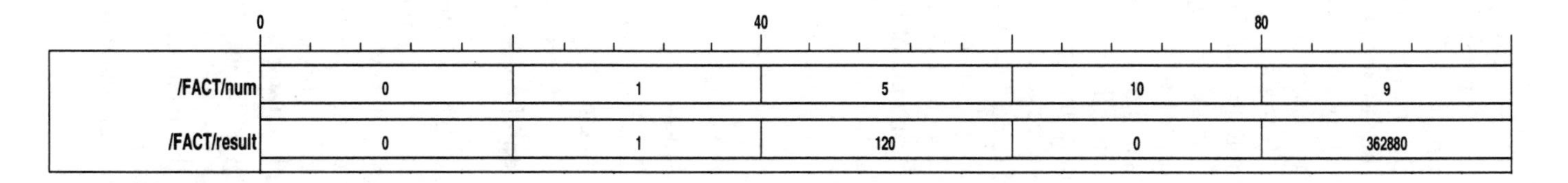

FIGURE 6.1 Simulation waveform for the factorial function.

```
1     entity FACT is
2     end FACT;
3
4     architecture RTL of FACT is
5        function FACTORIAL (constant N : in integer)
6           return integer is
7        begin
8           if (N = 1) then
9              return 1;
10          elsif (N > 1) and (N < 10) then
11             return N * FACTORIAL (N - 1);
12          else
13             assert FALSE report "0 < N < 10 is not true"
14                severity NOTE;
15             return 0;
16          end if;
17       end FACTORIAL;
18       signal NUM, RESULT : integer;
19    begin
20       RESULT <= FACTORIAL (NUM);
21    end RTL;
```

Note that a formal parameter may be an unconstrained array type. In this case, the bounds of the formal parameter are obtained from the actual parameter.

6.3 PACKAGE DECLARATION

We have briefly described package declaration and package body in Chapter 2. Here, we will discuss them in more detail.

package_declaration ::= **package** identifier **is**
    package_declarative_part
end [identifier] ;

If the identifier of the package name appears after keyword **end**, it must be the same as the identifier in the first line of the package declaration after the keyword **package**. The package declarative part may consist of subprogram declaration (function and procedure), type declaration, subtype declaration, constant declaration, signal declaration, file declaration, alias declaration, component declaration, attribute declaration, attribute specification, disconnect specification, and use clause. We will discuss attribute declaration and specification, disconnect specification, use clause, and file declaration later.

A constant is called a *deferred constant* if the constant declaration does not have the ":=" to define its value. The full constant declaration would exist in the package body.

A package declaration can be shared by a project design team to have the same type, subtype, and subprograms. As with the STANDARD package discussed in Chapter 2, basic types such as BIT and BOOLEAN are declared.

6.4 PACKAGE BODY

package_body ::= **package body** identifier **is**
package_body_declarative_part
end [identifier] ;

If the identifier after the keyword **end** exists, it should be the same as the identifier after the keywords **package body**. The package body declarative part can have the subprogram declaration, subprogram body, declaration of type (subtype, constant, file, alias), and use clause. The purpose of a package body is to define the bodies of subprograms or the values of deferred constants declared in the package declaration. Other declarative items are used to facilitate the definition of subprogram bodies and deferred constants. These declarative items cannot be made visible outside of the package body.

So far, we have been using predefined types such as BIT, BOOLEAN, and INTEGER. For simulation purposes, BIT type with two states of '0' and '1' is not enough. IEEE has defined the std_logic_1164 package. Inside the std_logic_1164 package, types std_ulogic, std_logic, std_ulogic_vector, and std_logic_vector are defined. The following VHDL code shows a portion of the IEEE std_logic_1164 package in a package PACK1164 for illustration:

```
1    package PACK1164 is
2      type std_ulogic is ( 'U',   -- Uninitialized
3                           'X',   -- Forcing  Unknown
4                           '0',   -- Forcing  0
5                           '1',   -- Forcing  1
6                           'Z',   -- High Impedance
7                           'W',   -- Weak     Unknown
8                           'L',   -- Weak     0
9                           'H',   -- Weak     1
10                          '-'); -- Don't care
11     type std_ulogic_vector is array ( NATURAL RANGE <> ) of std_ulogic;
12     function resolved ( s : std_ulogic_vector ) RETURN std_ulogic;
13     subtype UX01    is resolved std_ulogic RANGE 'U' TO '1';
14     subtype std_logic is resolved std_ulogic;
15     type std_logic_vector is array ( NATURAL RANGE <>) of std_logic;
16     function "and" ( l : std_ulogic; r : std_ulogic ) RETURN UX01;
17     function "and" ( l, r : std_logic_vector  ) RETURN std_logic_vector;
18     function "and" ( l, r : std_ulogic_vector ) RETURN std_ulogic_vector;
19   end PACK1164;
```

Lines 2 to 10 declare the type std_ulogic which has nine states. Note that each enumerated value is a character enclosed by two single quotes ('). The letters should be uppercase. 'U' indicates the uninitialized state. Since the default state (value) of an object is the left-most value of that type, 'U' would be the initial default value for the object of type std_ulogic. During simulation, if an object has value 'U', that may indicate the signal has never been updated. Values '0', '1' represent logic low and high. 'L' and 'H' are used for weak pull down and weak pull up. Value '-' means don't care condition. It is useful for better logic optimization by the synthesis tools. 'Z' indicates the high impedance state for the tristate interconnect. For example, the output of a tristate buffer would be 'Z' if the tristate buffer is not enabled. 'X' indicates the unknown value. It is usually caused by a bus contention when a signal has one driver driving '0' and one other driver driving '1'. 'W' represents the weak unknown. It can be caused by driving 'L' and 'H' by two different drivers for the same signal.

Line 11 declares type std_ulogic_vector as an unconstrained array of std_ulogic. Note that the range is NATURAL. Negative index is not allowed. Line 12 declares a resolution function, which we will discuss in more detail later in the next section. Line 13 declared a subtype which is based on an existing type std_ulogic. It is also a resolved type in that the value is resolved. Line 14 declares subtype std_logic which is resolved based on type std_ulogic. Note that std_logic and std_ulogic have the same nine states. Lines 16 to 18 show the function declaration of "and". Note that these three functions share the same name. However, the types of their formal parameters and the types of the function return values are not the same. We shall say that these functions are overloaded. The subprogram overloading will be discussed later in this chapter.

6.5 RESOLUTION FUNCTION

The purpose of a resolution function is to determine the result value for objects that are driven by more than one driver. For example, Figure 6.2 shows signal Y is driven by two tristate buffers.

```
1    library IEEE;
2    use IEEE.std_logic_1164.all;
3    entity DRIVE2 is
4       port (
5          A, B, A_ENn, B_ENn : in  std_logic;
6          Y                  : out std_logic);
7    end DRIVE2;
8
9    architecture RTL of DRIVE2 is
10   begin
11      Y <= A when A_ENn = '0' else 'Z';
12      Y <= B when B_ENn = '0' else 'Z';
13   end RTL;
```

Figure 6.2 schematic is obtained from synthesizing the preceding VHDL code. Note that all signals have the same type of std_logic which is declared in the library IEEE std_logic_1164 package. Line 1 declares the library IEEE to be referenced. Line 2 is a use clause to use everything (all) of the package std_logic_1164 in the library IEEE. The library and use clause will be discussed in more detail in Chapter 7. Lines 11 and 12 are two concurrent signal assignment statements. Each one of them would produce a driver for signal Y. In other words, signal Y has two drivers. This is how a resolution function comes into play.

The following VHDL code shows the package body of PACK1164 (portion of package body std_logic_1164). We have mentioned that the purpose of a package

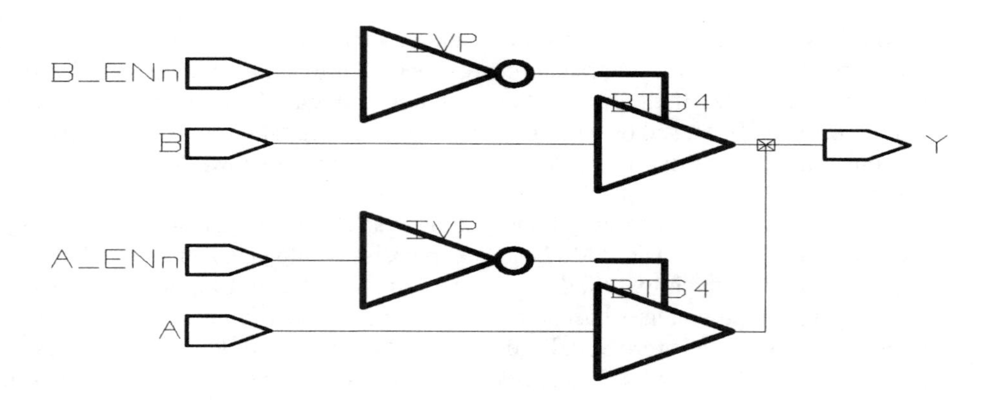

FIGURE 6.2 A signal driven by two sources.

body is to specify subprogram bodies and deferred constants. The other declaration items are to facilitate the specification of subprogram bodies and deferred constants. The examples of declaration items are shown in lines 2 to 18 where a one-dimension array, a two-dimension array, and a constant table are declared.

```
1     package body PACK1164 is
2        type stdlogic_1d is array (std_ulogic) of std_ulogic;
3        type stdlogic_table is array (std_ulogic, std_ulogic) of
std_ulogic;
4
5        constant resolution_table : stdlogic_table := (
6        --  ------------------------------------------------------
7        --  | U    X    0    1    Z    W    L    H    -        |   |
8        --  ------------------------------------------------------
9            ( 'U', 'U', 'U', 'U', 'U', 'U', 'U', 'U', 'U' ), -- | U |
10           ( 'U', 'X', 'X', 'X', 'X', 'X', 'X', 'X', 'X' ), -- | X |
11           ( 'U', 'X', '0', 'X', '0', '0', '0', '0', 'X' ), -- | 0 |
12           ( 'U', 'X', 'X', '1', '1', '1', '1', '1', 'X' ), -- | 1 |
13           ( 'U', 'X', '0', '1', 'Z', 'W', 'L', 'H', 'X' ), -- | Z |
14           ( 'U', 'X', '0', '1', 'W', 'W', 'W', 'W', 'X' ), -- | W |
15           ( 'U', 'X', '0', '1', 'L', 'W', 'L', 'W', 'X' ), -- | L |
16           ( 'U', 'X', '0', '1', 'H', 'W', 'W', 'H', 'X' ), -- | H |
17           ( 'U', 'X', 'X', 'X', 'X', 'X', 'X', 'X', 'X' )  -- | - |
18       );
19
20       function resolved ( s : std_ulogic_vector ) RETURN std_ulogic IS
21         variable result : std_ulogic := 'Z';  -- weakest state default
22       begin
23          if (s'LENGTH = 1) then
24             RETURN s(s'LOW);
25          else
26             for i in s'RANGE loop
```

```
27                result := resolution_table(result, s(i));
28             end loop;
29          end if;
30          RETURN result;
31       end resolved;
```

Lines 20 to 31 specify the function body of resolved. The formal parameter is an unconstrained array. The bounds will be determined by its actual parameter when the function is called. The function works for any length of the actual parameter array size. Line 23 checks—if the array has only one element (just one driver), the function returns that element value (no resolution is needed). Otherwise, the loop statement goes through each element and resolves the value with the variable result (which is initialized to 'Z') by the look up table as specified in the constant resolution_table (line 5). The simulation waveform in Figure 6.3 illustrates the resolution of signal Y in VHDL code for DRIVE2.

To simulate DRIVE2, the following commands are used :

trace A A_ENn B B_ENn Y
run 20
assign '0' A_ENn
run 20
assign '0' A
run 20
assign '1' A
run 20
assign '0' B
assign '0' B_ENn
run 20
assign 'L' B
run 20
assign 'H' A
run 20

At the start of the simulation, values of A, A_ENn, B, B_ENn are defaulted to 'U'. Each concurrent simulation process is executed until it is suspended. Both drivers of Y are 'Z' and they are resolved to 'Z'. At time 20, A_ENn is assigned to '0' that allows 'U' of A to come out of the three state from line 11, to be resolved with 'Z' to get 'U'. At time 40, A is assigned to '0', resolved with 'Z' (value of three state at line 12) to get '0'. At time 60, A is assigned to '1', resolved with 'Z' to get '1'. At time 80, both B and B_ENn are assigned '0', line 11 drives Y to '1' and line 12 drives Y to '0', unknown value 'X' is shown. At time 100, B is assigned to weak pull down 'L', '1' is the resolved value since '1' is stronger. At time 120, A is assigned to weak pull up 'H'. 'L' and 'H' would be resolved to 'W'. The waveform shows no difference of '0' and 'L' ('1' and 'H') due to black and white. The original waveform shows '0' and 'L'

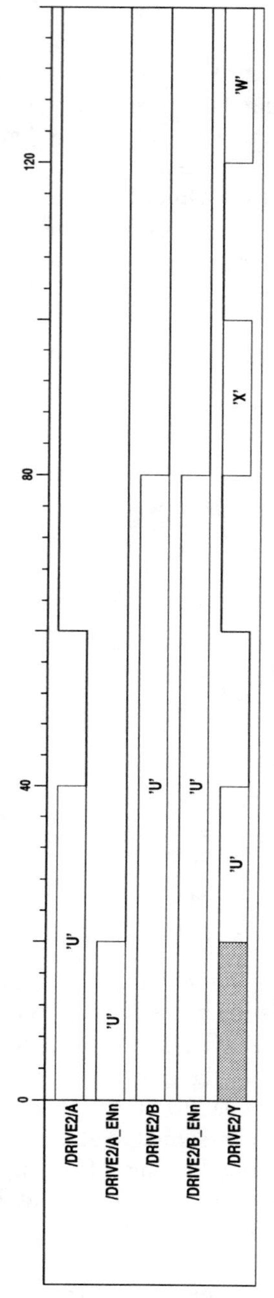

FIGURE 6.3 Simulation waveform of DRIVE2.

with different colors. Readers are encouraged to verify the two-dimension resolution_table to convince the resolution value for CMOS circuits.

In the DRIVE2 VHDL code, all signals use std_logic type. This is because signal Y has two drivers in lines 11 and 12. It would be an error if type BIT is used for those signals because BIT type is not a resolved type. It does not have an associated resolution function with the BIT type.

6.6 SUBPROGRAM OVERLOADING

A subprogram is overloaded if more than one subprogram specification is associated with the same function designator or procedure identifier. In PACK1164, lines 16 to 18 show function operator "and" is overloaded. Overloaded subprograms are differentiated by (1) the number of formal parameters, (2) the base type of the formal parameter, (3) the return type of the function. It would be an error if the simulator could not differentiate overloaded subprograms by these three factors.

The following VHDL code is the continuation of package body PACK1164. A constant table and_table is declared from lines 32 to 45. Lines 47 to 50 specify function operator with two formal parameters of std_ulogic type.

```
32      constant and_table : stdlogic_table := (
33      --  ----------------------------------------------------
34      --  | U    X    0    1    Z    W    L    H    -         |  |
35      --  ----------------------------------------------------
36          ( 'U', 'U', '0', 'U', 'U', 'U', '0', 'U', 'U' ),  -- | U |
37          ( 'U', 'X', '0', 'X', 'X', 'X', '0', 'X', 'X' ),  -- | X |
38          ( '0', '0', '0', '0', '0', '0', '0', '0', '0' ),  -- | 0 |
39          ( 'U', 'X', '0', '1', 'X', 'X', '0', '1', 'X' ),  -- | 1 |
40          ( 'U', 'X', '0', 'X', 'X', 'X', '0', 'X', 'X' ),  -- | Z |
41          ( 'U', 'X', '0', 'X', 'X', 'X', '0', 'X', 'X' ),  -- | W |
42          ( '0', '0', '0', '0', '0', '0', '0', '0', '0' ),  -- | L |
43          ( 'U', 'X', '0', '1', 'X', 'X', '0', '1', 'X' ),  -- | H |
44          ( 'U', 'X', '0', 'X', 'X', 'X', '0', 'X', 'X' )   -- | - |
45      );
46
47      function "and" ( l : std_ulogic; r : std_ulogic ) RETURN UX01 is
48      begin
49         RETURN (and_table(l, r));
50      end "and";
```

Lines 51 to 66 specify function operator "and" with two formal parameters and return value of type std_logic_vector (unconstrained). Lines 52 and 53 declare alias to normalize the array bounds since actual parameters passed in may be std_logic_vector(7 downto 4) and std_logic_vector(0 to 3). Line 54 declares variable *result* with the same type as alias *lv*. Lines 56 to 59 check whether both formal parameters have the same length. Otherwise, it loops through the normalized arrays *lv* and *rv* with the same index by a look up table. The function is returned in line 65. Lines 68 to 84 specify the function operator "and" with std_ulogic_vector type.

```
51      function "and" (l,r : std_logic_vector ) RETURN std_logic_vector
        is
52         alias lv : std_logic_vector ( 1 TO l'LENGTH ) is l;
53         alias rv : std_logic_vector ( 1 TO r'LENGTH ) is r;
54         variable result : std_logic_vector ( 1 TO l'LENGTH );
55      begin
56         if ( l'LENGTH /= r'LENGTH ) then
57            assert FALSE report
58              "arguments of 'and' operator are not of the same length"
59            severity FAILURE;
60         else
61            for i in result'RANGE loop
62               result(i) := and_table (lv(i), rv(i));
63            end loop;
64         end if;
65         RETURN result;
66      end "and";
67      ---------------------------------------------------------------
68      function "and"  ( l,r : std_ulogic_vector )
69         RETURN std_ulogic_vector is
70         alias lv : std_ulogic_vector ( 1 TO l'LENGTH ) is l;
71         alias rv : std_ulogic_vector ( 1 TO r'LENGTH ) is r;
72         variable result : std_ulogic_vector ( 1 TO l'LENGTH );
73      begin
74         if ( l'LENGTH /= r'LENGTH ) then
75            assert FALSE report
76              "arguments of 'and' operator are not of the same length"
77            severity FAILURE;
78         else
79            for i in result'RANGE loop
80                  result(i) := and_table (lv(i), rv(i));
81            end loop;
82         end if;
83         RETURN result;
84       end "and";
85   end PACK1164;
```

The preceding VHDL code shows function operator overloading with the function designator enclosed by double quotes ("). It is also possible to overload subprograms with the identifiers as subprogram designators. The following VHDL code shows overloading procedure MUX21:

```
1    package SUBPROG is
2       procedure MUX21(
3          signal SEL  : in  bit;
4          signal DIN0 : in  bit;
5          signal DIN1 : in  bit;
6          signal DOUT : out bit);
7       procedure MUX21(
```

```
8          signal SEL  : in  bit;
9          signal DIN0 : in  bit_vector;
10         signal DIN1 : in  bit_vector;
11         signal DOUT : out bit_vector);
12    end SUBPROG;
13
14    package body SUBPROG is
15       procedure MUX21(
16          signal SEL  : in  bit;
17          signal DIN0 : in  bit;
18          signal DIN1 : in  bit;
19          signal DOUT : out bit) is
20       begin
21          case SEL is
22             when '0' => DOUT <= DIN0;
23             when others => DOUT <= DIN1;
24          end case;
25       end MUX21;
26       --------------------------------------------------------------
27       procedure MUX21(
28          signal SEL  : in  bit;
29          signal DIN0 : in  bit_vector;
30          signal DIN1 : in  bit_vector;
31          signal DOUT : out bit_vector) is
32       begin
33          case SEL is
34             when '0' => DOUT <= DIN0;
35             when others => DOUT <= DIN1;
36          end case;
37       end MUX21;
38    end SUBPROG;
```

You may notice that alias is not used in the procedure MUX21 lines 27 to 37. There is no check on the length of DIN0, DIN1, and DOUT. Formal parameters are unconstrained, but they are not normalized with the alias as shown in function operator "and" earlier. This is because some synthesis tools do not accept alias declarations. To ensure that the code functions correctly and is synthesizable, users need to ensure the length consistency of DIN0, DIN1, and DOUT.

6.7 SUBPROGRAM RETURN VALUES AND TYPES

It is important to realize that the types of values returned from a subprogram depend on the return value inside the subprogram. The following VHDL code shows a simple example of a function INV that returns the inverted value of the argument being passed in. Note that the argument DIN is unconstrained. A local variable result is declared in line 9, which is constrained in the range from 1 to DIN'length. The function returns the value of result in line 12. The function return value is also unconstrained (line 8). Therefore, the value and the type of the function INV would be the same as result.

```
1     library IEEE;
2     use IEEE.std_logic_1164.all;
3     entity SUBRETN is
4        signal A : std_logic_vector(7 downto 4);
5        signal B : std_logic_vector(0 to     3);
6        signal C : std_logic_vector(2 downto 0);
7        signal LB, RB, LA, RA : std_logic;
8        function INV (DIN : std_logic_vector) return std_logic_vector is
9           variable result : std_logic_vector(1 to DIN'length);
10       begin
11          result := not DIN;
12          return result;
13       end INV;
14    end SUBRETN;
15    architecture RTL of SUBRETN is
16    begin
17       C  <= INV(A)(2 to 4);
18       LA <= C(C'left);
19       RA <= C(C'right);
20       RB <= INV(B)(4);
21       LB <= INV(B)(1);
22    end RTL;
23    architecture ORDER of SUBRETN is
24    begin
25       C  <= INV(A)(3 downto 1);
26    end ORDER;
27    architecture BOUND of SUBRETN is
28    begin
29       LB <= INV(A)(0);
30    end BOUND;
```

Architecture RTL shows examples of function calls. Line 17 gets a slice of the function and assigns to signal C. Lines 18 and 19 assign the left-most bit of C to LA and the right-most bit of C to RA. Lines 20 and 21 get the right-most bit and the left-most bit of inverted B to RB and LB, respectively. Note that function INV is called twice in lines 20 and 21 to the RB and LB. Lines 17, 18, and 19 only call function INV once, which is more efficient especially for the large function.

Line 25 of architecture ORDER (lines 23 to 26) shows an example of a function call with a slice reference (3 downto 1), not the same order of the function return value (1 to DIN'length). This results in an error. Line 29 of architecture BOUND shows an example of a function call with an index (0) that is not inside the range of the function return value (1 to DIN'length). This also results in a run-time error.

6.8 TYPE CASTING AND TYPE QUALIFICATION

A type conversion function can be used to convert types. Another way to change the type is through a type casting. A type casting is done with a type mark followed by an expression enclosed by a pair of parenthesis. Refer to the following VHDL code example for a 4-bit counter. In line 18, arithmetic operator "+" is used. This operator is overloaded in package std_logic_arith as referenced in line 3. The operator takes two operands of type UNSIGNED and returns a result of std_logic_vector type (even it can take other types of operands). The type UNSIGNED and operator "+" are declared in package std_logic_arith as follows :

type UNSIGNED is array (NATURAL range <>) of STD_LOGIC;
function "+"(L: UNSIGNED; R: UNSIGNED) return STD_LOGIC_VECTOR;

In line 10, signal CNTR4_FF is declared as std_logic_vector(3 downto 0). It cannot be used directly in the arithmetic operator "+" with UNSIGNED type operands expected. Therefore, CNTR4_FF (of type std_logic_vector) is cast to type UNSIGNED and then added with a literal "0001" after it is qualified to UNSIGNED type. This is because the literal "0001" can be bit_vector(3 downto 0), std_logic_vector(3 downto 0), a character string, or UNSIGNED type. Note that the type casting does not have the single quote "'" after the type mark as shown in line 18. The type qualification requires a single quote "'" after the type mark, as shown in line 16 (for "0000") and line 18 (for "0001").

```
1    library IEEE;
2    use IEEE.std_logic_1164.all;
3    use IEEE.std_logic_arith.all;
4    entity CONVTYPE is
5       port (
6          RSTn, CLK : in  std_logic;
7          CNTR4     : out std_logic_vector(3 downto 0));
8    end CONVTYPE;
9    architecture RTL of CONVTYPE is
10      signal CNTR4_FF : std_logic_vector(3 downto 0);
11   begin
12      CNTR4 <= CNTR4_FF;
13      p0 : process (RSTn, CLK)
14      begin
15         if (RSTn = '0') then
16            CNTR4_FF <= std_logic_vector'("0000");
17         elsif (CLK'event and CLK = '1') then
18            CNTR4_FF <= unsigned(CNTR4_FF) + unsigned'("0001");
19         end if;
20      end process;
21   end RTL;
```

6.9 EXERCISES

6.1 What is the reason that a function formal parameter cannot have the variable as the object class?

6.2 Write a function that takes two parameters and returns their Greatest Common Factor (GCF). For example, GCF of 8 and 12 is 4. Simulate your VHDL code.

6.3 Write a procedure that sorts an array of integers in nondecreasing order. Simulate your VHDL code.

6.4 Write procedures MUX31 and MUX41 (similar to MUX21) so that the code can be synthesizable with types std_logic and std_logic_vector.

6.5 What is the purpose of a package body? What is the purpose (and the restrictions) of the declaration items inside a package body?

6.6 Discuss whether a local variable, a signal, a constant, a type (subtype), and a subprogram can be declared in the package, package body, subprogram, process statement.

6.7 Sequential statements are used in the subprogram body. A wait statement is a sequential statement. Discuss the situations when a subprogram (function or procedure) with a wait statement is called inside a process statement with (1) an explicit wait statement, and (2) with a sensitivity list after the keyword **process**.

6.8 In package PACK1164, the object class of the formal parameter *S* of function resolved is not specified. Is *S* treated as a constant, a variable, or a signal?

6.9 If you have access to a VHDL simulator, find out whether you have the source code of the IEEE std_logic_1164 package. If you use SYNOPSYS, the file may be in $SYNOPSYS/packages/IEEE/src/std_logic_1164.vhd. Read through it and be sure to understand (1) what types are declared, (2) what functions are declared, (3) how the resolution function is specified and used, (4) what functions are overloaded, (5) why alias is used for subprograms.

6.10 If you use SYNOPSYS, read the VHDL code in file $SYNOPSYS/packages/IEEE/src/std_logic_arith.vhd. (1) what types are declared? (2) what functions are declared? (3) what functions are overloaded? (4) what is the purpose of this package std_logic_arith and how may you use it?

6.11 If you use SYNOPSYS, read the VHDL code in file $SYNOPSYS/packages/IEEE/src/std_logic_textio.vhd. (1) what types are declared? (2) what functions are declared? (3) what functions are overloaded? (4) what is the purpose of this package std_logic_textio and how may you use it?

6.12 The following VHDL code shows two architectures to assign signal Y. Is architecture SEQUENTIAL valid? What would be the value of signal Y? Is architecture CONCURRENT1 valid? If it is valid, what would be the value of signal Y. If it is not valid, why? Is architecture CONCURRENT2 valid? If it is valid, what would be the value of signal Y. If it is not valid, why?

```
1   entity SEQCON is
2   end SEQCON;
3   architecture SEQUENTIAL of SEQCON is
4      signal Y : bit;
5   begin
6      p0 : process
7      begin
8         Y <= '1';
9         Y <= '0';
10        wait for 20 ns;
11     end process;
12  end SEQUENTIAL;
13  architecture CONCURRENT1 of SEQCON is
14     signal Y : bit;
15  begin
```

```
16        Y <= '1';
17        Y <= '0';
18     end CONCURRENT1;
19     architecture CONCURRENT2 of SEQCON is
20        signal Y : bit;
21     begin
22        Y <= '1';
23        Y <= '1';
24     end CONCURRENT2;
```

6.13 Repeat the last problem if signal Y is of std_logic type.

6.14 In the following VHDL code, which procedure declaration is valid? Explain.

```
1      package CONIO is
2         procedure CON_INOUT(
3            constant DIN  : inout   bit);
4         procedure CON_BUFFER(
5            constant DIN  : buffer  bit);
6         procedure CON_LINKAGE(
7            constant DIN  : linkage bit);
8         procedure DEF_CON_INOUT(
9                     DIN  : inout   bit);
10        procedure DEF_CON_BUFFER(
11                    DIN  : buffer  bit);
12        procedure DEF_CON_LINKAGE(
13                    DIN  : linkage bit);
14     end CONIO;
```

Chapter 7

Design Unit, Library, and Configuration

7.1 ARCHITECTURE

We have presented several entity and architecture examples. In this chapter, we will discuss the concepts of entity and architecture in more detail.

architecture_body ::= **architecture** identifier **of** entity_name **is**
    architecture_declarative_part
begin
    [concurrent statements]
end [architecture_simple_name] ;

If the optional *architecture simple name* appears after the keyword **end**, it must be the same as shown in the identifier after the keyword **architecture**. Subprogram declarations, subprogram bodies, use clauses, disconnection specifications, configuration specifications, attribute specifications, declaration of types, subtypes, constants, signals, files, aliases, and components can be in the *architecture declarative part. Note that local variables cannot be declared in the architecture declarative part.* Refer to Appendix B for comparisons with other declaration parts. Only concurrent statements are allowed in the *architecture body* between keywords **begin** and **end.** They are executed asynchronously with respect to one another. Sequential statements can be inside a process statement or a subprogram. It is legal to have no concurrent statements inside the architecture body.

7.2 ENTITY DECLARATION

entity_declaration ::= **entity** identifier **is**
    [**generic** (generic_list) ;]
    [**port** (port_list) ;]

entity_declarative_part
[**begin**
entity_statement_part]
end [entity_simple_name] ;

If the optional *entity simple name* appears after the keyword **end**, it must be the same as the identifier after the keyword **entity**. Every item between keywords **is** and **end** is optional as we have seen in many examples in the previous chapters.

The format of the *generic list* and the *port list* is the same as for the interface list used in subprograms as discussed in the last chapter. The optional port clause interface list in the entity can only be the signal object class. It would be an error if an explicit object class other than a signal is specified in the port clause interface list. The optional generic clause interface list can only be a constant object class. It would be an error if an explicit object class other than constant is specified in the generic clause interface list. Generic interface list is a good way to pass parameters into the design entity such as the timing delay parameters, bus width, file name, and so on.

The entity declarative part can have subprogram declarations, subprogram bodies, use clauses, disconnection specifications, attribute specifications, declaration of types, subtypes, constants, signals, files, and aliases. Note that local variable, configuration, and component declaration is not allowed. Refer to Appendix B for the declaration part table to compare with other declaration parts.

The optional *entity statement part* can only have concurrent procedure calls, concurrent assertion statements, and process statements. All must be passive so that no signal is assigned (no signal assignment statement). The purpose is to monitor the entity operations and characteristics.

All declaration items, the *port list*, and the *generic list* are visible to the entity and the associated architectures. Note that the *generic clause* and the *port clause* are before the declarative part. The type (subtype) declared in the declarative part would not be visible to the generic and port clauses. If a type (subtype) not in the STANDARD package is to be used in the generic or port clause, the type (subtype) should be declared in a package and a reference to the package with a use clause should be made. The following VHDL code illustrates some of the concepts we have just discussed.

Lines 3 to 24 describe a D-type flipflop that is similar to what we have discussed in Chapter 5. All signals are of type std_logic or std_logic_vector rather than BIT or bit_vector. Type std_logic and std_logic_vector are declared in package std_logic_1164 in library IEEE. Therefore, lines 1 and 2 reference library IEEE and package std_logic_1164. Another difference is adding a generic constant PRESET_CLRn in the generic clause in the entity declaration. Based on the value of PRESET_CLRn, the D flipflop is preset to '1' or cleared to '0' in lines 15 to 19. Note that PRESET_CLRn is an INTEGER type. This is because some synthesis tools can only take INTEGER typed generic. For simulation purposes, it would be simpler to have PRESET_CLRn the same type as signal Q and lines 15 to 19 can be replaced by just Q <= PRESET_CLRn;.

```
1    library IEEE;
2    use IEEE.std_logic_1164.all;
3    entity DFF is
```

```
4       generic (
5          PRESET_CLRn   : in integer);
6       port (
7          RSTn, CLK, D : in  std_logic;
8          Q            : out std_logic);
9    end DFF;
10   architecture RTL of DFF is
11   begin
12      process (RSTn, CLK)
13      begin
14         if (RSTn = '0') then
15            if (PRESET_CLRn = 0) then
16               Q <= '0';
17            else
18               Q <= '1';
19            end if;
20         elsif (CLK'event and CLK = '1') then
21            Q <= D;
22         end if;
23      end process;
24   end RTL;
```

The 8-bit and 24-bit shift registers were discussed in Chapter 5. By using the generic, we can modify the code slightly so that the code can be used to model a variable length shift register. The modified code follows. A generic constant N of type INTEGER is used to specify the length of the shift register (number of D flipflops). After N is declared, it can be used as shown in line 34. Line 34 declares a signal T which is visible to (and used) both architectures RTL and RTL1. Lines 35 to 37 are for the optional concurrent assertion statement that checks the value of N. The synthesis tool ignores these lines.

```
25   library IEEE;
26   use IEEE.std_logic_1164.all;
27   entity SHIFTN is
28      generic (
29         PRESET_CLRn    : in  integer;
30         N              : in  integer);
31      port (
32         RSTn, CLK, SI : in  std_logic;
33         SO            : out std_logic);
34      signal T : std_logic_vector(N downto 0);
35      begin
36         assert (N > 3) and (N < 33) report
37            "N outside of range 3 to 32";
38   end SHIFTN;
```

Component declaration, defined as follows, is very similar to the entity declaration. Note that the keyword **is** after the identifier appears in the entity declaration but not in the component declaration. Keywords **end component** without an optional repeated identifier are used to close the component declaration.

component_declaration ::= **component** identifier
[**generic** (generic_list) ;]
[**port** (port_list) ;]
end component ;

The following architectures RTL and RTL1 declare component DFF in lines 40 to 46 and lines 68 to 74. This component declaration can be in a package to be referenced and shared in both architectures, but it cannot be in the entity as signal T in line 34. Note that the purpose of lines 34 to 37 is to illustrate what can be put inside the entity declaration. They do not suggest that it is better to have more things in the entity declaration. For example, line 34 has signal T of N+1 bits. However, architecture RTL needs only N-1 bits. If we would simulate architecture RTL of SHIFTN and trace the signal T, we would see 2 bits with value 'U'. Note also that N is used in the generate statements (lines 48, 49, 54, 78) and signal mapping (lines 52, 76).

```
39    architecture RTL of SHIFTN is
40       component DFF
41       generic (
42          PRESET_CLRn   : in integer);
43       port (
44          RSTn, CLK, D : in  std_logic;
45          Q            : out std_logic);
46       end component;
47    begin
48       g0 : for i in N-1 downto 0 generate
49          g1 : if (i = N-1) generate
50             bit7 : DFF
51                generic map (PRESET_CLRn => PRESET_CLRn)
52                port map (RSTn=> RSTn, CLK=> CLK, D=> SI, Q=> T(N-2));
53          end generate;
54          g2 : if (i > 0) and (i < N-1) generate
55             bitm : DFF
56                generic map (PRESET_CLRn => PRESET_CLRn)
57                port map (RSTn, CLK, T(i), T(i-1));
58          end generate;
59          g3 : if (i = 0) generate
60             bit0 : DFF
61                generic map (PRESET_CLRn => PRESET_CLRn)
62                port map (RSTn, CLK, T(0), SO);
63          end generate;
64       end generate;
65    end RTL;
66
67    architecture RTL1 of SHIFTN is
68       component DFF
69       generic (
70          PRESET_CLRn   : in integer);
71       port (
```

```
72          RSTn, CLK, D : in  std_logic;
73          Q            : out std_logic);
74       end component;
75    begin
76       T(N) <= SI;
77       SO   <= T(0);
78       g0 : for i in N-1 downto 0 generate
79          allbit : DFF
80             generic map (PRESET_CLRn=> PRESET_CLRn)
81             port map (RSTn=> RSTn, CLK=> CLK, D=> T(i+1), Q=> T(i));
82       end generate;
83    end RTL1;
```

The preceding VHDL code is synthesizable. The same code is used with various generic mappings that allows different designs to be generated. Figure 7.1 shows the synthesized schematic with N mapped to 5 and PRESET_CLRn mapped to 0. Figure 7.2 shows the synthesized schematic with N mapped to 3 and PRESET_CLRn mapped to 1.

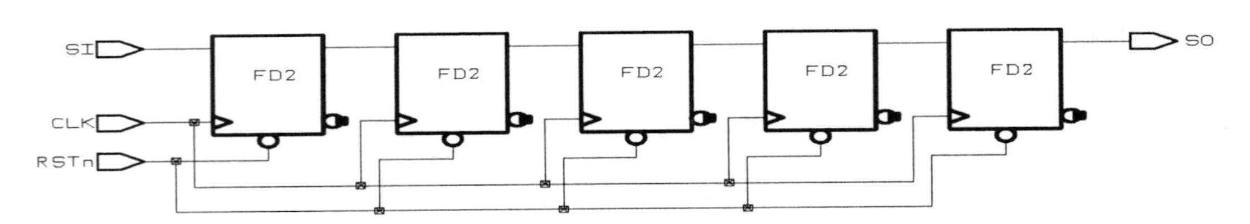

FIGURE 7.1 5-bit shift register with clear.

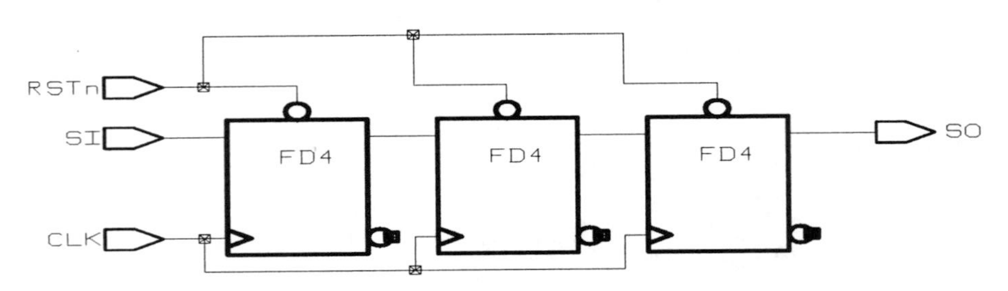

FIGURE 7.2 3-bit shift register with preset.

7.3 PORT MAP AND GENERIC MAP

The general rule for interface list parameter mapping (discussed in Chapter 6) applies to generic mapping (only constant object class) and the port mapping (only signal object class) of a component instantiation. For the port mapping, any locally declared signal can be used. For a hierarchical design, the lower-level component ports can be mapped to the higher-level entity port signals directly with the following restrictions. A port signal of mode **in** can be mapped with a port signal of mode **in**, **inout**, or **buffer**. A port signal of mode **out** can be mapped with a port signal of mode **out** or **inout**. A port signal of mode **inout** and **buffer** can only be mapped with a port signal of mode **inout** and **buffer** respectively. A port signal of mode **linkage** can be mapped with a port signal of any mode. A port of mode **in** can be unconnected (not mapped, or associated with the reserved word **open**) only if the port declaration includes a default expression. A port of mode other than **in** may be unconnected as long as its type is not an unconstrained array type.

From the hardware design point of view, mode **linkage** is not normally used. Mode **inout** is usually used for bidirectional pads which are used to communicate outside of the chip. Mode **buffer** is not recommended due to its restriction on the port mapping and allowing only one signal driver source. Using mode **buffer** or **inout** because a port signal of mode **out** cannot be read is not recommended. For an example, see the JKFF VHDL model in Chapter 4 exercise. Port Q is declared as mode **out**. From the JKFF entity point of view, Q is output only, rather than both input and output. The local variable or signal can be avoided inside the JKFF VHDL code if Q is declared as mode **inout** or **buffer**, as shown in the following BUF VHDL code. When entity BUF is instantiated as shown in BUFMAP VHDL code, lines 39 to 42, the mapping of Q (mode **buffer**) with DOUT (mode **out**) is not allowed. A temporary signal is necessary to be used in the port mapping. The DOUT is then assigned with the temporary signal. It is better to declare as mode **out** to match real hardware. This is the reason that mode **buffer** and **linkage** are not used throughout this book. VHDL'93 has a better solution for this problem with a new attribute 'DRIVING_VALUE, which will be discussed in Chapter 14.

```
1     library IEEE;
2     use IEEE.std_logic_1164.all;
3     entity BUF is
4        port (
5           RSTn, CLK, J, K : in  std_logic;
6           Q               : buffer std_logic);
7     end BUF;
8     architecture RTL of BUF is
9     begin
10       p0 : process (RSTn, CLK)
11       begin
12          if (RSTn = '0') then
13             Q <= '0';
14          elsif (CLK'event and CLK = '1') then
15             if (J = '1') and (K = '1') then
16                Q <= not Q;
17             elsif (J = '1') and (K = '0') then
```

```
18                Q <= '1';
19            elsif (J = '0') and (K = '1') then
20                Q <= '0';
21            end if;
22        end if;
23    end process;
24  end RTL;

25  library IEEE;
26  use IEEE.std_logic_1164.all;
27  entity BUFMAP is
28     port (
29        RSTn, CLK, DA, DB : in  std_logic;
30        DOUT              : out std_logic);
31  end BUFMAP;
32  architecture RTL of BUFMAP is
33     component BUF
34     port (
35        RSTn, CLK, J, K : in  std_logic;
36        Q               : buffer std_logic);
37     end component;
38  begin
39     buf0 : BUF
40     port map (
41        RSTn => RSTn, CLK => CLK, J => DA, K => DB,
42        Q => DOUT); -- ERROR!! map mode buffer with mode out
43  end RTL;
```

7.4 CONFIGURATION

Recall the 24-bit shift register (SHIFT24) and 8-bit shift register (SHIFT) examples in Chapter 5. Note that there are three architectures RTL1, RTL2, RTL3 associated with the entity SHIFT. Each architecture has DFF component instances. SHIFT24 instantiates component SHIFT three times. The question is which architecture of SHIFT is used in SHIFT24. There are two ways to bind components to entities. One is called *soft binding* and the other is called *hard binding*. The idea of the soft binding is not to fix the component selection in the entity and architecture VHDL code. The binding is deferred with a configuration declaration, which is defined as follows:

a. configurartion_declaration ::= **configuration** identifier **of** entity_name **is**
b. { use_clause | attribute_specification }
c. { block_configuration }
d. **end** [configuration_simple_name] ;

e. block_configuration ::= **for** block_specification
f. { use_clause }
g. { block_configuration | component_configuration }
h. **end for** ;

i. block_specification ::= architecture_name | block_statement_label |
j. generate_statement_label [(discrete_range | static_expression)]

k. component_configuration ::= **for** component_specification
l. [**use** binding_indication ;]
m. [block_configuration]
n. **end for** ;

o. binding_indication ::= **entity** entity_name [(architecture_identifier)] |
p. **configuration** configuration_name | **open**
q. [generic_map_aspect]
r. [port_map_aspect]

s. component_specification ::= instantiation_label { , instantiation_label } | others | all : component_name

VHDL configuration is complex, as shown in the preceding definitions. We can explain these definitions with the associated lowercase letter in the left and use the following VHDL code to show ways of configuring the SHIFT entity which has DFF as components. Lines 1 to 4 illustrate a configuration of DFF, which does not have any subcomponents. There is no need of a use clause and attribute specification (item b). Line 2 uses architecture name RTL as its block specification (item i) for the block configuration (item c). Line 3 closes with end for (item h) and line 4 closes with the configuration name after the keyword **end** (item d).

```
1     configuration CFG_RTL_DFF of DFF is
2        for RTL
3        end for;
4     end CFG_RTL_DFF;
```

The following VHDL code shows a configuration for the architecture RTL1 (line 6) of entity SHIFT. Line 6 uses the architecture name RTL1 and closes with keywords **end for** in line 10. Line 7 through 9 is the component configuration (item g expanded to items k, l, m, n). Line 7 is the component specification (item s) keyword **all** and component name DFF. Line 8 is binding indication (item o) followed by the keyword **use**. Note that the binding indication in line 8 has keyword **entity** (item o) while line 17 has keyword **configuration** (item p). The name WORK.DFF(RTL) indicates library WORK (to be discussed later in this chapter), entity DFF, and architecture RTL.

```
5     configuration CFG_RTL1_SHIFT of SHIFT is
6        for RTL1
7           for all : DFF
8                 use entity WORK.DFF(RTL);
```

```
9           end for;
10        end for;
11     end CFG_RTL1_SHIFT;
```

The following VHDL code shows a configuration for architecture RTL2 of entity SHIFT. Note that lines 14, 15, 20, 25 are the *generate statement labels* (item j) which become the block specification. Line 26 has the instantiation label (item s) bit0 rather than keyword **all**.

```
12     configuration CFG_RTL2_SHIFT of SHIFT is
13        for RTL2
14           for g0
15              for g1
16                 for all : DFF
17                    use configuration WORK.CFG_RTL_DFF;
18                 end for;
19              end for;
20              for g2
21                 for all : DFF
22                    use entity WORK.DFF(RTL);
23                 end for;
24              end for;
25              for g3
26                 for bit0 : DFF
27                    use entity WORK.DFF(RTL);
28                 end for;
29              end for;
30           end for;
31        end for;
32     end CFG_RTL2_SHIFT;
```

The following VHDL code shows a configuration of architecture RTL3 of entity SHIFT. Line 35 (40, 45) uses the generate statement label with generate range or static expression.

```
33     configuration CFG_RTL3_SHIFT of SHIFT is
34        for RTL3
35           for g0 (2 downto 0)
36              for allbit : DFF
37                 use configuration WORK.CFG_RTL_DFF;
38              end for;
39           end for;
40           for g0 (7 downto 4)
41              for all : DFF
42                 use configuration WORK.CFG_RTL_DFF;
43              end for;
44           end for;
45           for g0 (3)
46              for allbit : DFF
47                 use entity WORK.DFF(RTL);
48              end for;
49           end for;
```

```
50        end for;
51     end CFG_RTL3_SHIFT;
```

The following VHDL code shows a configuration of architecture RTL5 of entity SHIFT24. Lines 3, 11, and 14 have instantiation labels for the component instantiation statements. Line 12 and 15 show that they are using different configurations for *stage1* and *stage2* (each has 8 bits). Lines 4 through 9 show an example of hierarchical configuration that configures SHIFT (line 4) and then configures DFF inside entity SHIFT (lines 5 to 9). This hierarchical definition corresponds to item g and item e where a block configuration can be inside a block configuration for hierarchical configuration.

```
1      configuration CFG_RTL5_SHIFT24 of SHIFT24 is
2         for RTL5
3            for stage0 : SHIFT
4               use entity WORK.SHIFT(RTL1);
5               for RTL1
6                  for all : DFF
7                     use entity WORK.DFF(RTL);
8                  end for;
9               end for;
10           end for;
11           for stage1 : SHIFT
12              use configuration WORK.CFG_RTL2_SHIFT;
13           end for;
14           for stage2 : SHIFT
15              use configuration WORK.CFG_RTL3_SHIFT;
16           end for;
17        end for;
18     end CFG_RTL5_SHIFT24;
```

After SHIFT24 is configured as CFG_RTL5_SHIFT24, we can simulate the 24-bit shift register. Figure 7.3 shows the simulation results. Note that signal T2 (SO) is signal T1 (T2) delayed by eight clocks.

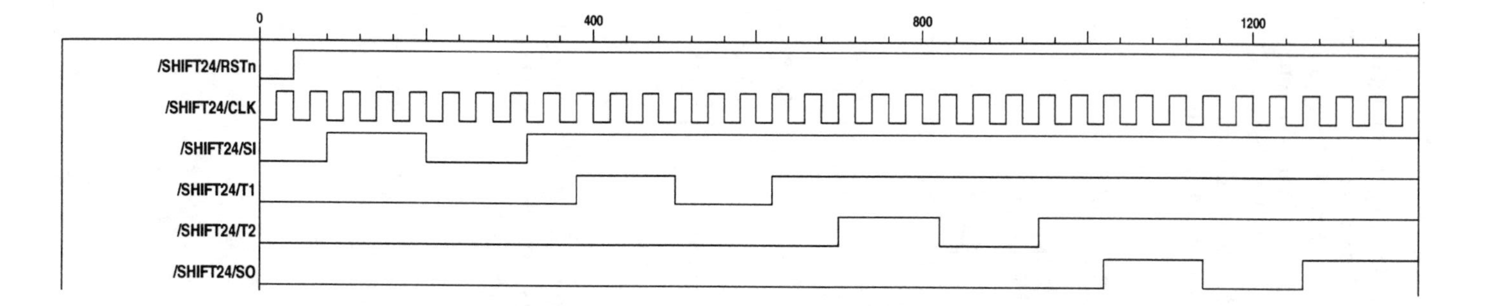

FIGURE 7.3 Simulation waveform for SHIFT24.

In Chapter 5, we discussed concurrent block statements with entity BLOCK-STMT as an example. We put *stage1* and *stage0* component instantiation statements inside a block statement labeled *block0*. To configure the entity BLOCKSTMT, the following VHDL code can be used. Note that line 6 uses the block label and then configures component instantiation labels *stage1* and *stage2* from line 7 to line 12. Line 13 corresponds to line 6 to close the block configuration.

```
1     configuration CFG_RTL5_BLOCKSTMT of BLOCKSTMT is
2        for RTL5
3           for stage2 : SHIFT
4              use configuration WORK.CFG_RTL1_SHIFT;
5           end for;
6           for block0
7              for stage1 : SHIFT
8                 use configuration WORK.CFG_RTL2_SHIFT;
9              end for;
10             for stage0 : SHIFT
11                use configuration WORK.CFG_RTL3_SHIFT;
12             end for;
13          end for;
14       end for;
15    end CFG_RTL5_BLOCKSTMT;
```

The preceding examples show various ways to soft (defer) bind components with entities by the configuration specification. The binding is not specified inside the SHIFT or SHIFT24 VHDL code. Another way is to use the hard binding such that the component configuration (items k to n) is specified with the component declaration inside the architecture declarative part, as shown in the following example.

In general, soft binding is preferred if several architectures are associated with the same entity. Separate configurations using different architecture bindings can be written and compiled once. During simulation, different configurations can be selected without recompiling the VHDL source code. If the hard-binding method is used, to change the binding of component architectures, the VHDL source code that is affected (may have hierarchical chain effect) needs to be modified and compiled before simulation can be performed.

Note that a generic mapping clause (item q) is optional in the binding indication (item o) to defer the mapping of generic. This carries the idea of soft binding to the next level. The following VHDL code shows the 8-bit shift register, SHIFT8, which uses DFF components with the generic PRESET_CLRn. Lines 11, 12, and 23 are commented out so that the PRESET_CLRn generic of entity DFF is not mapped.

```
1     library IEEE;
2     use IEEE.std_logic_1164.all;
3     entity SHIFT8 is
4        port (
5           RSTn, CLK, SI : in  std_logic;
6           SO            : out std_logic);
7        signal T : std_logic_vector(8 downto 0);
8     end SHIFT8;
9     architecture RTL1 of SHIFT8 is
```

```
10      component DFF
11   --   generic (
12   --      PRESET_CLRn   : in integer);
13      port (
14         RSTn, CLK, D : in  std_logic;
15         Q            : out std_logic);
16      end component;
17      constant N : integer := 8;
18   begin
19      T(N) <= SI;
20      SO   <= T(0);
21      g0 : for i in N-1 downto 0 generate
22         allbit : DFF
23   --          generic map (PRESET_CLRn=> 1)
24            port map (RSTn=> RSTn, CLK=> CLK, D=> T(i+1), Q=> T(i));
25      end generate;
26   end RTL1;
```

The following configuration VHDL code can be used to configure entity SHIFT8. Note that the entity SHIFT8 is reset to the initial value of "01010011" by using the generic mapping in the configuration (lines 6, 12, 18, 24, 30, 36). Figure 7.4 shows the simulation result. The shift register is reset to "01010011" or X"53" when RSTn is '0'. Signal T has 9 bits where the first bit is the same as SI.

```
1    configuration CFG_RTL1_SHIFT8 of SHIFT8 is
2       for RTL1
3          for g0 (1 downto 0)
4             for allbit : DFF
5                use entity WORK.DFF(RTL)
6                generic map (PRESET_CLRn => 1);
7             end for;
8          end for;
9          for g0(6)
10            for allbit : DFF
11               use entity WORK.DFF(RTL)
12               generic map (PRESET_CLRn => 1);
13            end for;
14         end for;
15         for g0(4)
16            for allbit : DFF
17               use entity WORK.DFF(RTL)
18               generic map (PRESET_CLRn => 1);
19            end for;
20         end for;
21         for g0 (3 downto 2)
22            for all : DFF
23               use entity WORK.DFF(RTL)
24               generic map (PRESET_CLRn => 0);
25            end for;
26         end for;
```

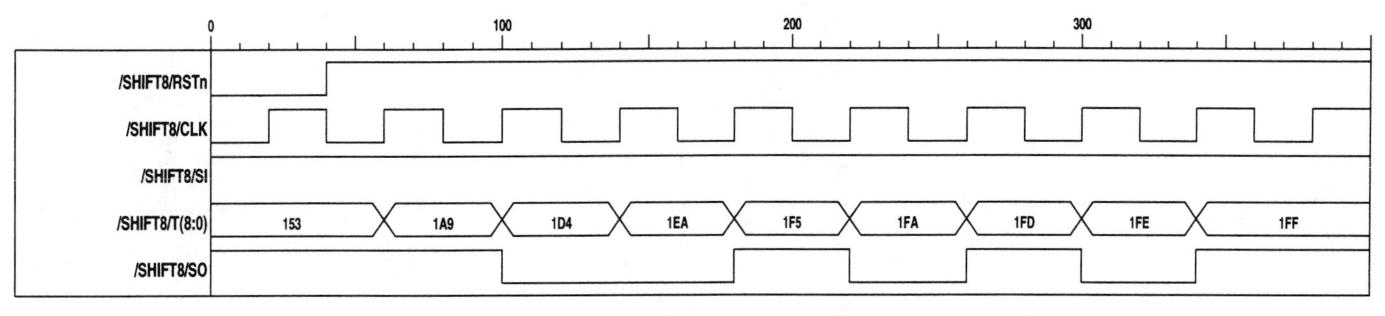

FIGURE 7.4 Simulation waveform of SHIFT8 with deferred generic mapping.

```
27         for g0(7)
28            for all : DFF
29               use entity WORK.DFF(RTL)
30               generic map (PRESET_CLRn => 0);
31            end for;
32         end for;
33         for g0(5)
34            for all : DFF
35               use entity WORK.DFF(RTL)
36               generic map (PRESET_CLRn => 0);
37            end for;
38         end for;
39      end for;
40   end CFG_RTL1_SHIFT8;
```

7.5 DESIGN UNIT

A design file can have one or many design units. There are primary design and secondary design units in VHDL. The *primary design unit* can be either entity declarations, configuration declarations, or package declarations. A VHDL primary design unit can be associated with many *secondary design units* that include an architecture body and a package body. Each primary unit in a given library must have a unique simple name. Each architecture body associated with a given entity declaration must be unique. For example, we have architecture names RTL1, RTL2, and RTL3 for entity SHIFT. Architecture body names associated with different entities can be the same, such as architecture RTL of entity DFF and architecture RTL of entity SHIFTN.

7.6 VHDL LIBRARY

A design library is an implementation dependent storage (usually a directory) facility for previously analyzed design units. A working library is the library in which the analysis of a design unit is placed. A resource library is a library to be referenced during the analysis of a given design unit. During the analysis, there can only be one working library. Any number of resource libraries can exist during the analysis, and a working library can also be a resource library. For example, library IEEE, a resource library is referenced (for std_logic_1164 package) in line 1 of entity DFF. The working library WORK is referenced in line 8 of configuration CFG_RTL1_SHIFT. Library logical name WORK denotes the current working library during a given analysis. Figure 7.5 shows relationship between VHDL source code files, the VHDL analyzer, the resource library, and the working library. The VHDL analyzer reads in VHDL source code files, references resource libraries (may include working library), and generates simulation databases into the working library.

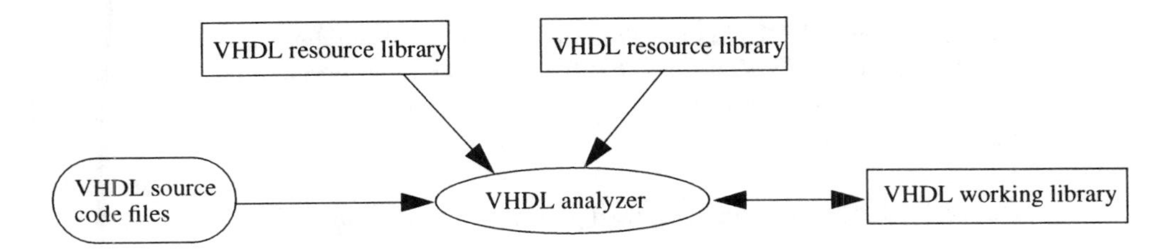

FIGURE 7.5 VHDL library and VHDL analyzer.

After VHDL source codes files are analyzed with a VHDL analyzer, VHDL simulation can be run with a VHDL simulator. Figure 7.6 shows the relationship between the resource library, the working library, text files, simulation commands, simulation waveforms, and assertion messages. To run VHDL simulation, a VHDL simulator selects a primary design unit from the working library. Resource libraries can be referenced. Simulation commands can be used to control the running of the VHDL simulator with regard to the amount of time to be run and the signals to be traced in the simulation waveform. Text files can be read when TEXTIO is used. The VHDL simulator generates simulation waveforms, assertion messages (if any), and text files (if TEXTIO is used). VHDL source codes can be shown on the screen when the simulator is running in the source code debugging mode for setting break points, single stepping, and so on.

Note that the library name (such as IEEE) after the keyword **library** in the library clause is a logical name. It can be independent in the VHDL source code. The logical name should be mapped into the actual storage facility (directory) known by the VHDL simulator. For example, in the SYNOPSYS VHDL simulator, file ".synopsys_vss.setup" (to specify the setup for the simulator) may have the following mapping :

```
WORK        > DEFAULT
DEFAULT      : ../work
SYNOPSYS     : $SYNOPSYS/$ARCH/packages/synopsys/lib
IEEE       : $SYNOPSYS/$ARCH/packages/IEEE/lib
```

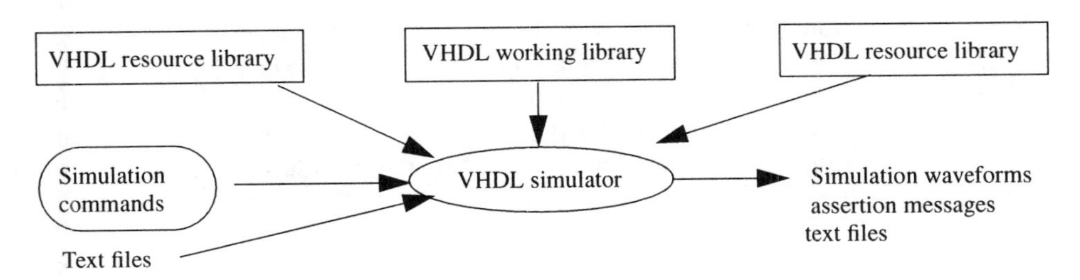

FIGURE 7.6 VHDL library and VHDL simulator.

Logical name WORK is the default logical name which is mapped to directory work. Logical names SYNOPSYS and IEEE are also mapped to corresponding directories. A library clause can specify one or more library logical names separated by comma(s) "," as follows:

library_clause ::= **library** logical_name { , logical_name } ;

The library logical name STD denotes the design library which includes package STANDARD and package TEXTIO. Every design unit is assumed to have the following implicit contexts:

library STD, WORK; **use** STD.STANDARD.**all**;

In other words, users do not need to specify STD and WORK library; everything inside package STANDARD and TEXTIO can be used without explicitly writing the preceding use clause again in the VHDL code.

Note that the scope of a library clause starts immediately after the library clause, and it extends to the end of the declarative region associated with the design unit in which the library clause appears. For example, in Section 7.2, entity DFF, architecture RTL of entity DFF, entity SHIFTN, architecture RTL of entity SHIFTN, and architecture RTL1 of entity SHIFTN are described in the same text file. The scope of library clause (library IEEE in line 1) extends from lines 2 to 24. Line 25 is necessary since a new primary design entity SHIFTN starts at line 27. Line 1 and line 25 are the same. The scope of library IEEE in line 25 extends from lines 26 to 83, which cover primary design unit SHIFTN and the two secondary design units architecture RTL and architecture RTL1.

The rules for the order of design units analysis are direct consequences of the visibility rules. A primary design unit whose name is referenced within a given design unit must be analyzed before the analysis of the given design unit. A primary design unit must be analyzed before the analysis of its corresponding secondary design unit. If a design unit in the library is changed (due to reanalysis), then all the library design units that are potentially affected by such change become obsolete and must be reanalyzed before they can be used again.

We have shown many VHDL examples where a library clause and a use clause are referenced before a design unit entity and architecture VHDL code. They can also be referenced for the configuration VHDL code. The following VHDL code shows such an example. Line 1 references a library COMP_LIB. The configuration is for the entity SHIFT and the architecture RTL1 (compared with the CFG_RTL1_SHIFT VHDL code). Note that lines 4 to 6 configure DFF with instance names (component instantiation statement labels) bit7, bit6, and bit4 with the DFF(RTL) in COMP_LIB library. Lines 7 to 9 configure DFF with instance name bit2 with the DFF(RTL) in the working library WORK. Library WORK is visible by default, and a library clause is not necessary. Lines 10 to 12 configure the rest of DFFs with the reserved word **others**.

```
1    library COMP_LIB;
2    configuration CFG1_RTL1_SHIFT of SHIFT is
```

```
3          for RTL1
4             for bit7, bit6, bit4 : DFF
5                   use entity COMP_LIB.DFF(RTL);
6             end for;
7             for bit2             : DFF
8                   use entity WORK.DFF(RTL);
9             end for;
10            for others           : DFF
11                  use configuration WORK.CFG_RTL_DFF;
12            end for;
13         end for;
14      end CFG1_RTL1_SHIFT;
```

7.7 BLOCK AND ARCHITECTURE ATTRIBUTES

A predefined attribute BEHAVIOR (ENT'behavior) is TRUE if there are no component instantiation statements inside the associated block statement or architecture. Another predefined attribute STRUCTURE (ENT'structure) is TRUE within the block or architecture if all process statements or equivalent process statements (concurrent statements) do not contain signal assignment statement.

7.8 EXERCISES

7.1 Modify SHIFTN VHDL code so that it can generate any N-bit shift register if N > 0.

7.2 Modify SHIFTN VHDL code so that it has a 2-bit MODE input, and a DATA input. When MODE is "00", the shift register remains the same value. When MODE is "01", it shifts right. When MODE is "10", it shifts left. When it is "11", it parallel loads the DATA. Verify your code by synthesis.

7.3 Verify the design in Exercise 7.2 by writing a configuration VHDL code and simulation.

7.4 Find out how your VHDL simulator maps logical library names to actual storage (directories).

7.5 The statement label for a component instantiation statement may be used in the configuration declaration. Can we safely claim that a statement label is required if a component instantiation statement can appear inside that statement? Explain.

7.6 Due to the scope of the library clause and the rule for the analysis order, is it a good practice to put all the design unit VHDL code in a text file? Explain your answer.

7.7 In the SHIFT8 VHDL code, the generic PRESET_CLRn is not mapped. It is mapped using configurations. Unfortunately, many synthesis tools do not accept (ignore) configuration declaration. To reset the register to "01010011" works for simulation but not for the synthesis. The synthesized schematic would not have the reset value of "01010011" (DFF is reset when the corresponding bit value is '0', preset when the corresponding bit value is '1'). Modify SHIFTN VHDL code to take a generic INIT of type INTEGER. D flipflops will be preset or reset based on the corresponding bit of INIT. Synthesize your VHDL code to generate a schematic. Write a configuration VHDL code and simulate.

7.8 In the CFG_RTL5_SHIFT24 VHDL code, one stage uses hierarchical configuration that configures SHIFT and then down to DFF. Write one configuration VHDL code that will configure all stages hierarchically. Discuss the advantages and disadvantages of having one large hierarchical configuration file rather than many configuration files where each configuration configures only its immediate subcomponents.

Chapter 8

Writing VHDL for Synthesis

8.1 GENERAL GUIDELINES OF VHDL SYNTHESIS

Synthesizable VHDL statements. VHDL was initially developed for the modeling (simulation) of digital circuits. It was not intended for synthesis. As we discussed in Chapter 1, synthesis is a process of translation and optimization based on a technology component library. There are VHDL constructs that can be used to generate hardware implementations. For example, a signal assignment statement "C <= A nand B;" can be translated into a two-input NAND gate easily. However, there are some constructs that do not correspond to actual hardware such as an assert statement, configuration declaration, and procedures for text I/O.

Time expressions. Another construct that is difficult to realize in hardware is the VHDL time expression—for example, a wait statement such as *wait for 20 ns*;. It is difficult to generate a circuit that will delay for exactly 20 ns. The same is true for a signal assignment statement *Z <= X xor Y after 3 ns*;. It is difficult to generate a circuit that would perform XOR function with exactly 3 ns timing delay. Therefore, most synthesis tools do not like these time expressions. A synthesis tool may generate a XOR gate and ignore the "after 3 ns" clause.

Initial values. In simulation, an object is initialized to its left-most value of its type. For example, a signal with BOOLEAN type is initialized to be FALSE. A signal with std_logic type is initialized to be 'U'. An object can be explicitly initialized when it is declared. For example, "*signal A : integer := 5;*" declares signal A of type INTEGER with an initial value 5. It is difficult (if not impossible) to implement the initial value for all kinds of objects in hardware. Therefore, most synthesis tools may ignore the initial value.

Synthesis is not defined by the IEEE standard. Various tools may have different capabilities, guidelines, and styles. In this chapter, we will present general use of VHDL for synthesis. The schematics are generated using the Synopsys Design Compiler. You are encouraged to check the restrictions and guidelines of your synthesis tools.

8.2 WRITING VHDL TO INFER FLIPFLOPS

A flipflop in general operates based on a clock signal. For example, the output of a D-type flipflop may be changed to the input data when a rising edge of the clock occurs. The rising edge of a signal CLOCK (of type bit) can be modeled in VHDL as *CLOCK'event and CLOCK = '1'*. The attribute event of a signal is of type BOOLEAN and it is TRUE when the signal changes value. Therefore, *CLOCK'event and CLOCK = '1'* indicates the rising edge of the signal CLOCK. The following VHDL code shows examples of inferring D-type flipflops. Note that IEEE std_logic_1164 package is referenced with lines 1 and 2. Type std_logic is used for all signals. We assume that *CLOCK'event and CLOCK = '1'* still represent the rising edge of the signal CLOCK, assuming the CLOCK signal would not have values other than '0' and '1'. Otherwise, *CLOCK'event and (CLOCK = '1') and (CLOCK'last_value = '0')* can be used.

```
1    library IEEE;
2    use IEEE.std_logic_1164.all;
3    entity FFS is
4       port(
5          CLOCK : in  std_logic;
6          ARSTn : in  std_logic;
7          SRSTn : in  std_logic;
8          DIN   : in  std_logic_vector(5 downto 0);
9          DOUT  : out std_logic_vector(5 downto 0));
10   end FFS;
```

The preceding VHDL defines the entity FFS with CLOCK, ARSTn, SRSTn as 1-bit input signals and DIN (DOUT) as a 6-bit input (output) signal. We shall use the signal CLOCK as the clock signal to connect to the D-type flipflops. Signal ARSTn is a low active signal to asynchronously reset the D flipflop to '0'. Signal SRSTn is a low active signal for synchronously resetting the D flipflop to '0'.

Lines 13 to 17 are a process statement. It does not have a sensitivity list after the keyword **process** in line 13. It has a wait statement in line 15 that is waiting for the rising edge of signal CLOCK. Then, signal DOUT(0) is assigned as *DIN(0) xor DIN(1)* in line 16. Figure 8.1 shows a XOR gate with a D flipflop corresponding to DOUT(0) are generated.

```
11   architecture RTL of FFS is
12   begin
13      seq0 : process
14      begin
15         wait until CLOCK'event and CLOCK = '1';
16         DOUT(0) <= DIN(0) xor DIN(1);
17      end process;
```

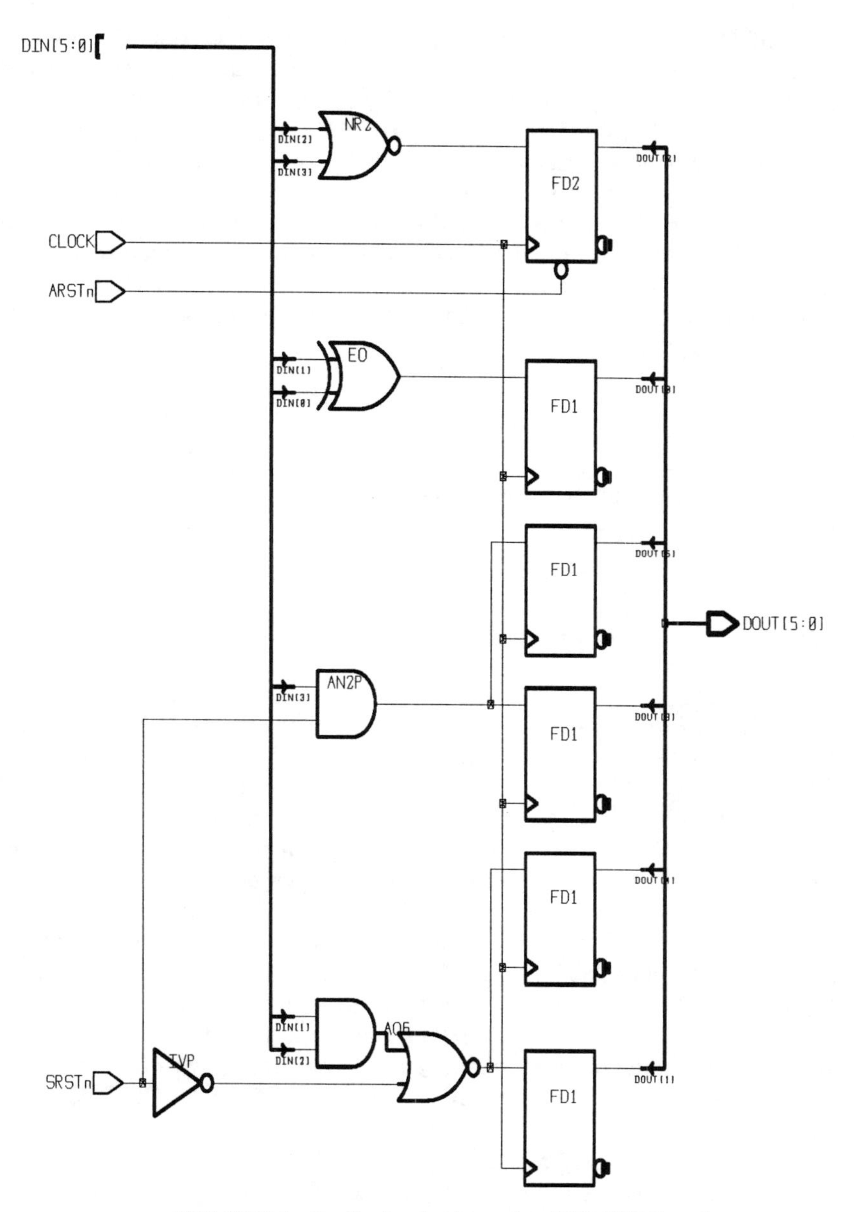

FIGURE 8.1 Synthesized schematic of FFS VHDL code.

Lines 18 to 27 define another process statement without a sensitivity list after the keyword **process** in line 18. Line 19 is again a wait statement waiting for the rising edge of the signal CLOCK. Lines 21 to 25 check whether signal SRSTn is '0'. If

SRSTn = '0', then signal DOUT(1) is reset to '0', otherwise, DOUT(1) gets the value of DIN(1) nand DIN(2). Note that SRSTn = '0' is checked right after line 20 (rising edge of CLOCK). We can say that DOUT(1) is synchronously reset to '0' when SRSTn is asserted '0'. Line 26 assigned DOUT(4) with an expression which is equivalent to the value assigned to DOUT(1) by the result of the if statement. Note that for any target signal (DOUT(1) and DOUT(4)) in the signal assignment statement after the wait statement of checking the rising edge of a signal, the target signal is synthesized as an output of a flipflop. The signal being tested as the rising edge is treated as a clock signal to that flipflop. Note that SRSTn is connected to an inverter and then an AND-NOR gate. The output of the AND-NOR gate is connected to the D-input of DOUT(1) and DOUT(4) in the bottom of the Figure 8.1 schematic. When SRSTn is asserted '0', both DOUT(1) and DOUT(4) will be reset to '0' after the rising edge of the signal CLOCK.

```
18        seq1 : process
19        begin
20           wait until CLOCK'event and CLOCK = '1';
21           if (SRSTn = '0') then
22              DOUT(1) <= '0';
23           else
24              DOUT(1) <= DIN(1) nand DIN(2);
25           end if;
26           DOUT(4) <= SRSTn and (DIN(1) nand DIN(2));
27        end process;
```

Lines 28 to 35 show a process statement with signals ARSTn and CLOCK as the sensitivity list. An if statement checks whether ARSTn is asserted '0'. If ARSTn = '0', DOUT(2) is reset to '0'. This is independent of the rising edge of the CLOCK signal because the rising edge test of signal CLOCK is performed only when ARSTn is not '0'. As soon as ARSTn is set to '0', DOUT(2) will be set to '0' without waiting for the rising edge of the CLOCK. This describes a flipflop behavior with asynchronous reset. Referring to the top of the Figure 8.1 schematic, the flipflop for DOUT(2) has an asynchronous reset pin connected to signal ARSTn. The D input of the flipflop connects to the output of a NOR gate. Note that all target signals will be synthesized with flipflop outputs, if the signal assignments are in the region between checking the rising edge (line 32) with keyword **elsif** and the keywords **end if** in line 34. In other words, more sequential statements can be used even though there is only one signal assignment statement enclosed by the checking of the CLOCK rising edge in this example.

Note also that the process has a sensitivity list of ARSTn and CLOCK, but not DIN. This is because DIN is only used when CLOCK is at the rising edge. Including DIN in the sensitivity list would not change the functional behavior. When DIN is changed, ARSTn = '1' and CLOCK is not at the rising edge, the process is evaluated with no effect. The evaluation of the process is a wasted effort.

```
28        seq2 : process (ARSTn, CLOCK)
29        begin
30           if (ARSTn = '0') then
31              DOUT(2) <= '0';
```

```
32          elsif (CLOCK'event and CLOCK = '1') then
33             DOUT(2) <= DIN(2) nor DIN(3);
34          end if;
35       end process;
```

Lines 36 to 46 show another process which is similar to the last process. The signal CLOCK is used as the process sensitivity list. Line 38 checks the rising edge of the signal CLOCK. The signal assignment statement and if statement are only executed when the CLOCK is at the rising edge. In this case, it is clear to see that DOUT(5) and DOUT(3) will be outputs of flipflops. SRSTn is treated as a synchronous reset. In Figure 8.1, signal SRSTn is connected to an AND gate with output connecting to inputs of both flipflops. DIN and SRSTn do not need to be included in the sensitivity for simulation efficiency.

```
36       seq3 : process (CLOCK)
37       begin
38          if (CLOCK'event and CLOCK = '1') then
39             DOUT(5) <= SRSTn and DIN(3);
40             if (SRSTn = '0') then
41                DOUT(3) <= '0';
42             else
43                DOUT(3) <= DIN(3);
44             end if;
45          end if;
46       end process;
47    end RTL;
```

Most synthesis tools treat the signal in the rising edge check as a clock signal to the flipflop being inferred. The testing of the rising edge of a signal for synthesis cannot be combined with other signals in the same expression. For example, *if (CLOCK'event and CLOCK = '1' and EN = '1')* may be used to model a flipflop with an enable pin EN. This is good for simulation. The synthesis tools, however, may not accept it and indicate the detection of an illegal edge. To achieve the same function for both simulation and synthesis, the checking of EN = '1' can be another if statement (lines 15 to 17) inside the "*if (CLOCK'event and CLOCK = '1') then*" as shown in the following VHDL code:

```
1     library IEEE;
2     use IEEE.std_logic_1164.all;
3     entity FFEN is
4        port(
5           CLOCK, ARSTn, EN, DIN   : in  std_logic;
6           DOUT                    : out std_logic);
7     end FFEN;
8     architecture RTL of FFEN is
9     begin
10       seq2 : process (ARSTn, CLOCK)
11       begin
12          if (ARSTn = '0') then
13             DOUT <= '0';
```

```
14         elsif (CLOCK'event and CLOCK = '1') then
15            if (EN = '1') then
16               DOUT <= DIN;
17            end if;
18         end if;
19      end process;
20   end RTL;
```

8.3 WRITING VHDL TO INFER LATCHES

A transparent latch usually has a data input signal (for example, D), an enable input signal (for example, LE for latch enable), and an output signal (for example, Q). When LE enables '1', the value of D is assigned to Q. When LE is not enabled (='0'), the value of Q is not changed. This behavior of a latch can be easily modeled with an if (or case) statement such that the target signal is not assigned in every path of the if (case) statement. In other words, there exists a condition that the target signal value is not assigned. The target signal needs to retain its old value. The target signal is not associated with an edge check to be synthesized as a flipflop. A latch is then inferred. The following VHDL code shows examples of inferring latches. Lines 1 and 2 reference IEEE package std_logic_1164. Lines 3 to 9 define the entity LATCHES.

```
1    library IEEE;
2    use IEEE.std_logic_1164.all;
3    entity LATCHES is
4       port(
5          RSTn   : in  std_logic;
6          LE     : in  std_logic;
7          DIN    : in  std_logic_vector(3 downto 0);
8          DOUT   : out std_logic_vector(3 downto 0));
9    end LATCHES;
```

Lines 12 to 17 show a process with signals DIN and LE as the sensitivity list. Lines 14 to 16 are an if statement. Note that the signal assignment in line 15 is only executed when LE = '1'. There is no **else** and **elsif** clauses in the if statement. DOUT(0) value will not be changed unless LE = '1'. Both DIN and LE are in the sensitivity list so that the process can be evaluated and DOUT(0) value may change when DIN is changed—even LE is stable at '1'. In the schematic in Figure 8.2, DOUT(0) is connected to the output of a latch where the enable pin is connected to LE with D pin connected to the output of a XOR gate.

```
10   architecture RTL of LATCHES is
11   begin
12      seq0 : process (DIN, LE)
13      begin
14         if (LE = '1') then
15            DOUT(0) <= DIN(0) xor DIN(1);
16         end if;
17      end process;
```

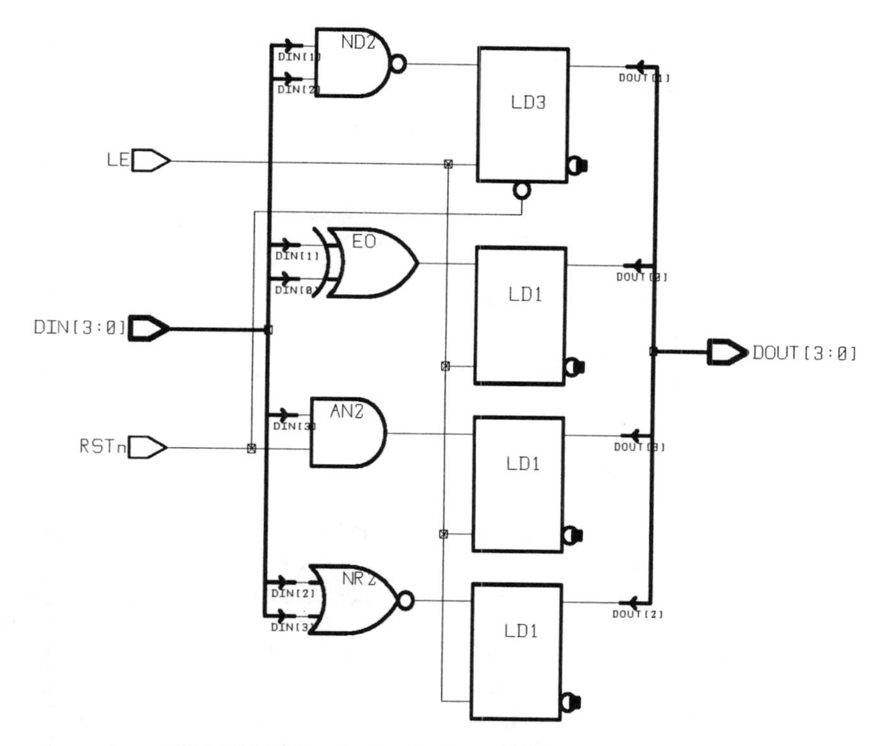

FIGURE 8.2 Synthesized schematic of LATCHES VHDL code.

Lines 18 to 25 have a process with sensitivity list of RSTn, LE, and DIN. RSTn = '0' is checked before checking LE = '1' in the if statement. The latch is an asynchronous reset latch (LD3). Lines 26 to 34 show another process statement with a case statement inside. DOUT(2) is not assigned a value when LE is not '1'. Lines 35 to 44 show another process statement. The checking of RSTn = '0' is inside the if statement when LE = '1'. The reset is performed only when LE = '1' and RSTn = '0'. The data input pin of the latch is the output of an AND gate with RSTn and DIN(3) as inputs.

```
18      seq1 : process (RSTn, LE, DIN)
19      begin
20         if (RSTn = '0') then
21            DOUT(1) <= '0';
22         elsif (LE = '1') then
23            DOUT(1) <= DIN(1) nand DIN(2);
24         end if;
25      end process;
26      seq2 : process (LE, DIN)
27      begin
28         case LE is
29            when '1' =>
30               DOUT(2) <= DIN(2) nor DIN(3);
31            when others =>
```

```
32                NULL;
33             end case;
34          end process;
35          seq3 : process (RSTn, LE, DIN)
36          begin
37             if (LE = '1') then
38                if (RSTn = '0') then
39                   DOUT(3) <= '0';
40                else
41                   DOUT(3) <= DIN(3);
42                end if;
43             end if;
44          end process;
45       end RTL;
```

The latch is inferred when a target object is not assigned a value in one of the execution paths of an if (case) statement. Extra latches may be synthesized due to mistakes in not assigning in every possible execution path. This mistake is easily made when there are many **elsif** clauses in an if statement or many choices in a case statement. Synthesis tools such as Synopsys will generate a report that summarizes which signal may be synthesized with flipflop, latch, or tristate buffer. Check the report to see if there are any extra latches. A good way to reduce this problem of extra latches is to use a sequential signal assignment statement to assign a default value. Later, another sequential signal assignment could replace the default value. If there is no sequential signal assignment value, the default value is used. No latch would be generated. For example, a default output value of a state machine can be set initially. During some states, the default value may be replaced. This reduces the burden to have one sequential signal assignment statement for every possible execution path. This also reduces the amount of VHDL code.

8.4 WRITING VHDL TO INFER TRISTATES

A tristate buffer has an input pin (for example, D), an enable pin (for example, EN), and an output pin (for example, Y). When EN is asserted '1', Y gets the value of D. When EN is not asserted, Y gets the high impedance state value 'Z'. Note that 'Z' is not a value of type bit. It is a value of type std_logic. This behavior can be easily described in VHDL. The following VHDL code shows examples of generating tristate buffers. Whenever a target signal can be assigned the value 'Z', it is synthesized as an output of a tristate buffer. Lines 3 to 8 define the entity STATE3. Type std_logic and std_logic_vector are used so that lines 1 and 2 reference the IEEE std_logic_1164 package.

```
1     library IEEE;
2     use IEEE.std_logic_1164.all;
3     entity STATE3 is
4        port(
5           EN    : in  std_logic;
6           DIN   : in  std_logic_vector(3 downto 0);
7           DOUT  : out std_logic_vector(3 downto 0));
8     end STATE3;
```

Lines 11 to 18 are a process statement with DIN and EN as the sensitivity list. DOUT(0) is assigned to 'Z' in line 16 when EN is not '1' in an if statement. Lines 19 to 27 are another process using a case statement where DOUT(1) is assigned to 'Z' in line 25. Line 28 is a concurrent conditional signal assignment statement. DOUT(2) gets the value 'Z' when EN is not '1'. Lines 29 to 31 are a concurrent selected signal assignment statement. DOUT(3) gets the value 'Z' in line 31 when EN is not '1'. Figure 8.3 shows the synthesized schematic.

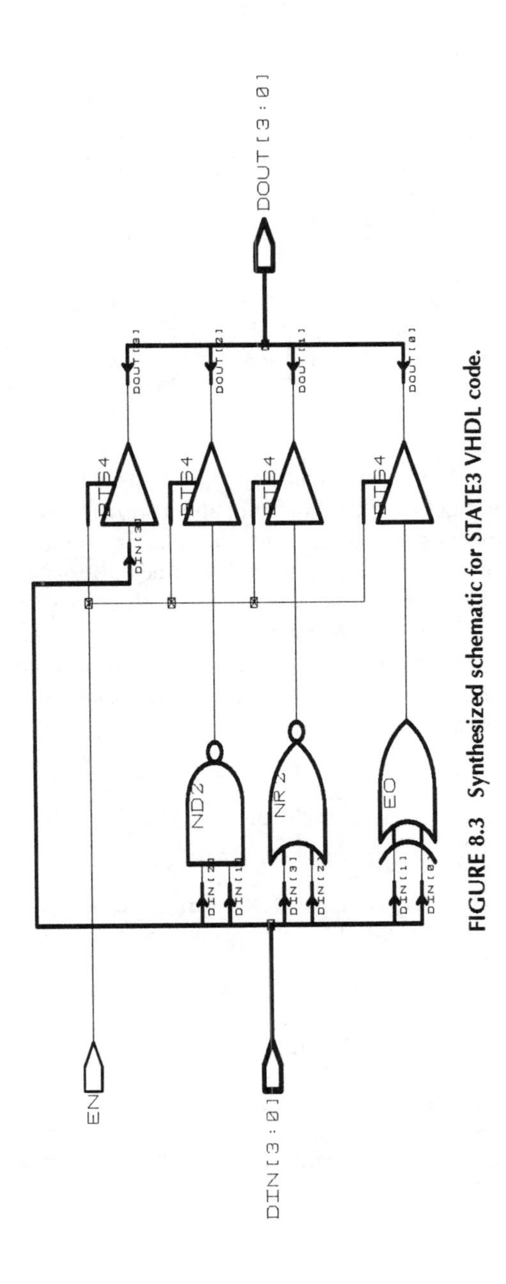

FIGURE 8.3 Synthesized schematic for STATE3 VHDL code.

```
9     architecture RTL of STATE3 is
10    begin
11       seq0 : process (DIN, EN)
12       begin
13          if (EN = '1') then
14             DOUT(0) <= DIN(0) xor DIN(1);
15          else
16             DOUT(0) <= 'Z';
17          end if;
18       end process;
19       seq1 : process (EN, DIN)
20       begin
21          case EN is
22             when '1' =>
23                DOUT(1) <= DIN(2) nor DIN(3);
24             when others =>
25                DOUT(1) <= 'Z';
26          end case;
27       end process;
28       DOUT(2) <= DIN(1) nand DIN(2) when EN = '1' else 'Z';
29       with EN select
30          DOUT(3) <= DIN(3) when '1',
31                     'Z'    when others;
32    end RTL;
```

It is important to know what the synthesis tools do if tristate buffers are inferred with flipflops or latches. For example, the following VHDL code shows a tristate buffer inferred in lines 13 and 21. They are also inside the rising edge check (line 15) and incomplete assigned (latch inferred) in lines 20 to 24. Figure 8.4 shows the synthesized schematic for TRIREG VHDL code. Note that when the output of a flipflop or a latch is tristated, the enable line of the tristate buffer is also flipflopped or latched, respectively. You are encouraged to check your synthesis tools to see what circuits are synthesized. You may want to keep the VHDL code simple for portability among VHDL simulation and synthesis tools.

```
1     library IEEE;
2     use IEEE.std_logic_1164.all;
3     entity TRIREG is
4        port (
5           CLOCK, LE, TRIEN, DIN : in  std_logic;
6           FFOUT, LATCHOUT       : out std_Logic);
7     end TRIREG;
8     architecture RTL of TRIREG is
9     begin
10       ff0 : process (TRIEN, CLOCK)
11       begin
12          if (TRIEN = '0') then
13             FFOUT <= 'Z';
14          elsif (CLOCK'event and CLOCK = '1') then
15             FFOUT <= DIN;
16          end if;
17       end process;
```

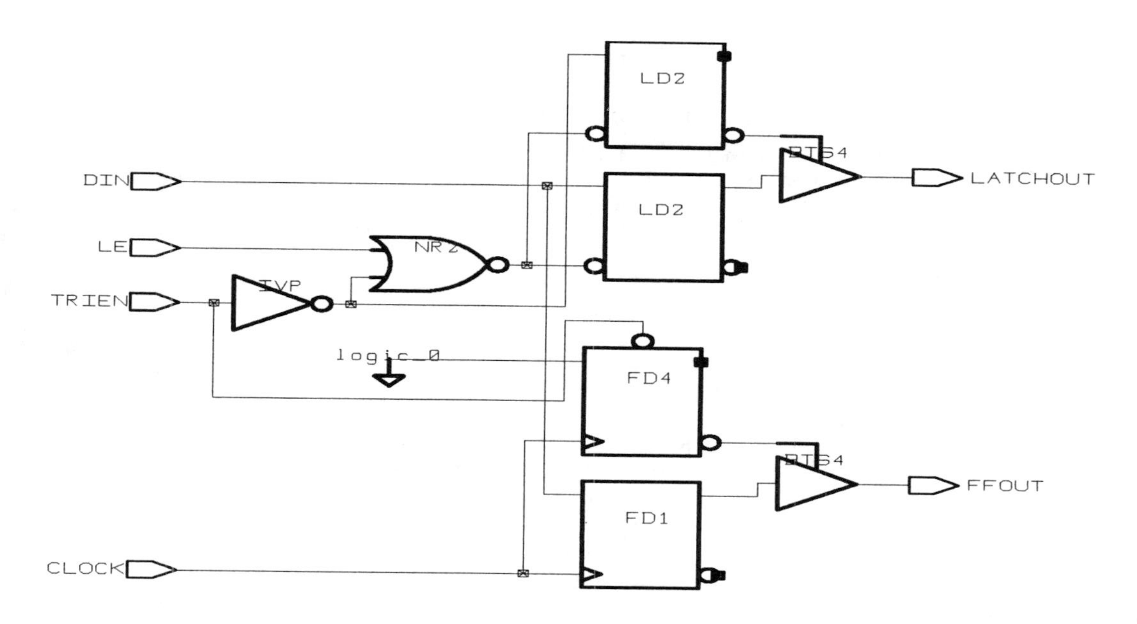

FIGURE 8.4 Tristate buffer with flipflop and latch inferred.

```
18      latch0 : process (LE, TRIEN, DIN)
19      begin
20         if (TRIEN = '0') then
21            LATCHOUT <= 'Z';
22         elsif (LE = '1') then
23            LATCHOUT <= DIN;
24         end if;
25      end process;
26   end RTL;
```

8.5 WRITING VHDL TO GENERATE COMBINATIONAL CIRCUITS

Combination circuits can be generated in several ways. Expressions with a BOOLEAN algebra operator as shown in the preceding examples such as XOR, NAND, AND gates are generated *before* going to flipflops, latches, and tristate buffers. When a target signal (variable) is not inferred as flipflops, latches, or tristate buffers, then it will be synthesized as a combinational circuit. For example, the IFCASE VHDL code as shown in Chapter 5 exercises, shows architectures that can be synthesized to combinational circuits without flipflops, latches, and tristates. Their corresponding concurrent (conditional or selected) signal assignment will be synthesized to combinational circuits too. The sequential if (case) statements and concurrent conditional (selected) signal assignments are usually synthesized with combinational circuits of MUX functions.

The following VHDL code shows entity PARGEN which generates an even parity for a 32-bit AD bus and 4-bit CBEn bus with a flipflop output PAR. Lines 13 to 22 show a function which returns the even parity value by passing in an unconstrained parameter DIN. The process statement in lines 24 to 31 is very similar to the process used to infer flipflop as discussed in Section 1. Note that signal PAR will correspond to the output of the flipflop while the combinational circuits for parity generation are connected to the D-input of the flipflop. Figure 8.5 shows the synthesized schematic. Note that the function calculates the parity with a for loop (lines 18 to 20). The schematic does not show a series of two-input XOR in a long chain. This is the result of the synthesis tool that understands its function and performs optimizations.

```
1    library IEEE;
2    use IEEE.std_logic_1164.all;
3    entity PARGEN is
4       port(
5          RSTn   : in  std_logic;
6          CLOCK  : in  std_logic;
7          AD     : in  std_logic_vector(31 downto 0);
8          CBEn   : in  std_logic_vector(3 downto 0);
9          PAR    : out std_logic);
10   end PARGEN;
11
12   architecture RTL of PARGEN is
13      function EVENPAR (constant DIN : in std_logic_vector)
14         return std_logic is
15         variable result : std_logic;
16      begin
17         result := '0';
18         for i in DIN'range loop
19            result := result xor DIN(i);
20         end loop;
21         return result;
22      end EVENPAR;
23   begin
24      delay : process (RSTn, CLOCK)
25      begin
26         if (RSTn = '0') then
27            PAR <= '0';
28         elsif (CLOCK'event and CLOCK = '1') then
29            PAR <= EVENPAR(AD) xor EVENPAR(CBEn);
30         end if;
31      end process;
32   end RTL;
```

It is important to think about the corresponding hardware when writing VHDL code for synthesis. Comparison operators such as =, >, < would imply an equal checking and comparator combinational circuits. Arithmetic operators such as +, -, and * would imply an adder, subtracter, and a multiplier. You are encouraged to check your synthesis tools to determine whether they would synthesize the comparison and the arithmetic operators. Some synthesis tools can take a division operator "/" if the divisor is a constant of 2 to an integer's power (2, 4, 8, and so on). An arithmetic shift is synthesized.

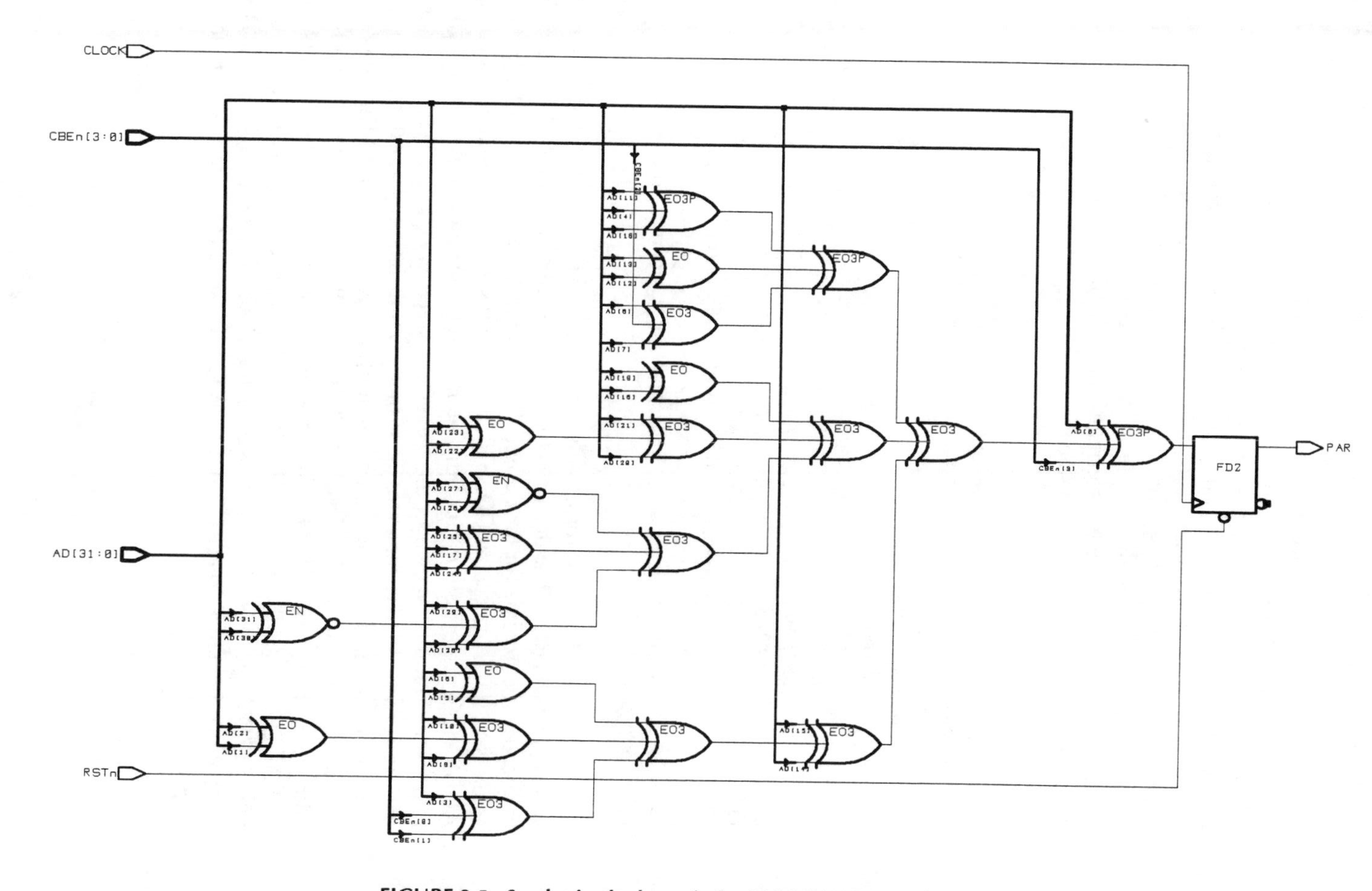

FIGURE 8.5 Synthesized schematic for PARGEN VHDL code.

8.6 PUTTING THEM TOGETHER

Let's consider a design case to illustrate how we can apply what we have discussed. Suppose we are to design a circuit called CONTROL that has inputs RSTn, CLOCK, AD, FRAMEn, IRDYn and output signals WR, WORD, DATA, and ADDR. RSTn is an asynchronous input to the flipflops. CLOCK is the clock signal. AD is an 8-bit address and data multiplexed input. FRAMEn is a low active signal. When it is asserted '0', the address appears at the AD bus. After that, whenever IRDYn is asserted low '0', a valid data is put in the AD bus. When FRAMEn is deasserted, it indicates the last data to be transferred when IRDYn is asserted low '0'. Lines 4 to 15 show the entity CONTROL VHDL code.

```
1     library IEEE;
2     use IEEE.std_logic_1164.all;
3     use IEEE.std_logic_arith.all;
4     entity CONTROL is
5        port(
6           RSTn   : in  std_logic;
7           CLOCK  : in  std_logic;
8           AD     : in  std_logic_vector(7 downto 0);
9           FRAMEn : in  std_logic;
10          IRDYn  : in  std_logic;
11          WR     : out std_logic;
12          WORD   : out std_logic_vector(3 downto 0);
13          DATA   : out std_logic_vector(7 downto 0);
14          ADDR   : out std_logic_vector(3 downto 0));
15    end CONTROL;
```

The simulation waveform shown in Figure 8.6 can be used to describe the operations of CONTROL. The design is to latch the address when FRAMEn goes '0', then to decode whether it is for the design CONTROL. If the address most significant 4-bit value is "1100", the design CONTROL should respond. The address and data output ADDR (least significant 4-bit of the latched address) and DATA are normally in high impedance states as shown in the waveform when WR is '0'. WR indicates writing data is active. WORD indicates how many valid data are transferred (assuming at most 15 data transfers for each transaction). For example, FRAMEn goes '0' at time 80 ns, address value X'C3" (B"11000011") is latched. The most significant 4-bit address is "1100". CONTROL should be active by enabling tristate buffers for DATA and ADDR. FRAMEn goes to '1' after one clock cycle. The last data is transferred when IRDYn is '0' at the rising edge of the clock. WR outputs '1' when DATA and ADDR is driven by CONTROL. When WR = '0', both DATA and ADDR are in a high impedance state.

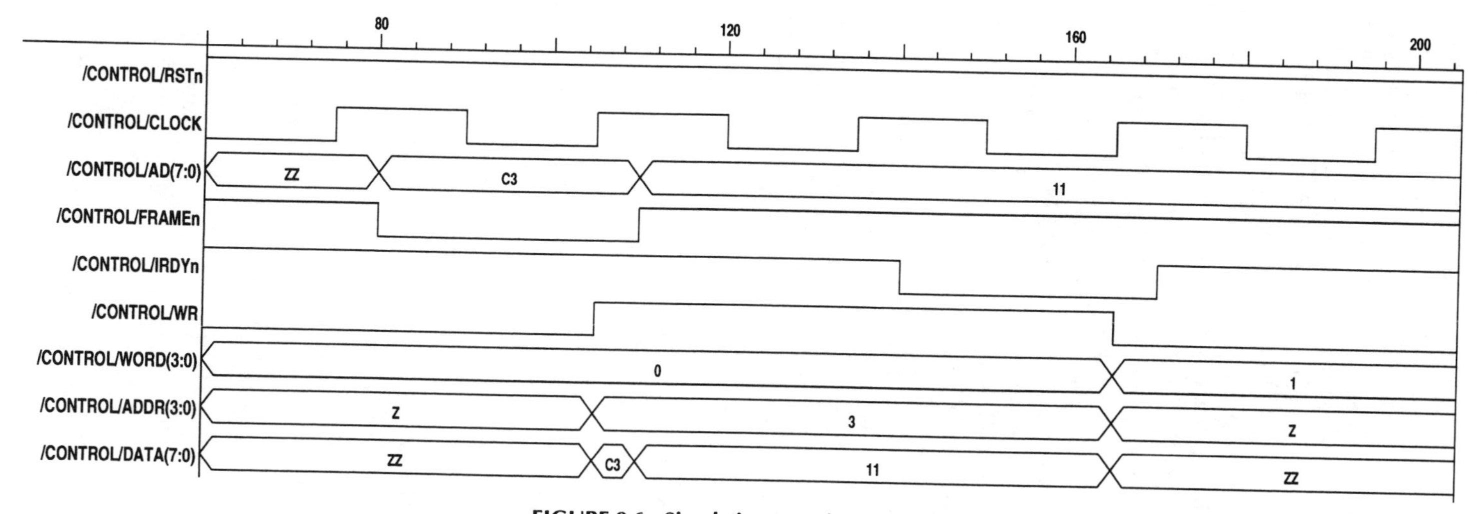

FIGURE 8.6 Simulation waveform for CONTROL.

To approach the design of CONTROL, we need to latch in the address when FRAMEn is asserted. Latches can be inferred and a latch enable signal should be generated. The most significant latched addresses are then decoded to compare with the value "1100" (combinational circuits). The counter for the WORD needs to have 4-bit flipflops inferred. The counter should be controlled regarding when to reset and when to enable the count. Output signals DATA and ADDR are outputs of tristate buffers. The enable signal to control tristate buffers needs to be generated also. The rough concept is shown in Figure 8.7. The signal ADDR_LATCH is declared in line 17. It is used to infer 8-bit latches. A 4-bit counter is required, and line 18 declares a signal WORD_CNTR. Lines 19 and 20 declare other single bit signals. The tristate buffers would also be needed. Other circuits include the address decoder and other control signals such as CNT_RESET and CNT_EN to synchronously reset the counter and to enable the counter, respectively. The circular blocks in the figure are close to VHDL simulation concurrent processes communicated with signals.

The latch enable signal LATCH_EN is generated as follows. A signal FRAMEn_DELAY is generated by delaying FRAMEn by one clock cycle using an inferred D-type flipflop, as shown in lines 22 to 29. LATCH_EN is obtained in line 30. Lines 31 to 36 use AD and LATCH_EN as the sensitivity list of the process to infer 8-bit latches associated with signal ADDR_LATCH.

```
16    architecture RTL of CONTROL is
17       signal ADDR_LATCH : std_logic_vector(7 downto 0);
18       signal WORD_CNTR  : std_logic_vector(3 downto 0);
19       signal LATCH_EN, CNT_RESET, CNT_EN, BUSYn     : std_logic;
20       signal ADDR_HIT, FRAMEn_DELAY                 : std_logic;
21    begin
22       delay : process (RSTn, CLOCK)
23       begin
24          if (RSTn = '0') then
```

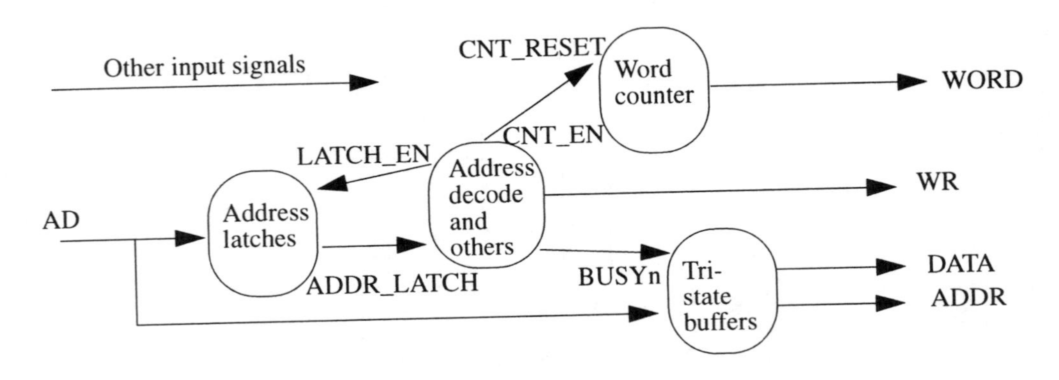

FIGURE 8.7 Rough concept for design CONTROL.

```
25             FRAMEn_DELAY <= '1';
26          elsif (CLOCK'event and CLOCK = '1') then
27             FRAMEn_DELAY <= FRAMEn;
28          end if;
29       end process;
30       LATCH_EN <= FRAMEn nor (not FRAMEn_DELAY);
31       latch : process (LATCH_EN, AD)
32       begin
33          if (LATCH_EN = '1') then
34             ADDR_LATCH <= AD;
35          end if;
36       end process;
37       ADDR_HIT  <= '1' when ADDR_LATCH(7 downto 4) =
38                                 "1100" else '0';
39       CNT_RESET <= ADDR_HIT and LATCH_EN;
40       CNT_EN    <= not (BUSYn or IRDYn);
41       WORD      <= WORD_CNTR;
42       WR        <= not BUSYn;
43       cnt_gen : process (RSTn, CLOCK)
44       begin
45          if (RSTn = '0') then
46             WORD_CNTR <= "0000";
47          elsif (CLOCK'event and CLOCK = '1') then
48             if (CNT_RESET = '1') then
49                WORD_CNTR <= "0000";
50             elsif (CNT_EN = '1') then
51                WORD_CNTR <= unsigned(WORD_CNTR) + 1;
52             end if;
53          end if;
54       end process;
55       bsy : process (RSTn, CLOCK)
56       begin
57          if (RSTn = '0') then
58             BUSYn <= '1';
59          elsif (CLOCK'event and CLOCK = '1') then
60             if (CNT_RESET = '1') then
61                BUSYn <= '0';
62             elsif (FRAMEn = '1') and (IRDYn = '0') then
63                BUSYn <= '1';
64             end if;
65          end if;
66       end process;
67       ADDR <= ADDR_LATCH(3 downto 0) when BUSYn = '0' else "ZZZZ";
68       DATA <= AD                     when BUSYn = '0' else (others => 'Z');
69    end RTL;
```

Lines 37 and 38 decode the address. ADDR_HIT signal is set to '1' when the right address "1100" is matched. The counter is described by lines 43 to 54. The counter is asynchronously reset to "0000" when RSTn = '0' (lines 45 and 46). At the rising edge of the CLOCK (line 47), the counter can be reset to "0000" (line 49) synchronously when CNT_RESET = '1' (line 48). Otherwise, the counter is increased by 1 when the counter is enabled by signal CNT_EN = '1'. The CNT_RESET signal is generated in line 39. The counter is reset to 0 in the beginning of each transaction. Line 41 puts the WORD_CNTR to the output. Note that WORD is of mode **out**. It cannot be read in line 51. Therefore, signal WORD_CNTR is declared and used. A signal BUSYn is asserted '0' to indicate the CONTROL circuit is active to drive ADDR and DATA. The BUSYn is modeled in lines 55 to 66. It is asynchronously preset to '1' when RSTn = '0'. It is asserted to '0' at the rising edge when CNT_RESET = '1' and deasserted to '1' when FRAMEn is '1' and IRDYn = '0'. Line 42 generates output signal WR. Lines 67 and 68 infer tristate buffers for ADDR and DATA output.

Note that line 51 uses an arithmetic operator "+". This operator usually takes both arguments of INTEGER type and returns an integer. This operator is overloaded in std_logic_arith package to take an argument of type unsigned (which is almost identical to std_logic_vector) and an argument of type INTEGER, and returns a value of type std_logic_vector. They are defined as:

```
type UNSIGNED is array (NATURAL range <>) of STD_LOGIC;
function "+"(L: UNSIGNED; R: INTEGER) return STD_LOGIC_VECTOR;
```

This is the reason that std_logic_arith package is referenced in line 3. An adder is inferred and optimized.

Figures 8.6, 8.8, 8.9, and 8.10 show the simulation waveform for CONTROL. Figure 8.6 shows a transfer of one data. Figure 8.8 shows a transfer of four data. Figure 8.9 shows no data being transferred because the address is not right for CONTROL. Figure 8.10 shows a transfer of three data. Note that data is not transferred when IRDYn is '1'. Readers are encouraged to verify the simulation waveforms and to understand the waveforms and VHDL code relationship.

The CONTROL VHDL code can be synthesized. Figure 8.11 shows the schematic. At the left side, a D-type flipflop with asynchronous preset corresponds to FRAMEn_DELAY. After the NOR gate, it is the LATCH_EN signal that goes to the enable pin of 8 latches. Four of the latches in the left side are for the most significant 4 bits of address that goes through an AND gate to check the value of "1100". It then goes through several gates to another asynchronous preset D-type flipflop which corresponds to BUSYn signal. The bottom part of the schematic is the 4-bit counter with outputs connected to output signal WORD. The least significant 4 bits of latched address go to four tristate buffers with ADDR as output signal. Another eight tristate buffers get the input directly from AD and output to DATA. WR then takes the inverted output of the BUSYn flipflop.

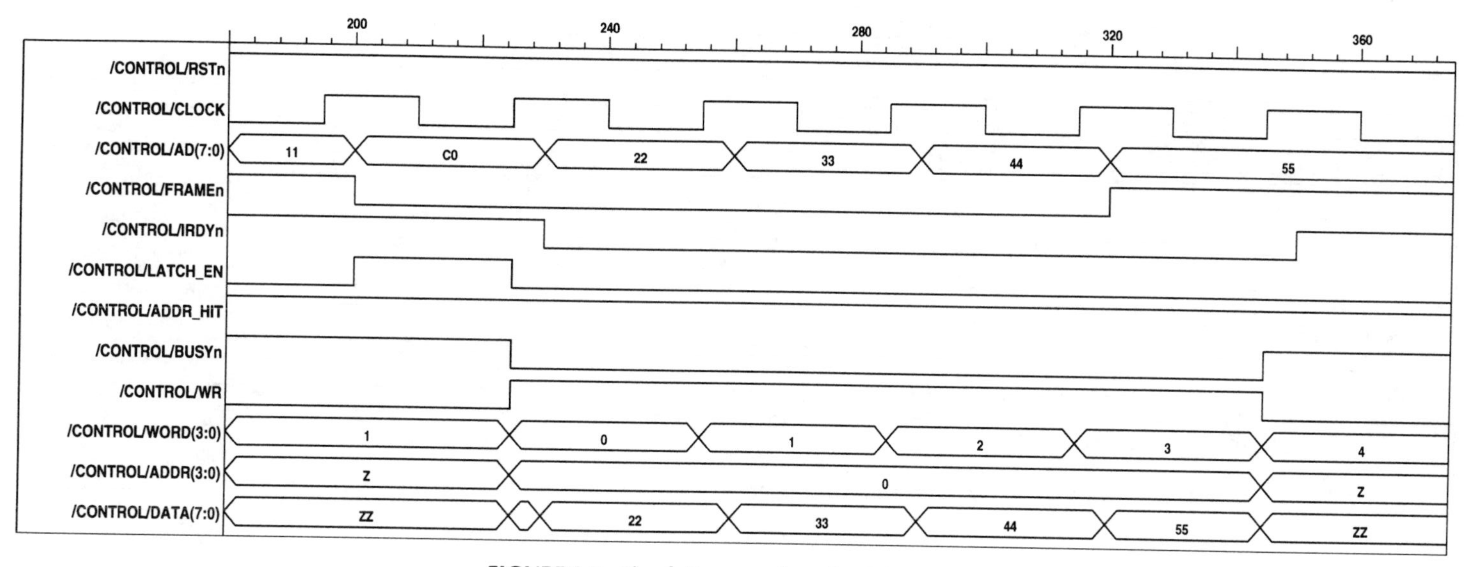

FIGURE 8.8 Simulation waveform for CONTROL.

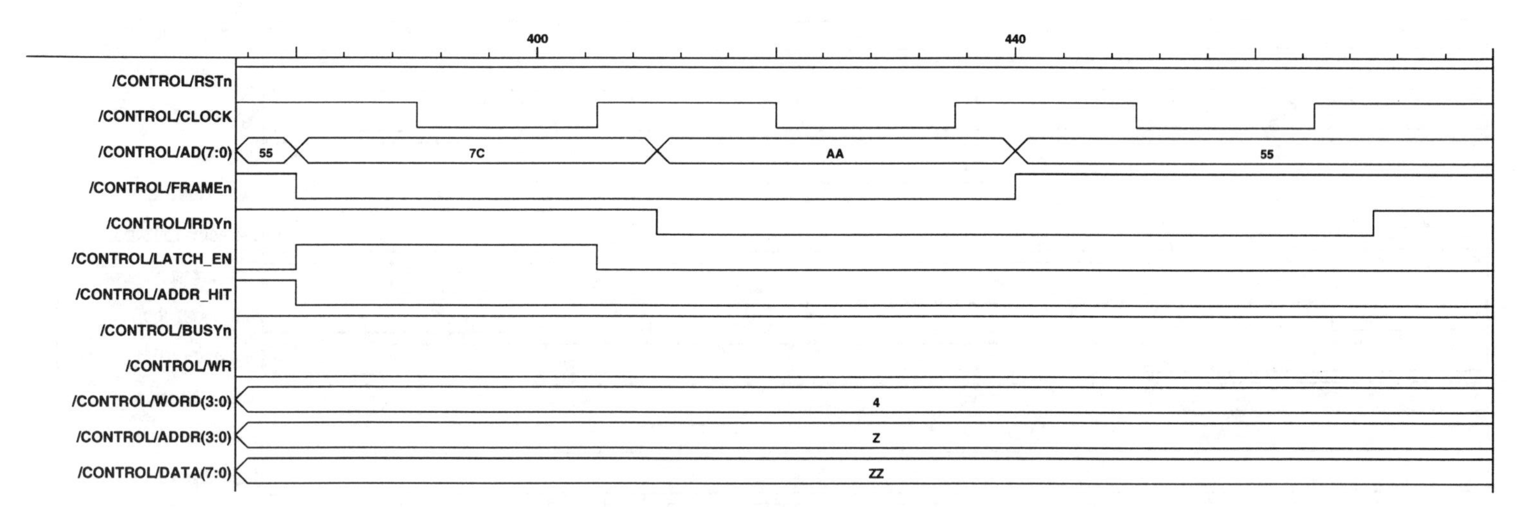

FIGURE 8.9 Simulation waveform for CONTROL.

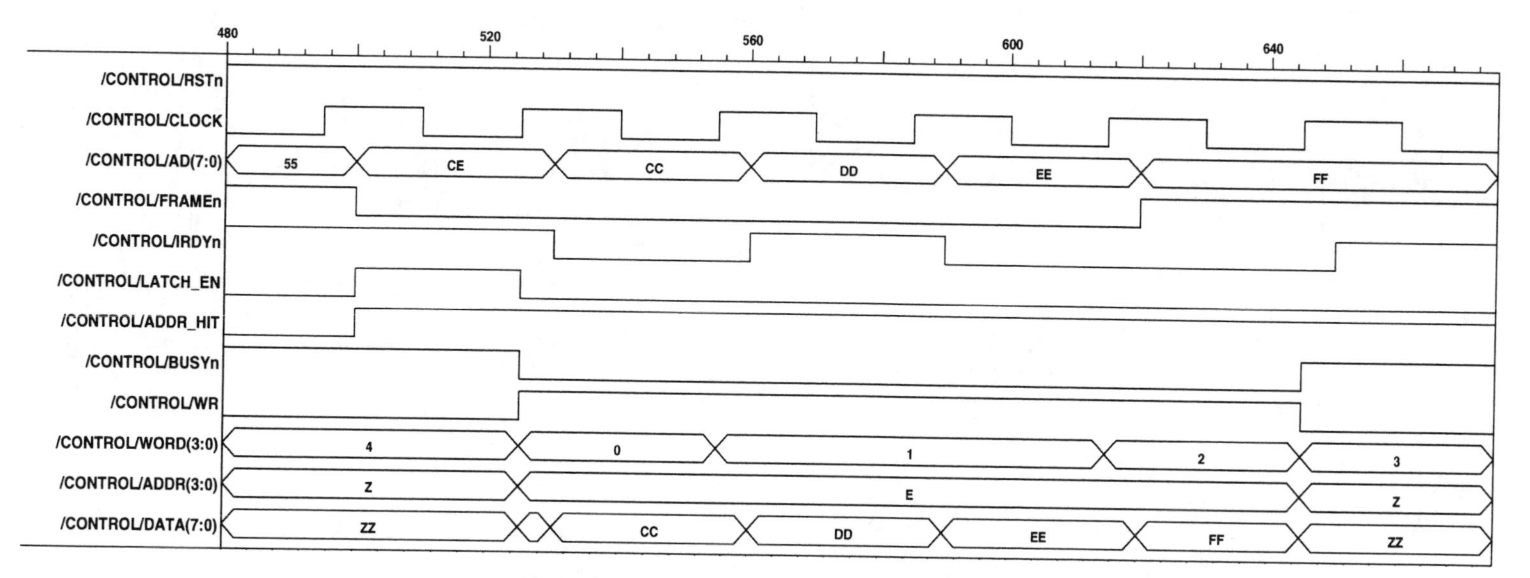

FIGURE 8.10 Simulation waveform for CONTROL.

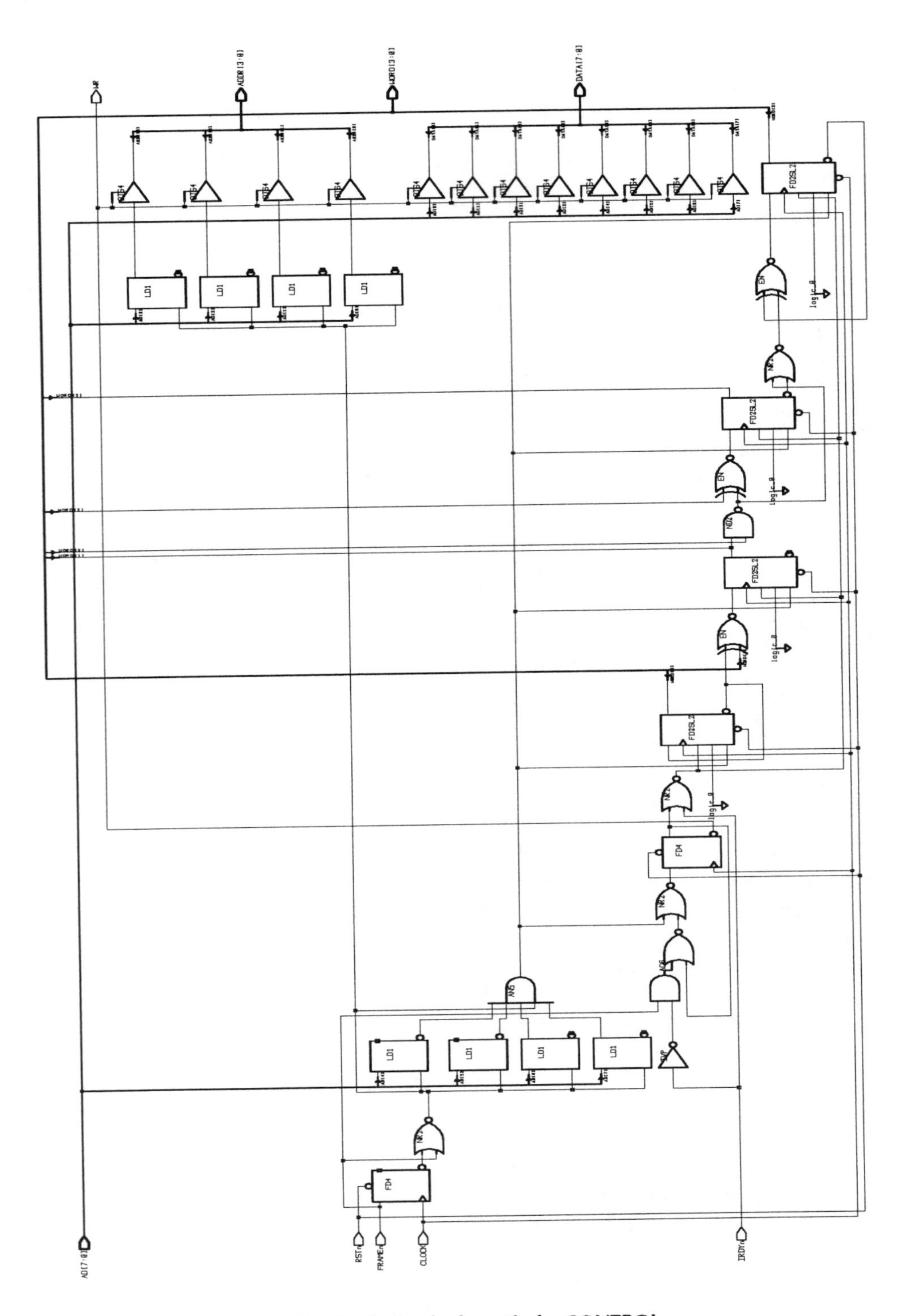

FIGURE 8.11 Synthesized schematic for CONTROL.

In real applications, it would be difficult (if not impossible) to check the schematic as described. The schematic may be too big. The correctness of the design is verified with the VHDL simulator in functional level. After it is synthesized, another structural VHDL code can be generated with an estimated pre-layout delay. A more detailed simulation can be performed. After the layout, the post-layout timing can be generated and included in the simulation. During the synthesis, it is recommended to review the synthesis summary such as the number of flipflops, latches, and tristate buffers. It will become an "act of faith" after you have more experience with VHDL and the synthesis tools. You will not have to review the schematic.

The architecture RTL of CONTROL has three processes to generate FRAMEn_DELAY, WORD_CNTR, and BUSYn. Note that they share the same sensitivity list (RSTn, CLOCK). The if statement constructs inside the processes are the same. They can be combined as one process as shown in lines 76 to 95. This saves 12 lines of VHDL code. It is also more efficient for simulation, since only one process is used. The amount of time to move the process among process states is reduced. RSTn and the rising edge check of CLOCK are done in one process rather than three separate processes.

```
70   architecture RTL1 of CONTROL is
71      signal ADDR_LATCH : std_logic_vector(7 downto 0);
72      signal WORD_CNTR  : std_logic_vector(3 downto 0);
73      signal LATCH_EN, CNT_RESET, CNT_EN, BUSYn   : std_logic;
74      signal ADDR_HIT, FRAMEn_DELAY               : std_logic;
75   begin
76      ff : process (RSTn, CLOCK)
77      begin
78         if (RSTn = '0') then
79            FRAMEn_DELAY <= '1';
80            WORD_CNTR    <= "0000";
81            BUSYn        <= '1';
82         elsif (CLOCK'event and CLOCK = '1') then
83            FRAMEn_DELAY <= FRAMEn;
84            if (CNT_RESET = '1') then
85               WORD_CNTR <= "0000";
86            elsif (CNT_EN = '1') then
87               WORD_CNTR <= unsigned(WORD_CNTR) + 1;
88            end if;
89            if (CNT_RESET = '1') then
90               BUSYn <= '0';
91            elsif (FRAMEn = '1') and (IRDYn = '0') then
92               BUSYn <= '1';
93            end if;
94         end if;
95      end process;
96      LATCH_EN <= FRAMEn nor (not FRAMEn_DELAY);
97      latch : process (LATCH_EN, AD)
98      begin
99         if (LATCH_EN = '1') then
100           ADDR_LATCH <= AD;
101        end if;
102     end process;
```

```
103      ADDR_HIT  <= '1' when ADDR_LATCH(7 downto 4) =
104                          "1100" else '0';
105      CNT_RESET <= ADDR_HIT and LATCH_EN;
106      CNT_EN    <= not (BUSYn or IRDYn);
107      WORD      <= WORD_CNTR;
108      WR        <= not BUSYn;
109      ADDR <= ADDR_LATCH(3 downto 0) when BUSYn = '0' else "ZZZZ";
110      DATA <= AD              when BUSYn = '0' else (others => 'Z');
111   end RTL1;
```

8.7 SIMULATION VERSUS SYNTHESIS DIFFERENCES

The process sensitivity list has a major impact on the preceding simulation of the corresponding process statement. However, the synthesis tools usually ignore the process sensitivity list and are able to generate corresponding hardware. For example, in the preceding VHDL code, lines 97 to 102 show a process for generating latches with LATCH_EN and AD as the sensitivity list. If either LATCH_EN or AD is missing from the process sensitivity list, the synthesis tools can generate the same latches. However, the simulation behavior would not be the same as the actual hardware behavior even if the actual hardware were correct. Missing a signal in a process sensitivity list in a process to generate a combinational circuit is usually a mistake. Sometimes it is difficult to check to see the differences through simulation, especially when the sensitivity list has many signals. A good way to approach this is to use the synthesis tool to parse the VHDL code. A warning message is generated if a signal is missing from the sensitivity list which is read inside a process statement corresponding to combinational circuits.

8.8 THINK ABOUT HARDWARE

When writing VHDL code for synthesis, it is important to keep in mind what corresponding hardware would be inferred or generated by the synthesis tools. Is it efficient in terms of area and speed? As an example, the following VHDL code shows an entity LE64 that inputs two 64-bit bus and one bit combinational circuit output LE. LE = '1' if DATA1 is less or equal to DATA2. Lines 1 to 9 describe the entity LE64 with references to library IEEE and packages std_logic_1164 and std_logic_arith. A quick, straightforward approach for the architecture is shown in lines 10 to 13. This is good for both simulation and synthesis.

```
1     library IEEE;
2     use IEEE.std_logic_1164.all;
3     use IEEE.std_logic_arith.all;
4     entity LE64 is
5        port (
6           DATA1  : in  std_logic_vector(63 downto 0);
7           DATA2  : in  std_logic_vector(63 downto 0);
8           LE     : out std_logic);
9     end LE64;
10    architecture RTL1 of LE64 is
```

```
11    begin
12       LE <= '1' when (data1 <= data2) else '0';
13    end RTL1;
```

The following VHDL code shows another architecture for LE64 to implement the same function. They would simulate exactly the same. More lines of VHDL code are used. However, architecture RTL2 implies a subtracter in line 21, while architecture RTL implies a comparator (line 12). From the hardware point of view, a subtracter (carry look ahead is available) is more efficient than a comparator. Both architectures are synthesized with the same conditions and constraints (LSI lca300k, B5X5 wire load model, and WCMIL operating conditions). RTL1 has an area of 480 combinational circuit units with 31 ns delay. RTL2 has an area of 386 combinational circuit units with about 15 ns delay.

```
14    architecture RTL2 of LE64 is
15    begin
16       process (data1, data2)
17          variable x, y, z : unsigned(64 downto 0);
18       begin
19          x := '0' & unsigned(DATA1);
20          y := '0' & unsigned(DATA2);
21          z := y - x;
22          if (z(64) = '0') then
23             LE   <= '1';
24          else
25             LE   <= '0';
26          end if;
27       end process;
28    end RTL2;
```

The following VHDL code shows an example of using an arithmetic operator "–". From the hardware point of view, the "–" operator corresponds to a subtracter or a decrementer. Input signal SEL selects either A-1, B-1, C-1, or D-1 to the register output DOUT. Architecture RTL uses four concurrent signal assignment statements (lines 14 to 17) and a process statement for the register output DOUT with a case statement inside. The synthesis tools may not optimize across several concurrent statements so that architecture RTL will have four separate 16-bit decrementers with output going through 16-bit 4-to-1 multiplexer, and then going to 16-bit flipflops. Architecture RTL1 uses only one concurrent process statement. All "–" operators are in the same process. The synthesis tool can implement the circuit with 16-bit 4-to-1 multiplexers to select A, B, C, or D, going through a single 16-bit decrementer, and then going to 16-bit flipflops. Three 16-bit decrementers are saved compared to architecture RTL with the same timing delay.

```
1     library IEEE;
2     use IEEE.std_logic_1164.all;
3     use IEEE.std_logic_arith.all;
4     entity RESOURCE is
5        port(
6           RSTn, CLOCK : in  std_logic;
7           SEL         : in  std_logic_vector(1 downto 0);
```

```
8          A, B, C, D  : in  std_logic_vector(15 downto 0);
9          DOUT        : out std_logic_vector(15 downto 0));
10     end RESOURCE;
11     architecture RTL of RESOURCE is
12        signal A1, B1, C1, D1 : std_logic_vector(15 downto 0);
13     begin
14        A1 <= unsigned(A) - 1;
15        B1 <= unsigned(B) - 1;
16        C1 <= unsigned(C) - 1;
17        D1 <= unsigned(D) - 1;
18        seq : process (RSTn, CLOCK)
19        begin
20           if (RSTn = '0') then
21              DOUT <= (others => '0');
22           elsif (CLOCK'event and CLOCK = '1') then
23              case SEL is
24                 when "00"    => DOUT <= A1;
25                 when "01"    => DOUT <= B1;
26                 when "10"    => DOUT <= C1;
27                 when others => DOUT <= D1;
28              end case;
29           end if;
30        end process;
31     end RTL;
32     architecture RTL1 of RESOURCE is
33     begin
34        seq : process (RSTn, CLOCK)
35        begin
36           if (RSTn = '0') then
37              DOUT <= (others => '0');
38           elsif (CLOCK'event and CLOCK = '1') then
39              case SEL is
40                 when "00"    => DOUT <= unsigned(A) - 1;
41                 when "01"    => DOUT <= unsigned(B) - 1;
42                 when "10"    => DOUT <= unsigned(C) - 1;
43                 when others => DOUT <= unsigned(D) - 1;
44              end case;
45           end if;
46        end process;
47     end RTL1;
```

8.9 USE OF SUBPROGRAM

We have shown a couple of examples of functions and procedures, such as EVENPAR (this chapter) and MUX21 (Chapter 6). Other useful procedures are MUX31, MUX41, LATCH, and BUFFER3 (for tristate buffers). They can be overloaded by the argument of a single bit std_logic or an unconstrained std_logic_vector. Other useful functions are COMPLEMENT2 (2's complement), REDUCE_OR, REDUCE_AND, and so on. They can be defined in a package to be shared. For example, the latches inferred in CONTROL can be just one line of "LATCH(AD, LATCH_EN, ADDR_LATCH);" as a concurrent procedure call statement. This increases the readability to indicate hardware latch(es) is (are) inferred. It reduces the amount of VHDL coding. Also, the sensitivity list is automatically taken care of. From synthesis point of view, the functions and procedures would be lumped with other VHDL constructs for optimization. Another level of hierarchy is reduced compared to using an entity and a component instantiation. Synthesis tools in general do not optimize beyond the boundary of the entity. Also, the overhead of extra VHDL coding for component declaration, port mapping, and configuration are saved. It is costly to make another level of hierarchy just to contain a MUX21 gate.

Synthesis tools may restrict the use of signal attributes inside subprograms, for example, the rising edge check of a clock signal may not be allowed by the synthesis tools. Therefore, a procedure for the flipflop inference can be implemented if your synthesis tools can handle it.

8.10 SYNTHESIS PROCESS

The general framework of the synthesis process is described in Figure 8.12. A VHDL synthesis tool reads in VHDL source code. The operating condition sets up the temperature and manufacturing process conditions. These conditions affect the timing delay of the circuits. The wire load model describes the statistics of the interconnect wire length based on the technology and the size of the circuits for the synthesis tool to estimate the timing delay due to the interconnects among components. The component library consists of the function of each component (such as NAND gate), gate timing delay, each output pin drive capability, and the loading of each input pin. The component library may also have large macro cells such as ROM and RAM. The component library is usually provided by the manufacturing vendor. The design constraints usually specify the clock frequency and input output timing requirements. The synthesis tool usually generates schematics, netlists (VHDL, EDIF, or other netlist format), timing files (such as Standard Delay Format, SDF), and results (area, number of gates, timing violations, and so on).

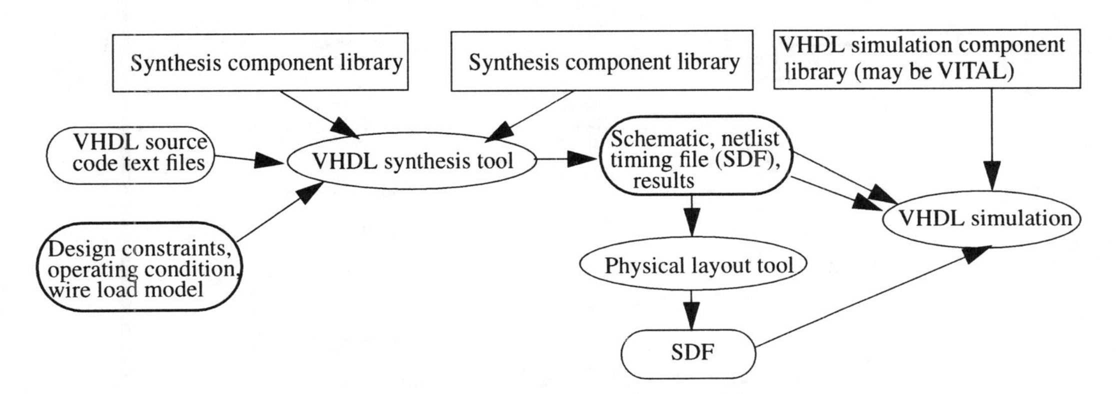

FIGURE 8.12 Synthesis framework.

The timing constraints are usually specified by the clock frequency and input output timing delays relative to the rising edge of the clock. Figure 8.13 shows the framework of the timing constraints for a design. There are two D-type flipflops which represent the registers in the design. The flipflops are clocked with the rising edge of the signal CLOCK. Combination circuits group A indicates the category of combination circuits that has inputs from the design input ports. Combination circuits group B indicates the category that has inputs from the design input ports and internal flipflops. Combinational circuits group C indicates the category of generating output signals. The output signals can also be directly from inputs (feed through) or directly from flipflops. The input delay for input ports and output delay for output ports are specified relative to the rising edge of the CLOCK. The input delay specifies the amount of time after the rising edge of the CLOCK that the input port signals are available. The output delay specifies the amount of time within which the output port signals should be ready before the rising edge of the clock. The period (frequency) of the CLOCK specifies the speed of the clock signal.

As an example, the following VHDL code shows a simple circuit with one flipflop and some combinational circuits. Output port signal FF_OUT is an output signal directly from the flipflop output, signal FF_COMB_OUT is an output from combinational circuits (with inputs from input ports and flipflop), and signal COMB_OUT is an output from combinational circuits (with inputs only from input ports).

```
1    library IEEE;
2    use IEEE.std_logic_1164.all;
3    entity CONSYN is
4       port (
5          RSTn, CLK                        : in  std_logic;
6          D0, D1, D2, D3, D4, D5, D6, D7 : in  std_logic;
```

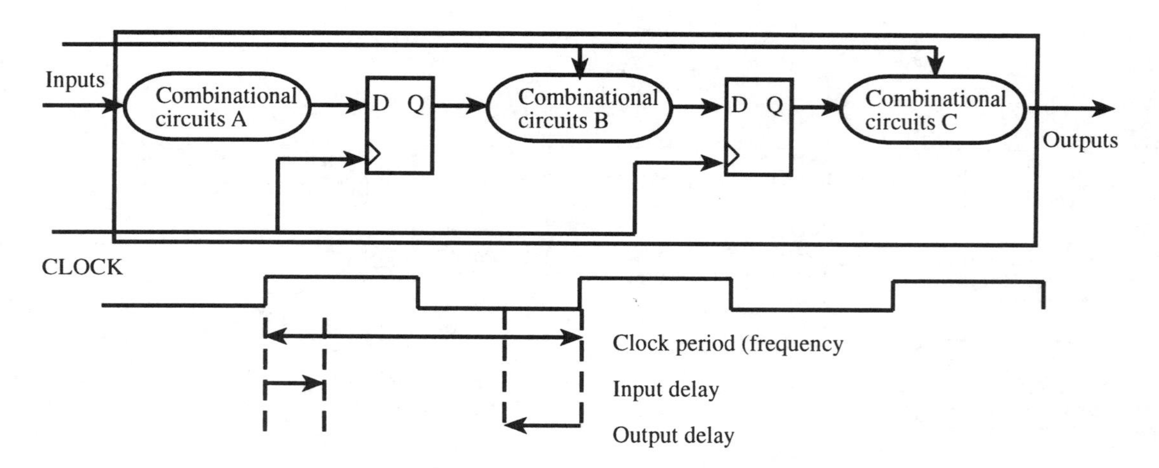

FIGURE 8.13 Timing constraints and the design.

```
7           FF_OUT, COMB_OUT, FF_COMB_OUT   : out std_logic);
8     end CONSYN;
9     architecture RTL of CONSYN is
10       signal XOR8, FF   : std_logic;
11    begin
12       XOR8        <= D7 xor D6 xor D5 xor D4 xor D3 xor D2 xor D1 xor D0;
13       COMB_OUT     <= XOR8;
14       FF_OUT       <= FF;
15       FF_COMB_OUT <= FF xor D5 xor D4 xor D3 xor D2 xor D1 xor D0;
16       p0 : process (RSTn, CLK)
17       begin
18          if (RSTn = '0') then
19             FF  <= '0';
20          elsif (CLK'event and CLK = '1') then
21             FF  <= XOR8;
22          end if;
23       end process;
24    end RTL;
```

The following design constraints are used to synthesize the circuit. Line 1 reads in the VHDL source code file. Line 2 includes another command file to specify the wire load model, operating conditions, component library (with physical path name in UNIX), and so on. Line 3 declares the clock period of 10 ns for the clock signal CLK. Line 4 specifies clock skew. Line 5 specifies input delay for input port signals. Lines 6 and 7 specify output delay for output signals. Line 8 tells the synthesis tool to connect the reset and clock signals RSTn and CLK directly to the flipflops without any gates. Line 10 starts the synthesis.

```
1     read -f vhdl "consyn.vhd"
2     include compile.common
3     create_clock "CLK" -name clk -period 10.0 -waveform {0.0 5.0}
4     set_clock_skew -uncertainty 1.0 clk
5     set_input_delay  1.0 -add_delay -clock clk {D*}
6     set_output_delay 3.0 -add_delay -clock clk {COMB_OUT}
7     set_output_delay 3.0 -add_delay -clock clk {FF_COMB_OUT}
8     set_dont_touch_network {CLK RSTn }
9     compile -map_effort medium
```

Figure 8.14 shows the synthesized schematic of CONSYN VHDL code. Note that the input delay of all inputs is set to be 1 ns after the rising edge of the clock. We can change the input delay and output delay as follows. Figure 8.15 shows the synthesized schematic. Note that input port D5 has input delay of 6.0 ns, which is slower than input port D0 of 1 ns input delay. D5 is connected to the last stage of the XOR gate to compensate for its late arrival.

```
1     set_input_delay  1.0 -add_delay -clock clk {D0}
2     set_input_delay  2.0 -add_delay -clock clk {D1}
3     set_input_delay  3.0 -add_delay -clock clk {D2}
4     set_input_delay  4.0 -add_delay -clock clk {D3}
5     set_input_delay  5.0 -add_delay -clock clk {D4}
```

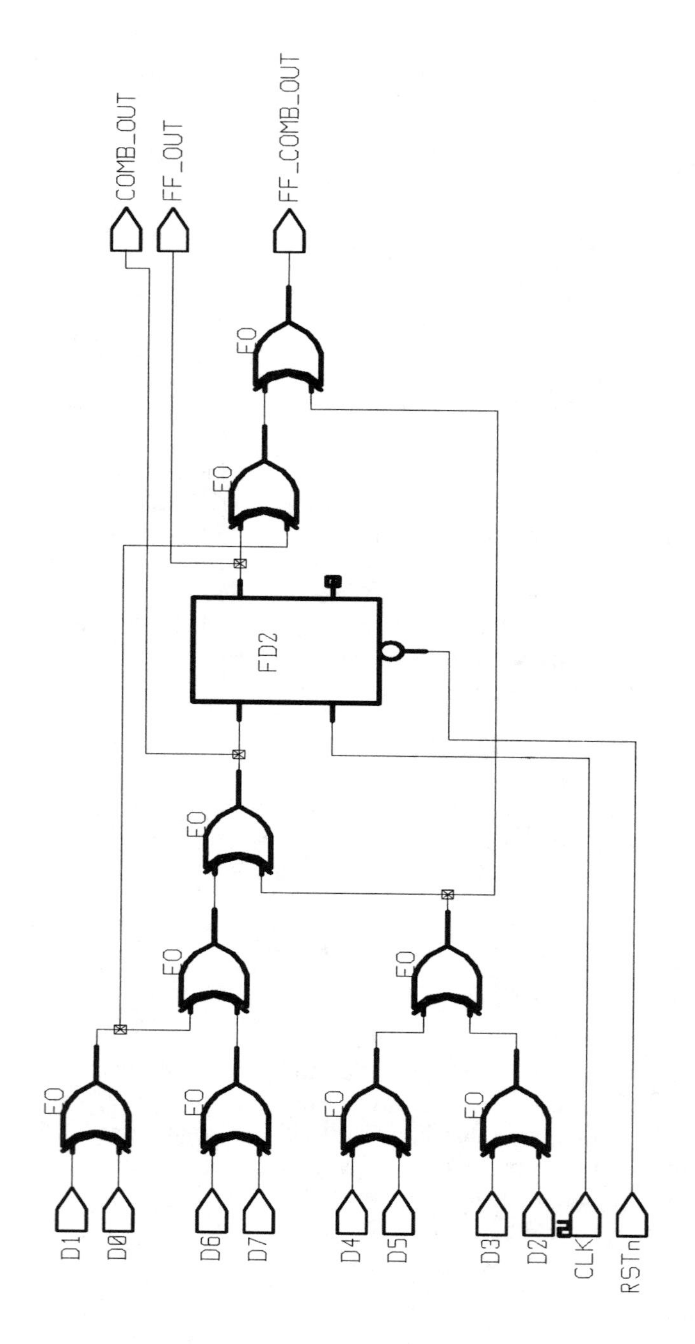

FIGURE 8.14 Synthesized schematic for CONSYN VHDL code.

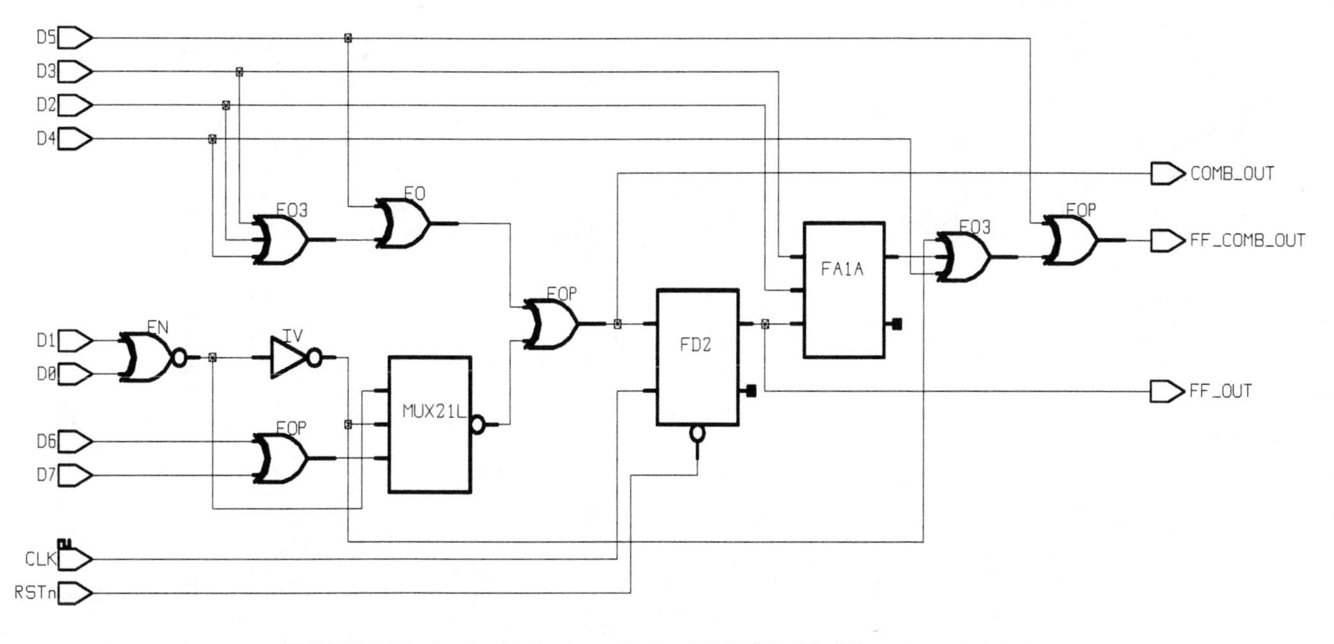

FIGURE 8.15 Synthesized schematic for CONSYN with different constraints.

```
6     set_input_delay  6.0 -add_delay -clock clk {D5 D6 D7}
7     set_output_delay 2.0 -add_delay -clock clk {COMB_OUT}
8     set_output_delay 3.0 -add_delay -clock clk {FF_COMB_OUT}
```

The preceding example shows that the same VHDL code can be synthesized to different schematics with different constraints. Before the VHDL code is synthesized, simulation is usually run to verify its functionality. Figure 8.16 shows a simulation waveform for CONSYN VHDL code. The inputs are changed at the falling edge of the clock. The simulation assumes that there is no timing delay. Output signal COMB_OUT changes at the same time as input signal D3 changes. FF_OUT changes right at the rising edge of the clock. To simulate the timing delay, the synthesis tool generates SDF (Standard Delay Format) as the industry standard delay file and the netlist in VHDL.

The VHDL netlist output from the synthesis tool follows. It consists of many component instantiation statements with library references (lines 1 to 4).

```
1     library IEEE;
2     library TC200G;
3     use IEEE.std_logic_1164.all;
4     use TC200G.components.all;
5
6     entity CONSYN is
7        port( RSTn,CLK,D0,D1,D2,D3,D4,D5,D6,D7 : in std_logic;  FF_OUT,
8              COMB_OUT, FF_COMB_OUT : out std_logic);
9     end CONSYN;
10
11    architecture structural of CONSYN is
12    signal XOR8,FF,n70,n71,n72,n73,n74,n75,n76,n67,n68,n69 : std_logic;
13    begin
14       FF_OUT <= FF;
15       COMB_OUT <= XOR8;
16
17      FF_reg : FD2 port map( Q=>FF, QN=>n75, D=>XOR8, CP=>CLK, CD=>RSTn);
18       U30 : MUX21L port map( Z => n71, A => n67, B => n68, S => n69);
19       U31 : EN port map( Z => n67, A => D1, B => D0);
20       U32 : IV port map( Z => n68, A => n67);
21       U33 : EOP port map( Z => n69, A => D6, B => D7);
22       U34 : EO3 port map( Z => n70, A => D3, B => D2, C => D4);
23       U35 : EO port map( Z => n72, A => D5, B => n70);
24       U36 : EOP port map( Z => XOR8, A => n72, B => n71);
25       U37 : FA1A port map( S=> n73, CO=> n76, CI=> D3, A=> D2, B=> FF);
26       U38 : EO3 port map( Z => n74, A => n68, B => n73, C => D4);
27       U39 : EOP port map( Z => FF_COMB_OUT, A => D5, B => n74);
28
29    end structural;
```

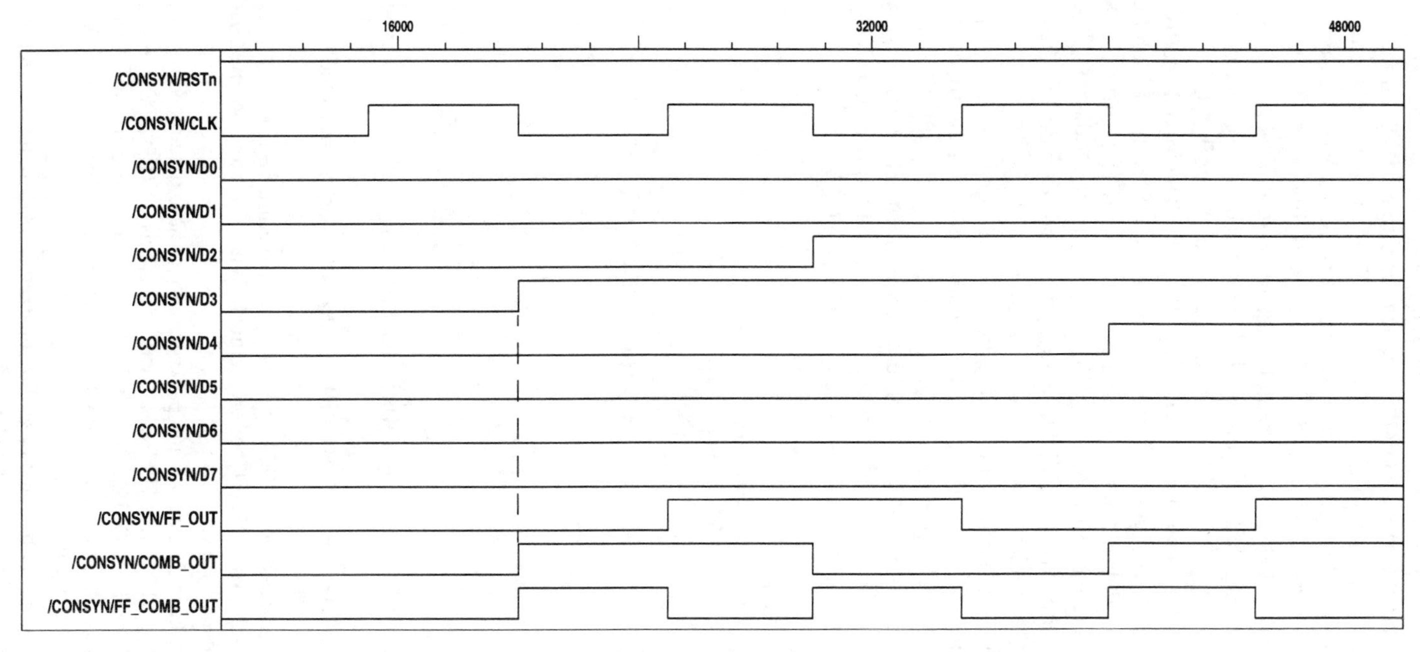

FIGURE 8.16 Simulation waveform of CONSYN VHDL code.

A small part of the SDF file follows. It indicates an XOR gate with delays (from low to high, and high to low) of minimum, typical, and worst case timing. The structural VHDL netlist and the SDF file can be read into a VHDL simulator. The SDF timing delays associated with the cell instance (U34) will be converted to generics (of type time) of the XOR gate simulation primitive in the VHDL component simulation library, which may be VITAL (VHDL Initiated Toward ASIC Library) compliant. VITAL is another IEEE 1076.4 standardization effort. This process is called overlaying the SDF timing delay. Figure 8.17 shows the simulation waveform after the SDF timing delay is overlaid. Note that output signal COMB_OUT changes with some timing delay after input port signal D3 is changed. The VHDL netlist can be used in physical layout. The layout tool can also generate a SDF file which has a much more accurate timing delay due to physical interconnect wire length. The SDF file can be used for post-layout verification. As shown in Figure 8.12, the VHDL simulation (gate level) can be done with the VHDL netlist generated from the synthesis tool. The SDF can be from the synthesis tool (before layout) or from the layout tool (after layout) for pre-layout or post-layout simulation. These simulations, in general, are much slower than simulating the original RTL VHDL code. It is important to verify the functionality with RTL VHDL simulation to achieve faster turnaround. The number of pre-layout and post-layout simulation should be minimized. They are usually used to check the detailed timing. The results can be compared with the original RTL VHDL simulation with a test bench (to be discussed in Chapter 11).

```
1    (DELAYFILE
2     ( CELL ( CELLTYPE "XOR" )
3            ( INSTANCE U34.Z_VTX )
4            ( DELAY ( INCREMENT
5                     ( DEVICE O1
6               (0.385090:0.385090:0.385090)(0.235177:0.235177:0.235177)
7                     )
8            )        )
9    )
```

8.11 EXERCISES

8.1 In the PARGEN VHDL code, what would be changed if PAR is a combinational circuit output?

8.2 Try to synthesize the TRIREG VHDL code with your synthesis tools to see what circuits are synthesized.

8.3 If you have access to SYNOPSYS tools, read the file $SYNOPSYS/packages/IEEE/src/std_logic_arith.vhd. What types are defined? What operators are overloaded and what are their parameter types?

8.4 Write a package declaration and package body to have the procedures (MUX21, MUX31, MUX41, LATCH, BUFFER3) and functions (REDUCE_OR, REDUCE_AND, REDUCE_XOR, COMPLEMENT2). Use the overloading—for example, procedure LATCH may have std_logic or std_logic_vector, with or without reset. Verify your code by simulation and synthesis.

8.5 After you have the package described in Exercise 8.4, write a barrel shifter VHDL code with a generic of N for the number of stages. Input signals are N-bit SHIFT, 2**N-bit DIN,

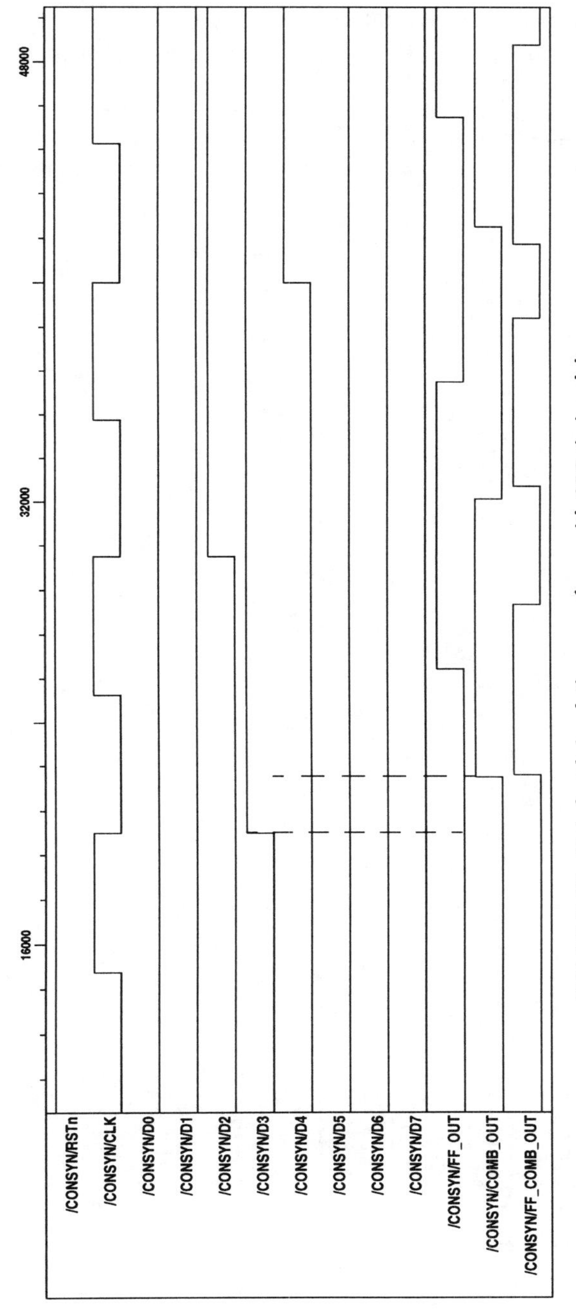

FIGURE 8.17 Gate level simulation waveform with SDF timing delay.

2-bit MODE, and output signal 2**N-bit DOUT. SHIFT indicates the number of bits to shift. DIN is the input data. MODE controls the mode of shift (left, right, rotate left, rotate right). DOUT is the output result. Simulate and synthesize your code. (Hint: a MUX21 procedure with generate statements may be handy.)

8.6 Synthesize architectures RTL and RTL1 of entity RESOURCE and compare the area and speed results.

8.7 Referring to architecture RTL1 of entity RESOURCE, write another architecture RTL2 so that only one "–" operator is used. Hint: multiplex A, B, C, D first, the output of the multiplexer is then used to minus 1. Discuss the advantage of this approach.

Chapter 9

Finite-State Machines

9.1 FINITE-STATE MACHINE BACKGROUND

Finite-state machines (FSM) are commonly used in digital design for control circuits. An FSM is usually described as a state transition diagram. For example, Figure 9.1 shows a state transition diagram for a read write controller (RWCNTL) FSM. RWCNTL starts at state IDLE, waiting for START to go to '1' and then changes to either state READING or state WRITING depending on the value of RW input. States READING and WRITING remain in the same state until signal LAST goes to '1' which changes to state WAITING. State WAITING always goes to state IDLE. The FSM has three outputs RDSIG, WRSIG, and DONE. They are '1' when they are in state READING, WRITING, and WAITING, respectively, otherwise they are '0'.

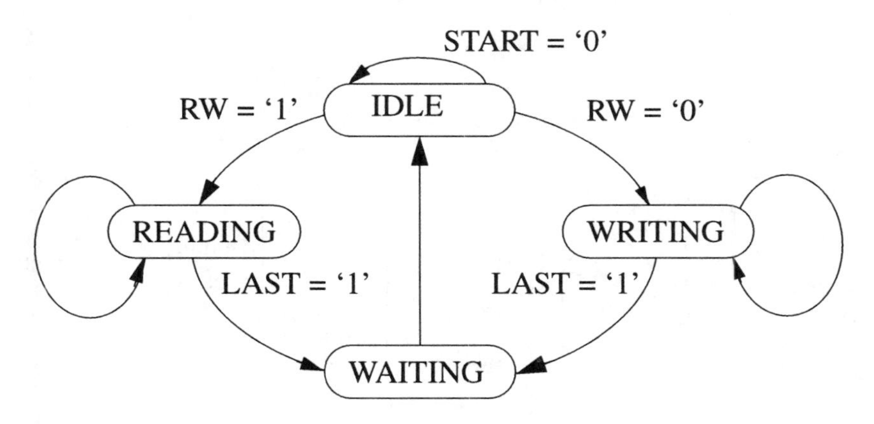

FIGURE 9.1 RWCNTL FSM state transition diagram.

From a hardware point of view, an FSM has two parts. One is a set of flipflops for remembering the states. The other part is the combinational circuits that control the state transition to go into the inputs of the flipflops and generate output signals. A state machine can be categorized as a MEALY FSM or a MOORE FSM. For a MEALY FSM, the output circuits are based on both inputs and the current state values. For a MOORE FSM, the output circuits depend only on the current state value. Figures 9.2 and 9.3 show the overall hardware implementation of a MOORE FSM and MEALY FSM, respectively. The flipflops would have the next_state value after the rising edge of the clock. The next state circuits and the output circuits can be considered combinational circuits in this case.

9.2 WRITING VHDL FOR A FSM

The RWCNTL FSM can be described as the following VHDL code. Lines 3 to 12 declare the entity. An enumerated type STATE_TYPE is declared in line 14 to include all possible states in the state transition diagram. Signals CURRENT_STATE and NEXT_STATE are declared in line 15.

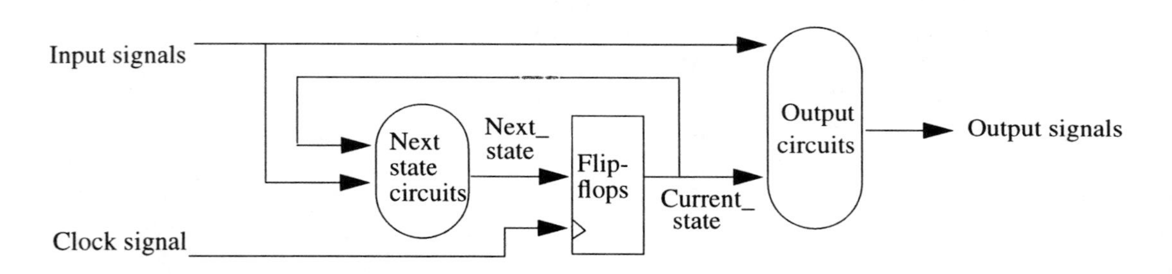

FIGURE 9.2 MEALY FSM hardware view.

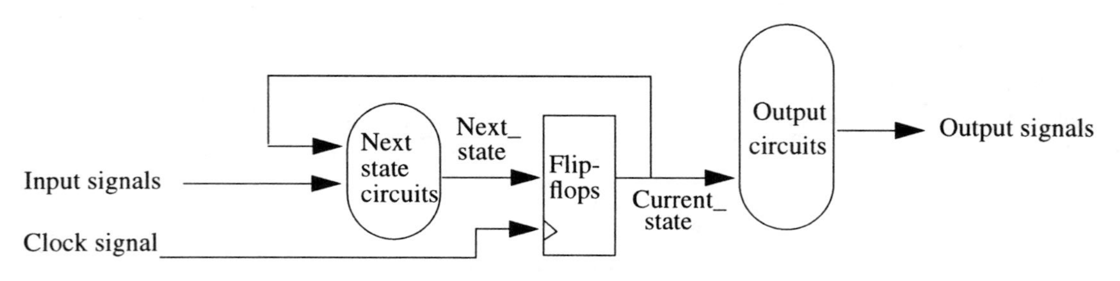

FIGURE 9.3 MOORE FSM hardware view.

```
1    library IEEE;
2    use IEEE.std_logic_1164.all;
3    entity RWCNTL is
4       port(
5          CLOCK : in  std_logic;
6          START : in  std_logic;
7          RW    : in  std_logic;
8          LAST  : in  std_logic;
9          RDSIG : out std_logic;
10         WRSIG : out std_logic;
11         DONE  : out std_logic);
12   end RWCNTL;
13   architecture RTL of RWCNTL is
14      type STATE_TYPE is (IDLE, READING, WRITING, WAITING);
15      signal CURRENT_STATE, NEXT_STATE: STATE_TYPE;
16   begin
17      comb : process(CURRENT_STATE, START, RW, LAST)
18      begin
19         DONE  <= '0';
20         RDSIG <= '0';
21         WRSIG <= '0';
22         case CURRENT_STATE is
23            when IDLE =>
24               if START = '0' then
25                  NEXT_STATE <= IDLE;
26               elsif RW = '1' then
27                  NEXT_STATE <= READING;
28               else
29                  NEXT_STATE <= WRITING;
30               end if;
31            when READING =>
32               RDSIG      <= '1';
33               if LAST = '0' then
34                  NEXT_STATE <= READING;
35               else
36                  NEXT_STATE <= WAITING;
37               end if;
38            when WRITING =>
39               WRSIG      <= '1';
40               if LAST = '0' then
41                  NEXT_STATE <= WRITING;
42               else
43                  NEXT_STATE <= WAITING;
44               end if;
45            when WAITING =>
46               DONE       <= '1';
47               NEXT_STATE <= IDLE;
48         end case;
49      end process;
50
51      seq : process
```

```
52       begin
53          wait until CLOCK'event and CLOCK = '1';
54          CURRENT_STATE <= NEXT_STATE;
55       end process;
56    end RTL;
```

Note that there are two concurrent processes, "comb" and "seq", inside the architecture body. Processes "comb" and "seq" describe the combinational and sequential circuits (to infer flipflops), respectively. The "comb" process has a sequential "case" statement to describe state transitions and the output signals. These signals are handled with sequential signal assignment statements. Note that output signals are assigned default values ('0' in lines 19 to 21) before the "case" statement. However, these default values may be changed by the VHDL sequential signal assignment statements later in the same process (such as line 32). Without setting up default values, we need to have a signal assignment statement for each output signal in every execution path of the case statement. This will increase the amount of VHDL code. Also, a latch will be inferred and synthesized if a signal is not completely specified in every execution path of a "case" or "if" statement. This technique is a good design practice, especially when the number of states is large. Note that signals CURRENT_STATE, START, RW, and LAST are included in the sensitivity list. The process assigns signals NEXT_STATE, RDSIG, WRSIG, and DONE. Signals RDSIG, WRSIG, and DONE are assigned in lines 19 to 21, 32, 39, and 46. Their values are dependent only on the value of CURRENT_STATE, not on the values of input signals. Therefore, RWCNTL is a MOORE FSM. A MEALY FSM can be described similarly.

Process "seq" has a "wait" statement that waits for the rising edge of the clock. CURRENT_STATE is assigned with the value of NEXT_STATE. CURRENT_STATE will infer flipflops since the signal assignment statement appears after the rising edge check in line 53. If the binary encoding is used, signal CURRENT_STATE will be synthesized with two flipflops (due to four states). Note that the process does not contain a sensitivity list because a "wait" statement is inside the process. Figure 9.4 shows the simulation waveform of RWCNTL VHDL code.

From the simulation waveform, it is clear that output signals RDSIG, WRSIG, and DONE are high when they are in states READING, WRITING, and WAITING, respectively.

The same VHDL code can be synthesized. The synthesized schematic in Figure 9.5 depicts two D-type flipflops. This assumes the "BINARY" encoding style such that states IDLE, READING, WRITING, and WAITING are encoded with binary numbers "00", "01", "10", and "11", respectively. Some synthesis tools also allow other encoding styles such as "GRAY CODE" and "ONE HOT". Note that output signals RDSIG, WRSIG, and DONE are outputs from combinational circuits since they are assigned in the process "comb" without a rising edge clock check.

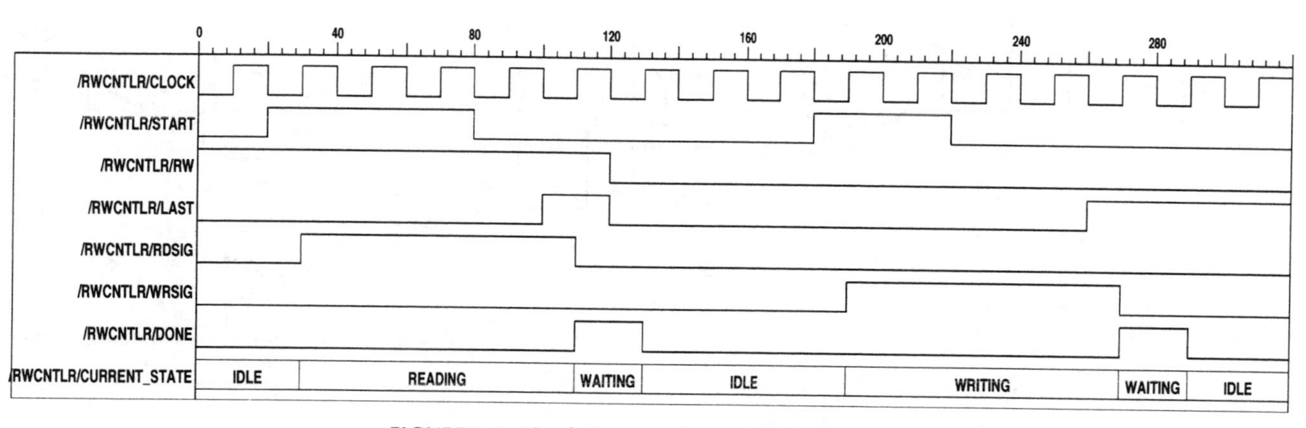

FIGURE 9.4 Simulation waveform of RWCNTL.

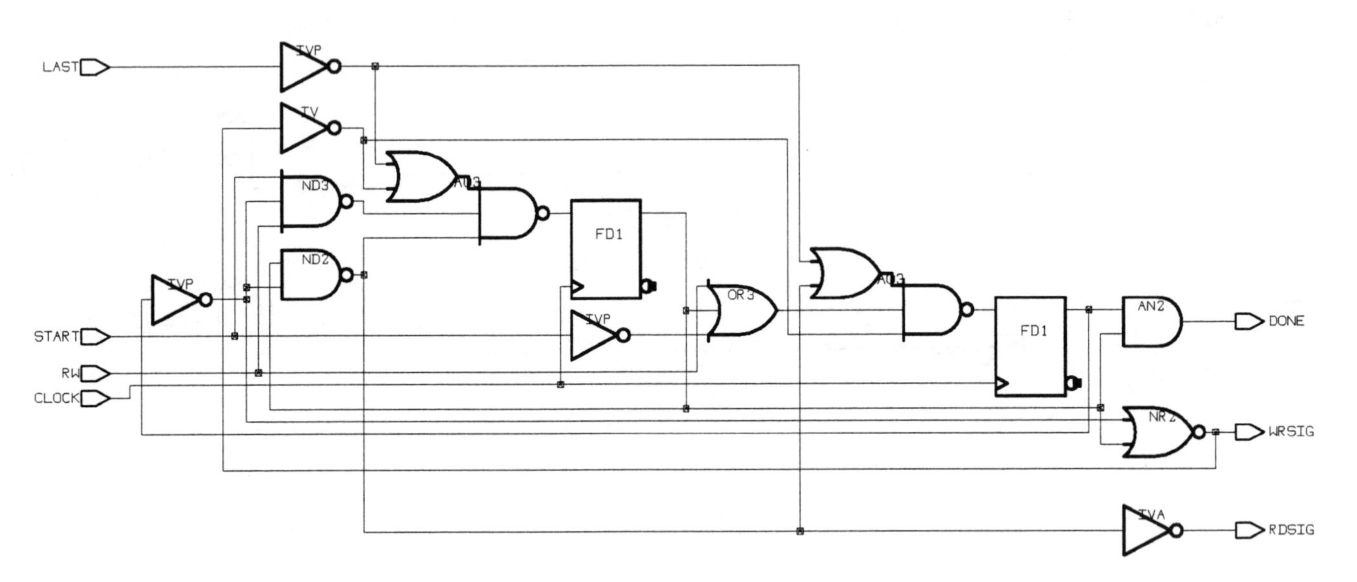

FIGURE 9.5 Synthesized schematic for RWCNTL.

9.3 FSM INITIALIZATION

A VHDL simulator assumes the default value of an object as its left-most value of its type. Hence, CURRENT_STATE and NEXT_STATE would have the initial default state as IDLE. The VHDL simulation would assume that the state machine starts at IDLE state as though it has been initialed to IDLE state. This may not be true in the real hardware. Therefore, it is common to have an asynchronous reset signal to reset (initialize) the flipflops. The following shows how we can write VHDL codes so that the synthesized schematic would add a reset signal to the flipflops.

Considering the following VHDL code, a signal RESETn is added in the entity port list (line 5). The process "comb" remains the same. Process "seq" is changed so that the CURRENT_STATE is set to IDLE when RESETn is '0', otherwise CURRENT_STATE is set to NEXT_STATE after the rising edge of the clock (lines 55 to 60). Note that the original "wait" statement is changed to an "if" statement and RESETn and CLOCK are added to the sensitivity list in the process statement. Note that NEXT_STATE does not have to be in the sensitivity list because the signal assignment for CURRENT_STATE is always after the rising edge of the CLOCK and signal CLOCK has been in the sensitivity list. This improves simulation efficiency.

```
1     library IEEE;
2     use IEEE.std_logic_1164.all;
3     entity RWCNTLR is
4        port(
5           RESETn : in  std_logic;
6           CLOCK  : in  std_logic;
7           START  : in  std_logic;
8           RW     : in  std_logic;
9           LAST   : in  std_logic;
10          RDSIG  : out std_logic;
11          WRSIG  : out std_logic;
12          DONE   : out std_logic);
13    end RWCNTLR;
14
15    architecture RTL of RWCNTLR is
16       type STATE_TYPE is (IDLE, READING, WRITING, WAITING);
17       signal CURRENT_STATE, NEXT_STATE: STATE_TYPE;
18    begin
19       comb : process(CURRENT_STATE, START, RW, LAST)
20       begin
21          DONE  <= '0';
22          RDSIG <= '0';
23          WRSIG <= '0';
24          case CURRENT_STATE is
25             when IDLE =>
26                if START = '0' then
27                   NEXT_STATE <= IDLE;
28                elsif RW = '1' then
29                   NEXT_STATE <= READING;
30                else
31                   NEXT_STATE <= WRITING;
32                end if;
```

```
33              when READING =>
34                 RDSIG      <= '1';
35                 if LAST = '0' then
36                    NEXT_STATE <= READING;
37                 else
38                    NEXT_STATE <= WAITING;
39                 end if;
40              when WRITING =>
41                 WRSIG      <= '1';
42                 if LAST = '0' then
43                    NEXT_STATE <= WRITING;
44                 else
45                    NEXT_STATE <= WAITING;
46                 end if;
47              when WAITING =>
48                 DONE       <= '1';
49                 NEXT_STATE <= IDLE;
50           end case;
51        end process;
52
53        seq : process (CLOCK, RESETn)
54        begin
55           if (RESETn = '0') then
56              CURRENT_STATE <= IDLE;
57           elsif (CLOCK'event and CLOCK = '1') then
58              CURRENT_STATE <= NEXT_STATE;
59           end if;
60        end process;
61     end RTL;
```

Figure 9.6 shows the synthesized schematic of the preceding VHDL code. Note that signal RESETn is connected to the active low asynchronous reset input of the flipflops.

Another way of resetting flipflops is through the synchronous reset. This means that the FSM is not reset until the rising edge of the clock signal occurs, even if the reset signal is asserted before the clock rising edge. This can be done by adding an "if" statement before the "case" statement in the process "comb" as shown in the following VHDL code (lines 24 to 26). Note that signal RESETn is added to the sensitivity list in line 19. The process "seq" remains the same as in RWCNTL so that the flipflops do not have asynchronous reset. Otherwise, we have reset in both the "comb" and "seq" processes as shown in the following VHDL code. The reset signal RESETn will connect to the flipflops asynchronous reset input and also to the combinational gates that feed to flipflops D inputs as shown in Figure 9.7 synthesized schematic. Compare the schematics in Figures 9.6 and 9.7—the amount of hardware (area or the number of gates) in Figure 9.7 is greater.

```
1      library IEEE;
2      use IEEE.std_logic_1164.all;
3      entity RWCNTLR1 is
4         port(
5            RESETn : in  std_logic;
6            CLOCK  : in  std_logic;
```

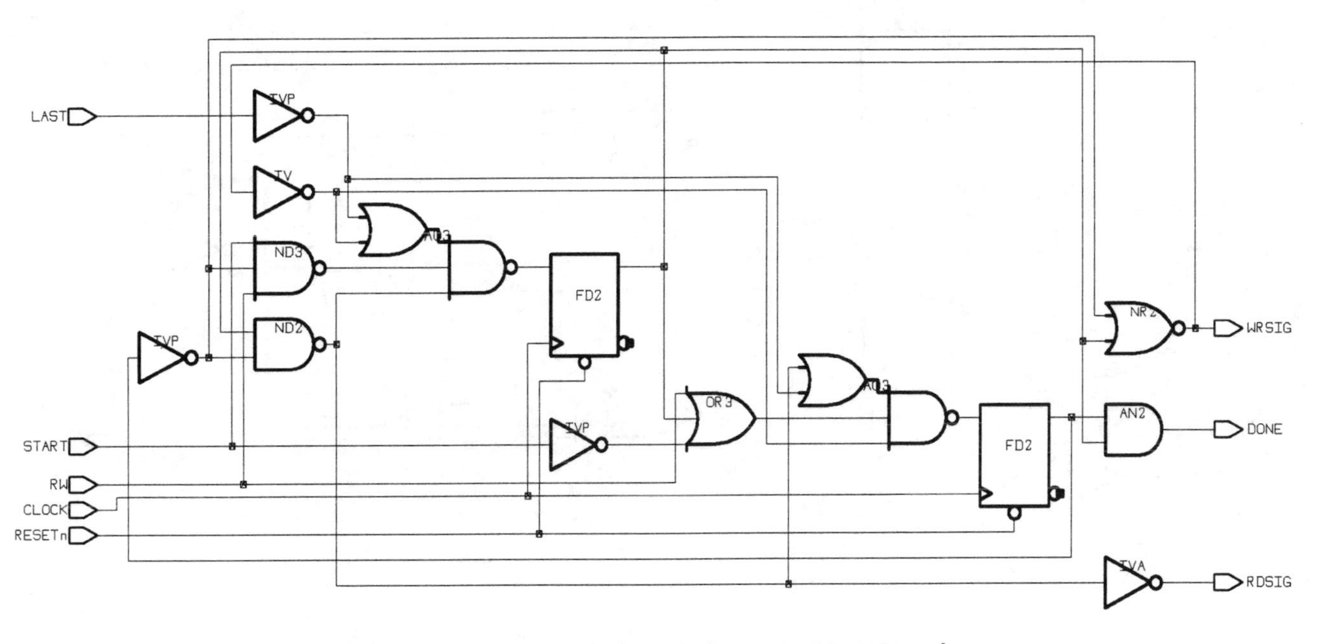

FIGURE 9.6 Synthesized schematic for RWCNTLR VHDL code.

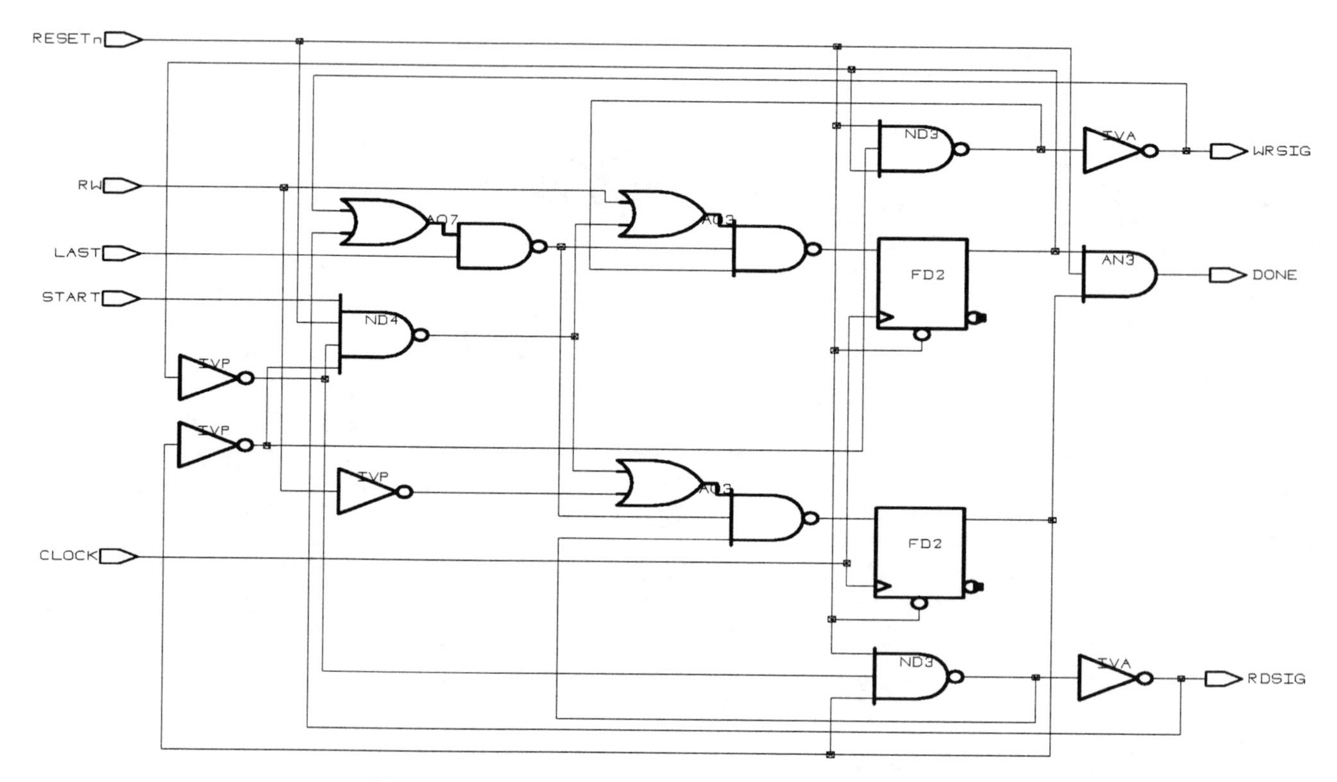

FIGURE 9.7 Synthesized schematic for RWCNTLR1 VHDL code.

```
7            START  : in  std_logic;
8            RW     : in  std_logic;
9            LAST   : in  std_logic;
10           RDSIG  : out std_logic;
11           WRSIG  : out std_logic;
12           DONE   : out std_logic);
13     end RWCNTLR1;
14
15     architecture RTL of RWCNTLR1 is
16        type STATE_TYPE is (IDLE, READING, WRITING, WAITING);
17        signal CURRENT_STATE, NEXT_STATE: STATE_TYPE;
18     begin
19        comb : process(RESETn, CURRENT_STATE, START, RW, LAST)
20        begin
21           DONE  <= '0';
22           RDSIG <= '0';
23           WRSIG <= '0';
24           if (RESETn = '0') then
25              NEXT_STATE <= IDLE;
26           else
27              case CURRENT_STATE is
28                 when IDLE =>
29                    if START = '0' then
30                       NEXT_STATE <= IDLE;
31                    elsif RW = '1' then
32                       NEXT_STATE <= READING;
33                    else
34                       NEXT_STATE <= WRITING;
35                    end if;
36                 when READING =>
37                    RDSIG      <= '1';
38                    if LAST = '0' then
39                       NEXT_STATE <= READING;
40                    else
41                       NEXT_STATE <= WAITING;
42                    end if;
43                 when WRITING =>
44                    WRSIG      <= '1';
45                    if LAST = '0' then
46                       NEXT_STATE <= WRITING;
47                    else
48                       NEXT_STATE <= WAITING;
49                    end if;
50                 when WAITING =>
51                    DONE       <= '1';
52                    NEXT_STATE <= IDLE;
53              end case;
54           end if;
55        end process;
56
57        seq : process (CLOCK, RESETn)
58        begin
```

```
59          if (RESETn = '0') then
60             CURRENT_STATE <= IDLE;
61          elsif (CLOCK'event and CLOCK = '1') then
62             CURRENT_STATE <= NEXT_STATE;
63          end if;
64       end process;
65    end RTL;
```

9.4 FSM FLIPFLOP OUTPUT SIGNAL

It is also common to have the output signals from the flipflops instead of combination circuit outputs. This can be done by adding the signal assignment for output signals in the process "seq" as shown in lines 58 to 60 and lines 63 to 65. Temporary signals DONE_comb, RDSIG_comb, and WRSIG_comb are declared in line 18 and are assigned in the process "comb".

```
1     library IEEE;
2     use IEEE.std_logic_1164.all;
3     entity RWCNTLRF is
4        port(
5           RESETn : in  std_logic;
6           CLOCK  : in  std_logic;
7           START  : in  std_logic;
8           RW     : in  std_logic;
9           LAST   : in  std_logic;
10          RDSIG  : out std_logic;
11          WRSIG  : out std_logic;
12          DONE   : out std_logic);
13    end RWCNTLRF;
14
15    architecture RTL of RWCNTLRF is
16       type STATE_TYPE is (IDLE, READING, WRITING, WAITING);
17       signal CURRENT_STATE, NEXT_STATE: STATE_TYPE;
18       signal DONE_comb, RDSIG_comb, WRSIG_comb : std_logic;
19    begin
20       comb : process(CURRENT_STATE, START, RW, LAST)
21       begin
22          DONE_comb  <= '0';
23          RDSIG_comb <= '0';
24          WRSIG_comb <= '0';
25          case CURRENT_STATE is
26             when IDLE =>
27                if START = '0' then
28                   NEXT_STATE <= IDLE;
29                elsif RW = '1' then
30                   NEXT_STATE <= READING;
31                else
32                   NEXT_STATE <= WRITING;
33                end if;
34             when READING =>
```

```
35                  RDSIG_comb <= '1';
36                  if LAST = '0' then
37                     NEXT_STATE <= READING;
38                  else
39                     NEXT_STATE <= WAITING;
40                  end if;
41               when WRITING =>
42                  WRSIG_comb     <= '1';
43                  if LAST = '0' then
44                     NEXT_STATE <= WRITING;
45                  else
46                     NEXT_STATE <= WAITING;
47                  end if;
48               when WAITING =>
49                  DONE_comb  <= '1';
50                  NEXT_STATE <= IDLE;
51            end case;
52         end process;
53
54         seq : process (CLOCK, RESETn)
55         begin
56            if (RESETn = '0') then
57               CURRENT_STATE <= IDLE;
58               DONE          <= '0';
59               RDSIG         <= '0';
60               WRSIG         <= '0';
61            elsif (CLOCK'event and CLOCK = '1') then
62               CURRENT_STATE <= NEXT_STATE;
63               DONE          <= DONE_comb;
64               RDSIG         <= RDSIG_comb;
65               WRSIG         <= WRSIG_comb;
66            end if;
67         end process;
68      end RTL;
```

The preceding VHDL code is synthesized in the schematic shown in Figure 9.8. Note that three more flipflops are generated in the schematic because DONE, RDSIG, and WRSIG are assigned in the process "seq", which has a rising edge check of the CLOCK signal. Here, the output timing of DONE, RDSIG, and WRSIG would be different from the timing in RWCNTLR VHDL code, (Figure 9.4) as shown in the Figure 9.9 simulation waveform.

If all the output signals are flipflop outputs, the case statement inside the process "comb" can be included inside the process "seq" after the rising edge clock check. The process "comb" can then be deleted. Signals DONE_comb, RDSIG_comb, WRSIG_comb are not necessary. The NEXT_STATE signal appearing in the case statement can be replaced by signal CURRENT_STATE. Line 62 of assigning NEXT_STATE to CURRENT_STATE can be deleted. The following shows the resulting VHDL code. There is only one process statement inside the architecture COMBINED. Signals CURRENT_STATE, RDSIG, WRSIG, and DONE infer flipflops since a clock edge check is done in line 80.

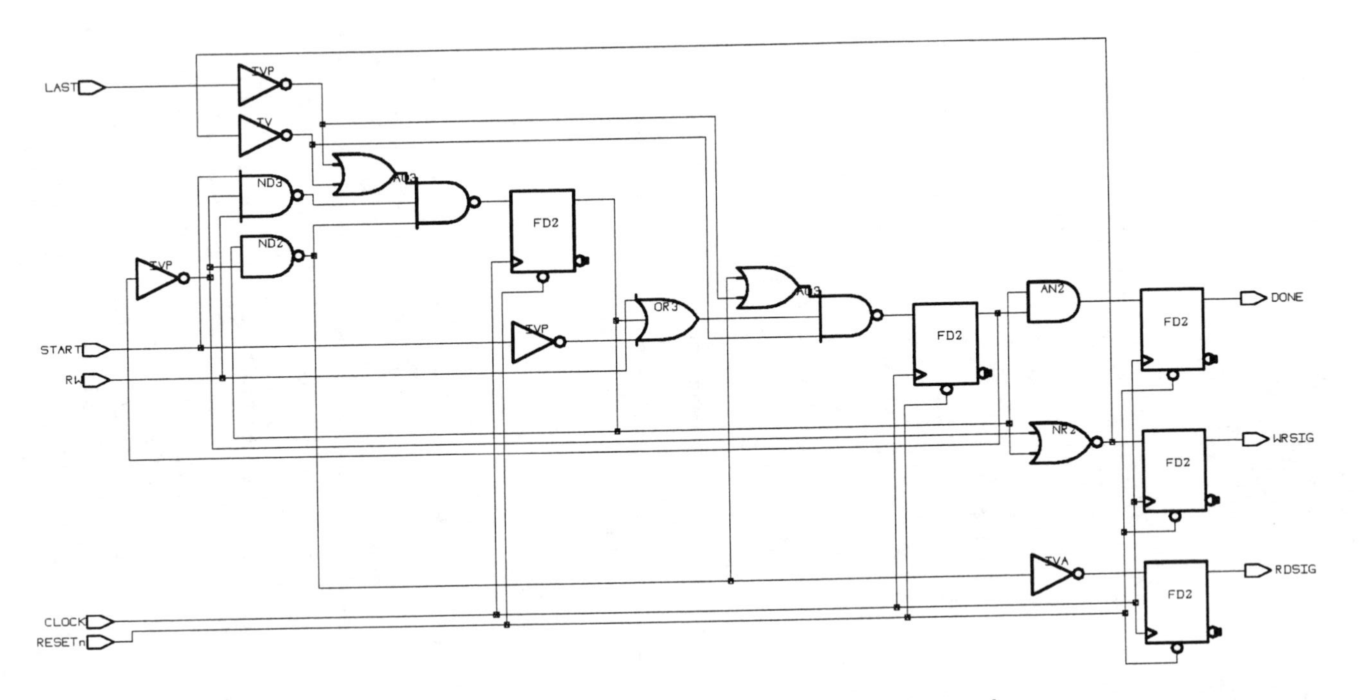

FIGURE 9.8 Synthesized schematic for RWCNTLRF VHDL code.

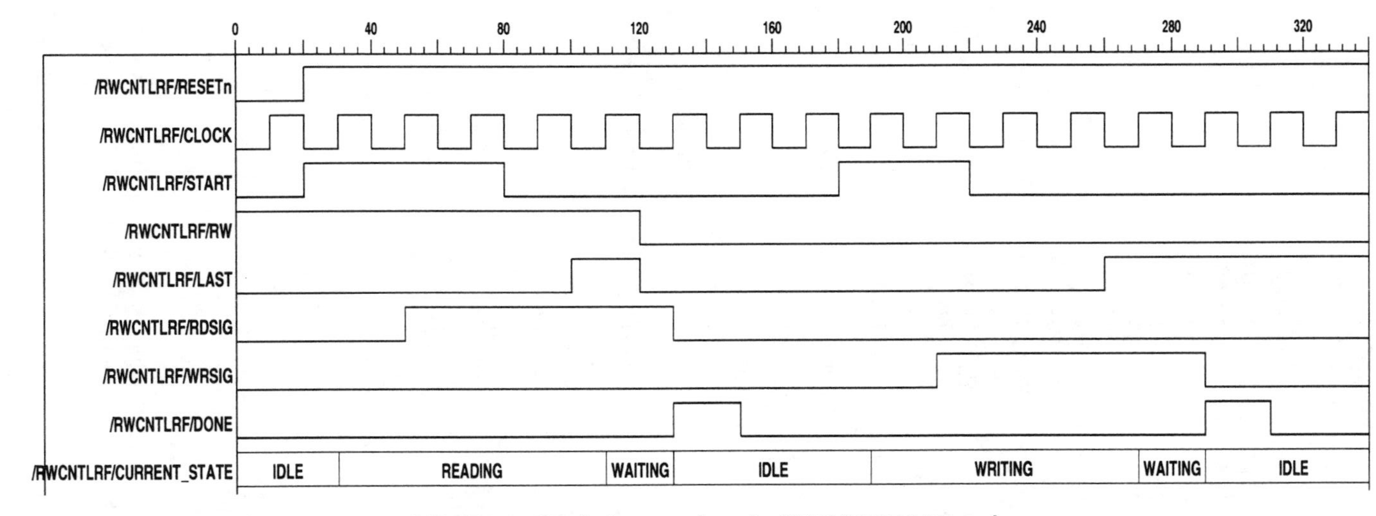

FIGURE 9.9 Simulation waveform for RWCNTLRF VHDL code.

```
69   architecture COMBINED of RWCNTLRF is
70      type STATE_TYPE is (IDLE, READING, WRITING, WAITING);
71      signal CURRENT_STATE : STATE_TYPE;
72   begin
73      seq : process (CLOCK, RESETn)
74      begin
75         if (RESETn = '0') then
76            CURRENT_STATE <= IDLE;
77            DONE          <= '0';
78            RDSIG         <= '0';
79            WRSIG         <= '0';
80         elsif (CLOCK'event and CLOCK = '1') then
81            DONE          <= '0';
82            RDSIG         <= '0';
83            WRSIG         <= '0';
84            case CURRENT_STATE is
85               when IDLE =>
86                  if START = '0' then
87                     CURRENT_STATE <= IDLE;
88                  elsif RW = '1' then
89                     CURRENT_STATE <= READING;
90                  else
91                     CURRENT_STATE <= WRITING;
92                  end if;
93               when READING =>
94                  RDSIG <= '1';
95                  if LAST = '0' then
96                     CURRENT_STATE <= READING;
97                  else
98                     CURRENT_STATE <= WAITING;
99                  end if;
100              when WRITING =>
101                 WRSIG   <= '1';
102                 if LAST = '0' then
103                    CURRENT_STATE <= WRITING;
104                 else
105                    CURRENT_STATE <= WAITING;
106                 end if;
107              when WAITING =>
108                 DONE  <= '1';
109                 CURRENT_STATE <= IDLE;
110           end case;
111        end if;
112     end process;
113  end COMBINED;
```

This approach could reduce a couple of lines of the VHDL code and thereby improve the simulation speed since the process is only sensitive to the reset and clock signals (line 73). The event of other signals would not trigger the evaluation of the process as it would for the process "comb" (in architecture RTL). However, using two processes makes modifying the design easier if an output signal is changed between the combination output and the flipflop output.

It is possible that the number of states of an FSM is not two to the power of an integer N. For example, an FSM may have five states. We know that the VHDL simulator always initializes the object as its left-most value of the type. If the binary encoding is used and the flipflops are not reset synchronously or asynchronously, the VHDL simulator initializes both the NEXT_STATE and CURRENT_STATE with value IDLE. The VHDL simulation runs nicely. However, in the actual hardware, it could be any value in power up. For example, an FSM with five states may have state encoding with "000", "001", "010", "011", and "100". When the hardware is powered up, the three flipflops may have encoding "111". If the state transition is not included, the FSM may not transition to the initial state of "000" to function correctly. Three extra unused states can be added to have full eight states (2 to the power of 3). The following two lines of VHDL code can be added to assign the NEXT_STATE to the initial state in the case statement as the last choice.

```
1       when others =>
2          NEXT_STATE <= IDLE;
```

9.5 FSM SYNTHESIS

Some synthesis tools have included special algorithms for synthesizing FSMs since FSM synthesis algorithms have been much more developed. They usually generate better results than a general synthesis algorithm for an FSM. To take advantage of the special FSM synthesis algorithms, it is better for an FSM to be an independent block (VHDL code). This makes the synthesis process easier, especially in writing synthesis constraints, encoding states, and synthesis commands.

9.6 EXERCISES

9.1 Refer to RWCNTLR and TWCNTLRF VHDL code and simulation waveforms in Figure 9.4 and Figure 9.9. Explain the timing difference for the output signals DONE, RDSIG, WRSIG.

9.2 In RWCNTL FSM, READING, WRITING, and WAITING are the names for the states. Can we change their names to READ, WRITE, and WAIT, respectively? Justify your conclusions.

9.3 What needs to be changed if RESETn input is used as a synchronous reset in the RWCNTLRF VHDL code (both architectures RTL and COMBINED).

9.4 In some applications, we take the encoded state vector as output signals to input to other modules. For example, an FSM state vector may represent the address to a ROM which takes an INTEGER type as its address input. FSM STATE_TYPE does not match with the INTEGER type. Type mismatch would result if we tie them together directly. How do we solve this type mismatch problem? Use RWCNTLR as an example of another output with type std_logic_vector(1 downto 0) that takes the CURRENT_STATE. What is the hardware implication of your solution? Synthesize your VHDL code to see whether extra hardware is used.

9.5 Write a VHDL code to model the PCI (peripheral component interconnect) protocol target state machine. Simulate and synthesize the VHDL code. See the PCI specification document.

9.6 Write a VHDL code to model the PCI protocol master state machine. Simulate and synthesize the VHDL code. See the PCI specification document.

9.7 Try your synthesis tool on RWCNTLR VHDL code to see whether the ONE-HOT or GRAY code encoding for state machine is available. Are you allowed to specify your own encoding?

Chapter 10

More on Behavioral Modeling

10.1 FILE TYPES AND FILE I/O

File types, file objects, and file I/O provide a way for a VHDL design to communicate through files that may be supplied externally. *File type* is defined as follows:

type file_type_identifier **is file of** type_mark;

Note that a type mark is used after the keywords **is file of**. The following file type declaration is not valid since bit_vector(7 downto 0) is not a type mark.

type BIT_F is FILE of bit_vector(7 downto 0); -- not valid, syntax error

File declaration is defined as follows:

file file_identifier : file_type_identifier **is** [**in** | **out**] file_logical_name;

The following VHDL code shows examples of file type declarations and file objects declarations. Lines 5 to 7 show file type declarations with predefined types such as INTEGER, REAL, and STRING. Line 8 defines a subtype BYTE to be used as a type mark in line 9. Lines 12 to 15 show examples of file object declarations. File type TEXT is defined in the TEXTIO package of library STD.

```
1     use STD.TEXTIO.all;
2     entity FILES is
3     end FILES;
4     architecture BEH of FILES is
5        type INT_FILE  is FILE of INTEGER;
6        type REAL_FILE is FILE of REAL;
7        type STR_FILE  is FILE of STRING;
8        subtype BYTE is bit_vector(7 downto 0);
9        type BIT_FILE  is FILE of BYTE;
10       constant FILE_NAME : string := "input_file";
```

```
11      constant PATH       : string := "design/";
12      file MEMIN  : text is in  "vector.in";
13      file MEMOUT : text is out "vector.out";
14      file INF1   : INT_FILE is FILE_NAME;
15      file STRF   : STR_FILE is out PATH & "str.out";
16   begin
17   end BEH;
```

Part of the TEXTIO package VHDL code follows. Line 3 of package TEXTIO defines file type TEXT as a file of STRING. Package TEXTIO is referenced in line 1 of entity FILES. If the mode **in** or **out** is not specified (as in line 13), **in** is assumed, as in line 14. Note that the file_logical_name should be a string expression as shown in lines 14 and 15. These string constants such as PATH and FILE_NAME can also be passed in as generics.

A file with mode **in** can only be read but is not to be updated. A file with mode **out** can only be updated but is not to be read. If multiple file objects of mode **out** have the same file logical name, the output will go to the same file. The VHDL standard does not specify the order in which the values are written.

```
1    package TEXTIO is
2        type LINE is access STRING;
3        type TEXT is file of STRING;
4        type SIDE is (RIGHT, LEFT);
5        subtype WIDTH is NATURAL;
6        file INPUT:  TEXT is in  "STD_INPUT";
7        file OUTPUT: TEXT is out "STD_OUTPUT";
8        procedure READLINE(variable F: in TEXT; L: inout LINE);
9        procedure READ(L: inout LINE; VALUE: out BIT);
10       procedure READ(L: inout LINE; VALUE: out BIT_VECTOR);
11       procedure READ(L: inout LINE; VALUE: out BIT; GOOD: out
                        BOOLEAN);
12       procedure READ(L: inout LINE; VALUE: out BIT_VECTOR; GOOD: out
                        BOOLEAN);
13       procedure READ(L: inout LINE; VALUE: out INTEGER);
14       procedure READ(L: inout LINE; VALUE: out TIME);
15
16       procedure WRITELINE(F: out TEXT; L: inout LINE);
17       procedure WRITE(L: inout LINE; VALUE: in BIT;
18          JUSTIFIED: in SIDE := RIGHT; FIELD: in WIDTH := 0);
19       procedure WRITE(L: inout LINE; VALUE: in BIT_VECTOR;
20          JUSTIFIED: in SIDE := RIGHT; FIELD: in WIDTH := 0);
21       procedure WRITE(L: inout LINE; VALUE: in INTEGER;
22          JUSTIFIED: in SIDE := RIGHT; FIELD: in WIDTH := 0);
23       procedure WRITE(L: inout LINE; VALUE: in TIME;
24          JUSTIFIED: in SIDE := RIGHT; FIELD: in WIDTH := 0;
25          UNIT: in TIME := ns);
26   end TEXTIO;
```

The TEXTIO package defines many overloaded READ and WRITE procedures. Type SIDE is defined (line 4) to be used in the WRITE procedure to specify either right or left justified field. Note also a type LINE is defined as an access type to the

STRING type in line 2. An object of type LINE is used as a buffer for the READLINE or WRITELINE procedure. Procedure READLINE reads a line from a file and puts the string in the buffer of type LINE. The actual values are then extracted from the buffer with procedure READ. The writing sequence is almost the reverse of the reading sequence. Procedure WRITE is used to write values to the line buffer. Then, procedure WRITELINE outputs the buffer of LINE type to the output file. Suppose we have a file with each line consisting of three integers (VAL, ADD, MINUS). The output file will have each line as three integers too with the values (VAL, VAL + ADD, VAL - MINUS). The following VHDL code shows a way to accomplish this.

```
1     use STD.TEXTIO.all;
2     entity CALCULATE is
3     end CALCULATE;
4     architecture BEH of CALCULATE is
5     begin
6        add_sub : process
7           file INFILE  : text is in  "calculate.in";
8           file OUTFILE : text is out "calculate.out";
9           variable VAL, ADD, MINUS     : integer;
10          variable READ_BUF, WRITE_BUF : line;
11       begin
12          while (not ENDFILE(INFILE)) loop
13             readline(INFILE, READ_BUF);
14             read(READ_BUF, VAL);
15             read(READ_BUF, ADD);
16             read(READ_BUF, MINUS);
17             write(WRITE_BUF,VAL, right, 10);
18             write(WRITE_BUF,VAL+ADD, right, 10);
19             write(WRITE_BUF,VAL-MINUS, right, 10);
20             writeline(OUTFILE,WRITE_BUF);
21          end loop;
22          wait;
23       end process;
24    end BEH;
```

Lines 7 and 8 declare input and output file objects with file names. Line 10 declares the line buffer for reading and writing. Lines 12 to 21 are a while loop statement which is to be completed when the end of file function ENDFILE returns TRUE. Line 13 reads a line from the input file into the line buffer READ_BUF. Three integers are extracted from the line buffer READ_BUF in lines 14 to 16. Lines 17 to 19 write values to a line buffer WRITE_BUF with field width of 10 and right justified. Line 20 outputs the line buffer WRITE_BUF to the output file. The input file "calculate.in" is as follows with the line numbers on the left which are not part of the file.

```
1     21 20 18
2     0     3   6
3     1E2  5_4 23
4     3e4  1  10
5     123_000 100_0 3E5
```

The output file "calculate.out" is as follows with the line numbers on the left which are not part of the file.

```
6          21         41         3
7           0          3        -6
8         100        154        77
9       30000      30001     29990
10     123000     124000   -177000
```

10.2 ROM MODEL

Read-only memory (ROM) is frequently used in the design. The behavior of a ROM can be modeled in VHDL. The following VHDL code shows a simple ROM VHDL model. The entity ROMRF has an input signal ADDR and an output signal DATA. Their widths are defined with the values of generics M and N, respectively (lines 10 and 11). The entity generic WORDS indicates the number of words in the ROM. Generic T_DELAY of type TIME is defined in line 15 to indicate the timing delay from the address ADDR change to the DATA output. Line 13 is a generic for the file name with the ROM binary code image. The code image can be read into the ROM during the initialization. The same file can also be used for the foundry (or layout tools such as silicon compilers) to create the equivalent ROM block. This approach in general is better than defining a constant table in the VHDL code. When the ROM image requires modifications, the ROM image file can be changed without changing the VHDL code. Line 14 defines a trace file name so that the access to ROM is recorded into a text file. The benefit is seeing the values of ADDR and DATA much easier than viewing the traces of the VHDL simulator on the screen.

```
1     library IEEE;
2     use IEEE.std_logic_1164.all;
3     use IEEE.std_logic_arith.all;
4     use IEEE.std_logic_textio.all;
5     use STD.TEXTIO.all;
6     use work.txtpack.all;
7
8     entity ROMRF is
9        generic (
10          N          : integer := 13 ;    -- number of bits per word
11          WORDS      : integer := 16 ;    -- number of words
12          M          : integer := 4 ;     -- address bus width
13          infile     : string  := "rom.in";
14          tracefile  : string  := "rom.trace";
15          T_DELAY    : time    := 4.0 ns);  -- address to data delay
16       port (
17          ADDR : in  std_logic_vector((M-1) downto 0);
18          DATA : out std_logic_vector((N-1) downto 0));
19    end ROMRF;
```

Lines 23 and 24 declare input and output files. Lines 25 and 26 define the type and subtype for the ROM as one-dimensional array of one-dimensional array (line 30) based on the generic values. Other local variables are defined in lines 27 (address), line 28 (output line buffer), line 29 (input line buffer), and line 31 (datatemp). Line 32 declares a BOOLEAN variable READDONE which is used as a switch to control the reading of the ROM image file. It is initialized to FALSE.

Lines 34 to 41 are an if statement to ensure that reading the ROM image file only during the initialization phase by assigning READDONE to TRUE in line 40 after the image file is read WORDS lines (MEM'range is used). This allows flexibility of reading the initial portion of the file (not the whole file) so that the MEM array is small enough for quick testing in the early stage of the verification.

```
20    architecture BEH of ROMRF is
21    begin
22       check : process(ADDR)
23          file MEMIN   : text is in  infile;
24          file memout : text is out tracefile;
25          subtype ARRAYWORD is bit_vector((N - 1) downto 0);
26          type    ARRAYTYPE is array(Natural range <>) of ARRAYWORD;
27          variable address   : integer;
28          variable OUTLINE   : line;
29          variable INLINE    : line;
30          variable MEM       : ARRAYTYPE (0 to (WORDS - 1));
31          variable datatemp : bit_vector(N-1 downto 0);
32          variable READDONE : boolean := FALSE;
33       begin
34          if (not READDONE) then
35             for i in MEM'range loop
36                readline(MEMIN, INLINE);
37                read(INLINE, datatemp);
38                MEM(i) := datatemp;
39             end loop;
40             READDONE := TRUE;
41          end if;
42          if (IS_X(ADDR)) then
43             DATA <= (others => 'X') after T_DELAY;
44             assert FALSE report "ADDR not valid" severity NOTE;
45             write(OUTLINE, string'(" -- READ address not valid : "));
46             write(OUTLINE, NOW);
47             writeline(memout, OUTLINE);
48          else
49             address := TO_INTEGER(ADDR);
50             DATA    <= To_stdlogicvector(MEM(address)) after T_DELAY;
51             write(OUTLINE, string'(" Address: "));
52             write(OUTLINE, TO_HEX_STRING(ADDR));
53             write(OUTLINE, string'(" Value: "));
54             write(OUTLINE,
               TO_HEX_STRING(to_stdlogicvector(MEM(address))));
55             write(OUTLINE, string'("  -- Memory READ  at time: "));
56             write(OUTLINE, NOW);
57             writeline(MEMOUT, OUTLINE);
58          end if;
```

```
59        end process;
60    end BEH;
```

Line 42 checks whether the address is valid so that each bit of the ADDR consists of only '0' or '1' since std_logic type can have other values. 'X' values are output to DATA, and assert statements are used to generate messages. Line 45 puts a string in the output line buffer. Line 46 puts the current simulation time to the output line buffer. Line 47 puts the output line buffer to the output file.

If the address is valid, ADDR is converted to an integer (line 49) address to index to the MEM since the array index of MEM is of INTEGER type. The DATA is obtained by converting MEM(address) of bit_vector type to std_logic_vector type in line 50. Type bit_vector is used for the MEM to save memory space for the VHDL simulator, especially if a large ROM is used (two values rather than nine values for each bit). Lines 51 to 57 record the ROM access ADDR, DATA, and simulation time in the output file.

The following shows the ROM image file:

```
1     0000000000000
2     0000100010001
3     0001000100010
4     0001100110011
5     0010001000100
6     0010101010101
7     0011001100110
8     0011101110111
9     0100010001000
10    0100110011001
11    0101010101010
12    0101110111011
13    0110011001100
14    0110111011101
15    0111011101110
16    0111111111111
```

Figure 10.1 shows the simulation waveform of the ROMRF VHDL code. Note that the output DATA is not initialized with value "UUUU" between time 0 ns and 4 ns. At time 40 ns, ADDR is set to "ZZZZ". The output DATA shows the value of "XXXX". The output file "rom.trace" is as follows:

```
Address: 0 Value: 000   -- Memory READ  at time: 0  NS
Address: 2 Value: 222   -- Memory READ  at time: 20 NS
-- READ address not valid  : 40 NS
Address: A Value: AAA   -- Memory READ  at time: 60 NS
Address: F Value: FFF   -- Memory READ  at time: 80 NS
```

You may be wondering where functions IS_X (line 42), TO_INTEGER (line 49), TO_STDLOGICVECTOR (line 50), TO_HEX_STRING (line 52) are located and how they are implemented. Function TO_HEX_STRING only outputs three hexa

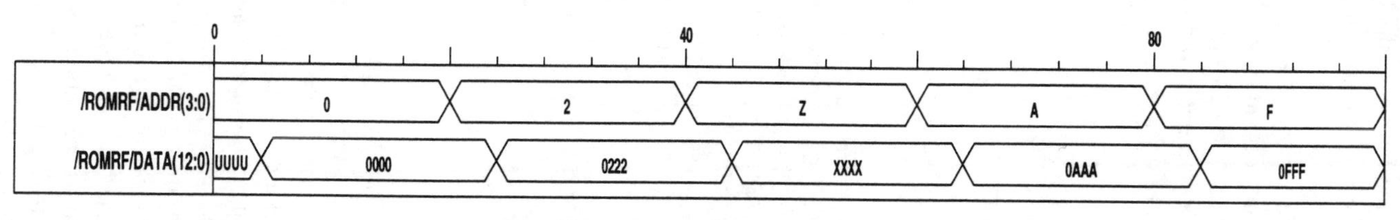

FIGURE 10.1 Simulation waveform of ROMRF.

digits because it does not generate a hexa digit if there are not exactly 4 bits (only bits 11 downto 0 are used). These are left as an exercise!

Note that type std_logic_vector is used for both ADDR and DATA so that the same base type is used to interface between blocks. Other types are used only internally to a block. This reduces the type mismatch among blocks, especially if written by different people in the same project.

The preceding VHDL code for the ROM reads the content of the ROM from a text file. It is possible to define the content of the ROM with a constant table. The following VHDL code shows an example of implementing the seven-segment decoder. Note that the VHDL code can be synthesized so that the ROM is implemented with gates. Figure 10.2 shows the synthesized schematic.

```
1     library IEEE;
2     use IEEE.std_logic_1164.all;
3     use IEEE.std_logic_arith.all;
4     entity ROM4SYN is
5        port (
6           ADDR : in  std_logic_vector(3 downto 0);
7           DOUT : out std_logic_vector(6 downto 0));
8     end ROM4SYN;
9     architecture RTL of ROM4SYN is
10       subtype WORD is std_logic_vector(6 downto 0);
11       type ROM_TBL is array(0 to 15) of WORD;
12       constant ROM_VAL : ROM_TBL :=  ROM_TBL'(
13          "1111110", "1100000", "1011011", "1110011",
14          "1100101", "0110111", "0111111", "1100010",
15          "1111111", "1110111", "0111001", "0111101",
16          "0011001", "1111001", "1011111", "0001111");
17    begin
18       DOUT <= ROM_VAL(conv_integer(unsigned(ADDR)));
19    end RTL;
```

10.3 BIDIRECTIONAL PAD MODEL

Pads are necessary in each integrated circuit. Pads can be categorized into (1) input, (2) output, and (3) bidirectional. A pad may also have tristate, pull up, or pull down options. In this section, the functional behaviors of pads are discussed. Most silicon foundry vendors provide the detailed simulation pad models. Therefore, the detailed simulation pad models involving diode and transistors are not discussed here. The functional input and output pad models are easy. They can be modeled as straight connections (perhaps with some timing delay modeling). Here, we will concentrate on the functional bidirectional pad modeling. Figure 10.3 shows the relationship between a bidirectional pad, the core of the chip, and the outside of an integrated circuit chip. A bidirectional pad can be viewed as the schematic shaded rectangle in Figure 10.3. It consists of a tristate buffer, controlled by signal OUTEN. It may have a weak pull up (down) option. A bidirectional pad can be functionally modeled as the following VHDL code:

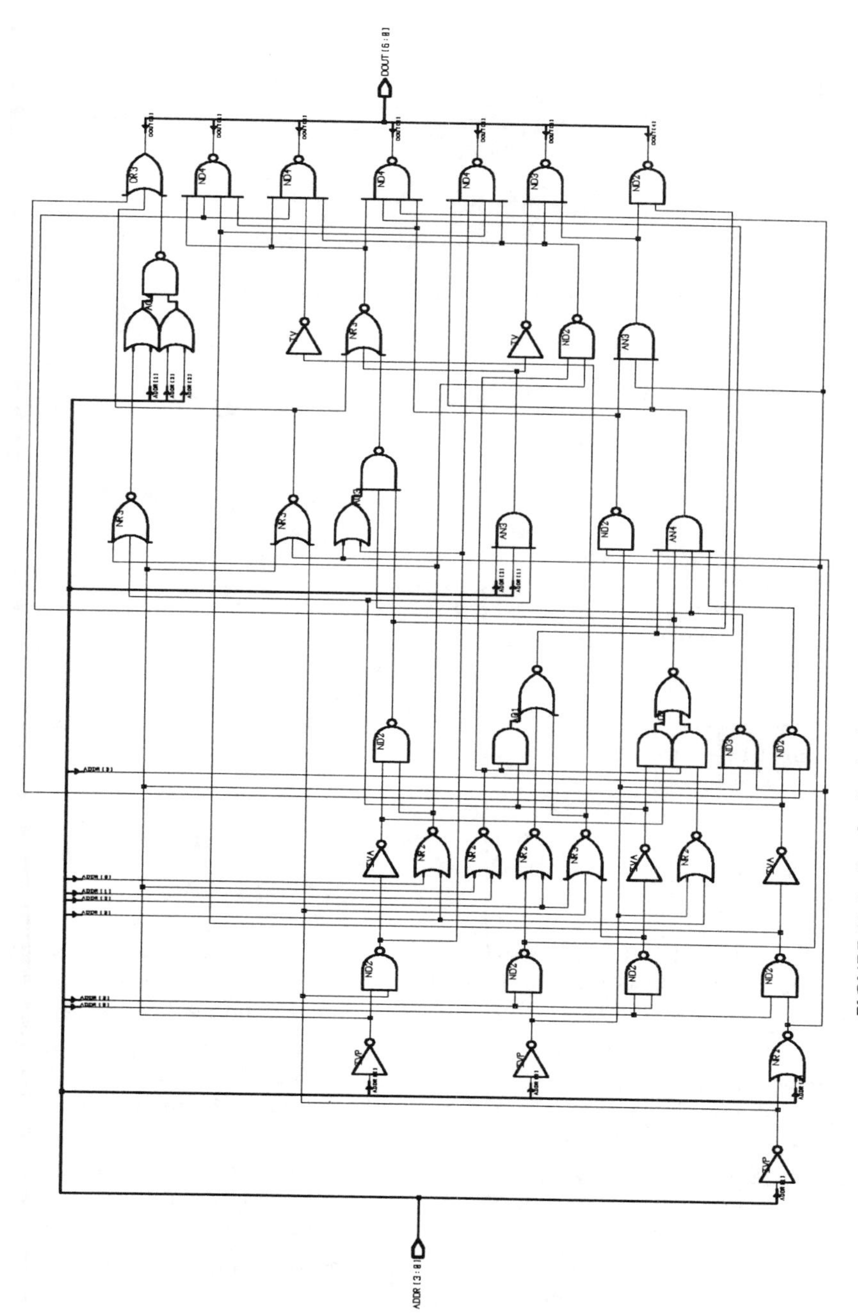

FIGURE 10.2 Synthesized schematic for ROM4SYN VHDL code.

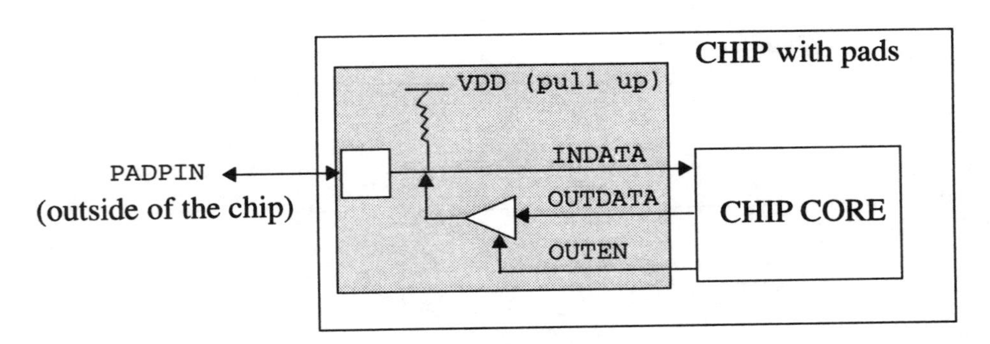

FIGURE 10.3 Bidirectional pad functional schematic.

```
1    library IEEE;
2    use IEEE.std_logic_1164.all;
3    use work.CONVERT.all;
4    entity BIPAD is
5       generic (
6          STRONG : in boolean;
7          PULL   : in std_logic);
8       port (
9          PADPIN  : inout std_logic_vector;
10         OUTEN   : in    std_logic;
11         OUTDATA : in    std_logic_vector;
12         INDATA  : out   std_logic_vector);
13   end BIPAD;
14   architecture BEH of BIPAD is
15   begin
16      PADPIN <= (PADPIN'range => PULL);
17      PADPIN <= OUTDATA when OUTEN = '1' else (PADPIN'range => 'Z');
18      INDATA <= TO_STRONG(PADPIN) when STRONG else PADPIN;
19   end BEH;
```

Lines 9 to 12 of BIPAD VHDL show that the bidirectional pad has two ports OUTEN and OUTDATA of mode **in**. They are connected to the outputs of the chip core logic. Port INDATA of mode **out** is considered to be the input to the core logic. Port PADPIN of mode **inout** is used to communicate to outside of the chip. There are two generics (lines 6 and 7) STRONG and PULL. PULL is of type std_logic to indicate the options for pull up (mapped with 'H'), pull down (mapped with 'L'), or none (mapped with 'Z'). STRONG is of type BOOLEAN to determine whether the pad converts weak values ('H' and 'L') to strong values ('1' and '0') or not. Function TO_STRONG is in the package CONVERT as shown in the following VHDL code. Package CONVERT is referenced in line 3 of BIPAD VHDL code. In BIPAD VHDL code, line 16 assigns PADPIN to the PULL generic. Line 17 assigns PADPIN to the value of OUTDATA when the tristate buffer is enabled by the OUTEN signal, otherwise, PADPIN is tristated with all 'Z's. Line 18 assigns the INDATA which is the same as the value of PADPIN or the value after the value has been converted to strong values when generic STRONG is TRUE.

```
1     library IEEE;
2     use IEEE.std_logic_1164.all;
3     package CONVERT is
4        function TO_STRONG (D : in std_logic_vector)
5           return std_logic_vector;
6     end CONVERT;
7     package body CONVERT is
8        function TO_STRONG (D : in std_logic_vector)
9           return std_logic_vector is
10          variable result : std_logic_vector(D'length-1 downto 0);
11          variable index  : integer;
12       begin
13          index := D'length - 1;
14          for i in D'range loop
15             if (D(i) = 'H') then
16                result(index) := '1';
17             elsif (D(i) = 'L') then
18                result(index) := '0';
19             else
20                result(index) := D(i);
21             end if;
22             index := index - 1;
23          end loop;
24          return result;
25       end TO_STRONG;
26    end CONVERT;
```

The following VHDL code shows examples of using the BIPAD bidirectional pad VHDL model. Lines 7 to 16 declare the BIPAD component, and line 17 specifies the hard binding configuration. Lines 23 to 25 instantiate four bidirectional pads which have weak pull up options (PULL is mapped to 'H'), and the value of INDATA should be converted to strong values (STRONG is mapped to TRUE). Lines 26 to 28 are similar to lines 23 to 25, except that the pads are not pull up or down, and the INDATA values are not converted to strong values. Lines 29 to 31 instantiate another four pads, the same as lines 26 to 28, except that the pads have weak pull up outside the pads, as modeled in line 32.

```
1     library IEEE;
2     use IEEE.std_logic_1164.all;
3     use IEEE.std_logic_arith.all;
4     entity PADEXAM is
5     end PADEXAM;
6     architecture BEH of PADEXAM is
7        component BIPAD
8        generic (
9           STRONG : in boolean;
10          PULL   : in std_logic);
11       port (
12          PADPIN  : inout std_logic_vector;
13          OUTEN   : in    std_logic;
14          OUTDATA : in    std_logic_vector;
```

```
15          INDATA   : out    std_logic_vector);
16       end component;
17       for all : BIPAD use entity work.BIPAD(BEH);
18       signal PADHZ, PADZZ, PADZH     : std_logic_vector(3 downto 0);
19       signal PADHZi, PADZZi, PADZHi    : std_logic_vector(3 downto 0);
20       signal PADo    : std_logic_vector(3 downto 0);
21       signal PAD_EN, SIG0, SIG1, SIG2, SIG3   : std_logic;
22    begin
23       pada : BIPAD
24          generic map (STRONG => TRUE, PULL => 'H')
25          port map (PADHZ, PAD_EN, PADo, PADHZi);
26       padb : BIPAD
27          generic map (STRONG => FALSE, PULL => 'Z')
28          port map (PADZZ, PAD_EN, PADo, PADZZi);
29       padc : BIPAD
30          generic map (STRONG => FALSE, PULL => 'Z')
31          port map (PADZH, PAD_EN, PADo, PADZHi);
32       PADZH    <= "HHHH";
33       ---------------------------------------------
34       dataout : process
35          variable index : integer := 0;
36       begin
37          index := (index + 1) mod 16;
38          PADo   <= conv_std_logic_vector(index, 4);
39          if (index >= 3) and (index <= 7) then
40             PAD_EN <= '1';
41          else
42             PAD_EN <= '0';
43          end if;
44          wait for 20 ns;
45       end process;
46       ---------------------------------------------
47       SIG0 <= '1' when PADHZi(1 downto 0) = "11" else '0';
48       SIG1 <= PADHZi(0) and PADHZi(1);
49       SIG2 <= '1' when PADZHi(1 downto 0) = "11" else '0';
50       SIG3 <= PADZHi(0) and PADZHi(1);
51    end BEH;
```

Lines 34 to 45 set up signal values to illustrate the pad functions. Lines 47 to 50 generate four signals SIG0, SIG1, SIG2, and SIG3 using different INDATA values to show the effect of TO_STRONG function. These four signals illustrate signals that can be generated inside the core of the chip. Line 47 assigns SIG0 to '1' when PADHZi(1 downto 0) is "11", otherwise '0'. Line 48 assigns SIG1 as the BOOLEAN AND operation of signals PADHZi(0) and PADHZi(1). Lines 49 and 50 are the same as lines 47 and 48 for the value PADZHi(1 downto 0). The waveform in Figure 10.4 shows the simulation of PADEXAM VHDL code. The tristate buffers of the bidirectional pads are enabled (PAD_EN = '1') when PADo values are 3 to 7. When PAD_EN is '0', the tristate buffers are disabled, and the values of PADPIN (outside of the chip) PADHZ, PADZZ, and PADZH are "HHHH", "ZZZZ", and "HHHH",

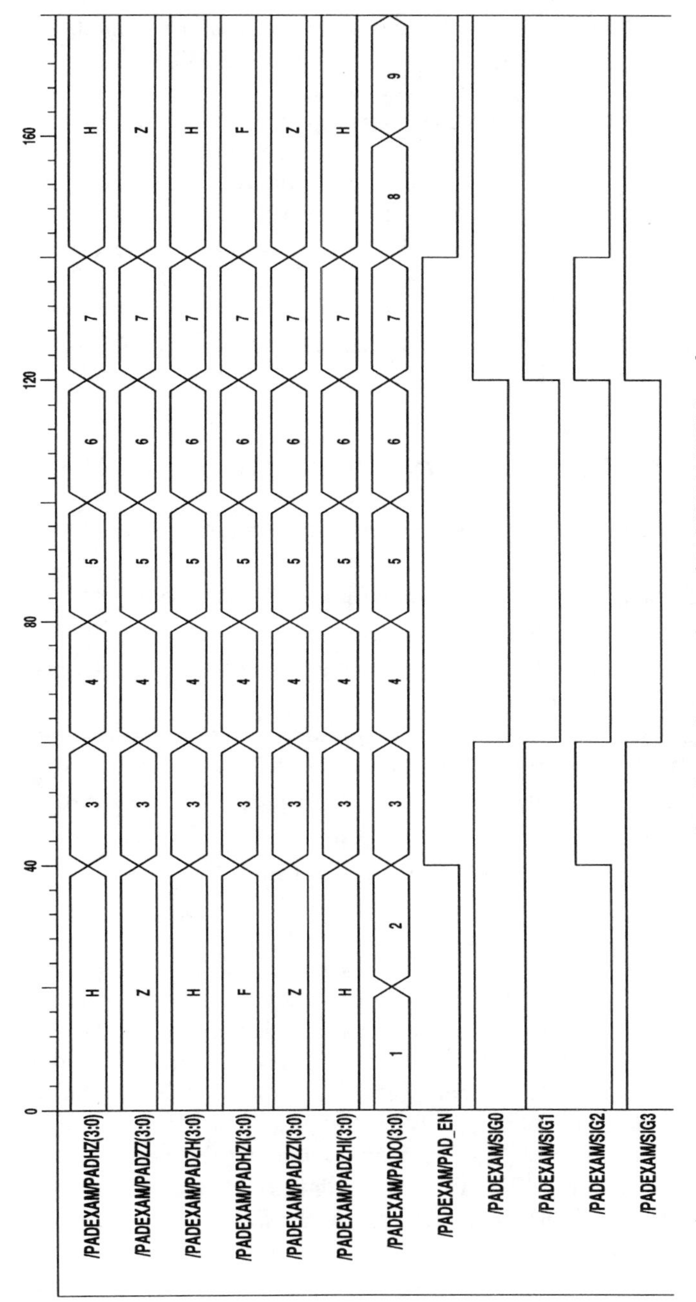

FIGURE 10.4 Simulation waveform of PADEXAM VHDL code.

respectively. The values of INDATA (inside of the chip) PADHZi, PADZZi, and PADZHi are "1111", "ZZZZ", and "HHHH", respectively. Note that SIG0 and SIG1 have the same waveform since the value of PADHZi(1 downto 0) cannot be "HH". Signals SIG2 and SIG3 are different since PADZHi(1 downto 0) = "HH", which is not the same as "11". SIG2 has the value '0', while SIG3 has the value '1', since 'H' and 'H' is '1' (refers to Chapter 6 and_table of std_logic_1164 package). These differences can only appear in VHDL simulation. From the hardware point of view (after synthesis), SIG2 and SIG3 would produce the same functional hardware implementations. To simplify the VHDL modeling of internal chip logic, converting the pad values to strong values is recommended to avoid the differences of VHDL simulation results and the synthesized results. Otherwise, line 48 may need to be written as follows:

```
SIG2 <= '1' when (PADZHi(1 downto 0) = "11") or (PADZHi(1 downto
0) = "HH") else '0';
```

10.4 ATTRIBUTE DECLARATION AND ATTRIBUTE SPECIFICATION

Attribute declaration and attribute specification syntax are as follows:

attribute_declaration ::= **attribute** identifier : type_mark ;
attribute_specification ::=
  attribute attribute_designator **of** entity_name_list : entity_class **is** expression ;
entity_class ::= **entity** | **architecture** | **configuration** | **procedure** |
  function | **package** |
  type | **subtype** | **constant** | **signal** | **variable** | **component** | **label**
entity_name_list ::= entity_designator { , entity_designator } | **others** | **all**
entity_designator ::= simple_name | opertor_symbol

An attribute specification associates a user-defined attribute (information) with one or more entities (of entity class which is not just entity). Reserved word **others** (**all**) must be the last specification of that given entity class and that attribute; if it is used, it refers to the immediate enclosing declaration of the specified entity class. Attribute specification for an attribute of a design unit (entity, architecture, configuration, and package) must appear immediately within the declarative part of that design unit. The rest of the entity class must appear within the declarative part in which the entity class is declared. It is an error when an attribute has more than one association within a given entity. Two different attributes with the same simple name cannot both be associated with the same given entity class.

The attached user-defined attributes (information) can be used to pass information among various computer-aided design (CAD) tools. The following VHDL code shows attribute declaration and attribute specification examples. Package ATTRPACK declares attributes CELL_LOC in line 3, PAD_KIND in line 5, and INFO in line 6. They are of type COORXY, PAD_TYPE, and string, respectively. Line 7 is an attribute specification for package (entity class) ATTRPACK. The package ATTR-

PACK is referenced in line 11 for VHDL entity ATTR. Lines 24 to 30 are more attribute specification examples. These attributes can be passed on to other CAD tools such as a layout tool. These attributes can also be referenced in the same way as the predefined attributes. Line 36 shows an example of referencing the INFO attribute of component ND2 as ND2'INFO.

```
1    package ATTRPACK is
2       type COORXY is record X, Y : integer; end record;
3       attribute CELL_LOC : COORXY;
4       type PAD_TYPE is (INPUT, OUTPUT, BIDIRECTIONAL);
5       attribute PAD_KIND : PAD_TYPE;
6       attribute INFO     : string;
7       attribute INFO of ATTRPACK : package is "package for attribute";
8    end ATTRPACK;
9    library IEEE;
10   use IEEE.std_logic_1164.all;
11   use work.ATTRPACK.all;
12   entity ATTR is
13      port (
14         IA, IB, IC, ID : in  std_logic;
15         IX, IY         : out std_logic);
16   end ATTR;
17   architecture BEH of ATTR is
18      signal SIG0, SIG1 : std_logic;
19      component ND2
20      port (
21         A, B : in  std_logic;
22         Z    : out std_logic);
23      end component;
24      attribute INFO of ND2 : component is "from XXX library";
25      attribute INFO of all : signal is "net weight = 1";
26      attribute CELL_LOC of cell0  : label is (10, 15);
27      attribute CELL_LOC of cell1  : label is (12, 30);
28      attribute CELL_LOC of others : label is (30, 50);
29      attribute PAD_KIND of IA, IB, IC, ID : signal is INPUT;
30      attribute PAD_KIND of others         : signal is OUTPUT;
31   begin
32      cell0 : ND2 port map (IA, IB, SIG0);
33      cell1 : ND2 port map (IC, ID, SIG1);
34      cell2 : ND2 port map (SIG0, SIG1, IX);
35      IY    <= IB xor IC xor ID;
36      assert FALSE report "ND2 is " & ND2'INFO severity NOTE;
37   end BEH;
```

10.5 ACCESS AND RECORD TYPE

As with many other high-level programming languages, VHDL has constructs for defining record and access (pointer) types. The access (pointer) type allows dynamic memory allocation. The following VHDL code shows an example of creating a binary search tree (BST). Lines 6 to 10 declare a record type NODE. The NODE record has a left pointer and a right pointer of type POINTER. The NODE record has a field to store the value of type CHARACTER. This NODE type is forward referenced in line 4 so that a POINTER type can be declared in line 5. Line 11 declares an array type of nine characters. Line 12 defines a constant with nine characters. Lines 14 to 22 are a procedure to dynamically get a new record NODE (line 18) and return a pointer to the NODE record. Lines 19 and 20 initialize the left and right pointer field to NULL.

Procedure BSTINSERT inserts a value into the BST. It is a recursive procedure that calls itself to find the bottom of the BST to insert. The insertion is done by getting a new node in line 30 and setting up the pointer (line 31). Depending on the value to be inserted, it will go through the left pointer (LPTR in line 33) or the right pointer (RPTR in line 40).

```
1     entity BSTREE is
2     end BSTREE;
3     architecture BEH of BSTREE is
4        type NODE;
5        type POINTER is access NODE;
6        type NODE is record
7           LPTR  : POINTER;
8           VALUE : CHARACTER;
9           RPTR  : POINTER;
10       end record;
11       type INTARRAY is array (1 to 9) of CHARACTER;
12       constant ITEMS : INTARRAY :=
13          ('3','2','6','4','8','9','1','7','5');
14       procedure GETNODE (
15          variable Z   : inout POINTER;
16          constant VAL : in    CHARACTER) is
17       begin
18          Z        := new NODE;
19          Z.LPTR   := NULL;
20          Z.RPTR   := NULL;
21          Z.VALUE  := VAL;
22       end GETNODE;
23       procedure BSTINSERT (
24          variable PARENT : in POINTER;
25          constant VAL    : in CHARACTER) is
26          variable Z : POINTER;
27       begin
28          if (VAL < PARENT.VALUE) then
29             if (PARENT.LPTR = NULL) then
30                GETNODE(Z, VAL);
31                PARENT.LPTR := Z;
32             else
33                BSTINSERT(PARENT.LPTR, VAL);
34             end if;
35          else
36             if (PARENT.RPTR = NULL) then
37                GETNODE(Z, VAL);
38                PARENT.RPTR := Z;
39             else
```

```
40               BSTINSERT(PARENT.RPTR, VAL);
41            end if;
42         end if;
43      end BSTINSERT;
```

Procedure INORDER (lines 44 to 53) traverses the BST in order. The values in the left subtree (pointed by the left pointer) of a node are visited before the node. The values in the right subtree (pointed by the right pointer) of a node are visited after the node. Line 50 is an assert statement to print out the message to indicate the node being visited. Procedure PREORDER (lines 54 to 63) is similar to INORDER. In the PREORDER traversing, the node is visited before the left subtree and then the right subtree. Procedure FIND (lines 64 to 80) returns a BOOLEAN value to indicate whether a value is in the BST or not.

```
44      procedure INORDER (variable P : in POINTER) is
45      begin
46         if (P.LPTR /= NULL) then
47            INORDER(P.LPTR);
48         end if;
49         assert FALSE report "Visit node " & P.VALUE severity NOTE;
50         if (P.RPTR /= NULL) then
51            INORDER(P.RPTR);
52         end if;
53      end INORDER;
54      procedure PREORDER (variable P : in POINTER) is
55      begin
56         assert FALSE report "Visit node " & P.VALUE severity NOTE;
57         if (P.LPTR /= NULL) then
58            PREORDER(P.LPTR);
59         end if;
60         if (P.RPTR /= NULL) then
61            PREORDER(P.RPTR);
62         end if;
63      end PREORDER;
64      procedure FIND (
65         variable P     : in  POINTER;
66         constant VAL   : in  CHARACTER;
67         variable MATCH : out BOOLEAN) is
68      begin
69         MATCH := FALSE;
70         if (P /= NULL) then
71            if (P.VALUE = VAL) then
72               MATCH := TRUE;
73               assert FALSE report "Found " & VAL severity NOTE;
74            elsif (P.VALUE < VAL) and (P.RPTR /= NULL) then
75               FIND(P.RPTR, VAL, MATCH);
76            elsif (P.VALUE > VAL) and (P.LPTR /= NULL) then
77               FIND(P.LPTR, VAL, MATCH);
78            end if;
79         end if;
80      end FIND;
81   begin
```

```
82      bst : process
83         variable HEAD   : POINTER;
84         variable FOUND : BOOLEAN;
85      begin
86         GETNODE(HEAD, '0');
87         for k in ITEMS'range loop
88            BSTINSERT(HEAD, ITEMS(k));
89            wait for 20 ns;
90         end loop;
91         INORDER(HEAD);  wait for 20 ns;
92         PREORDER(HEAD); wait for 20 ns;
93         FIND(HEAD, '9', FOUND);
94         assert FOUND report "9 not found" severity NOTE;
95         wait for 20 ns;
96         FIND(HEAD, 'A', FOUND);
97         assert FOUND report "A not found" severity NOTE;
98         deallocate(HEAD);
99         wait;
100     end process;
101  end BEH;
```

Lines 82 to 100 are a process statement with sequential statements inside to test the procedures. Line 86 gets the node for the root of the BST with value '0'. Lines 86 to 90 insert a value from the constant array every 20 ns. Line 91 calls procedure INORDER to traverse the BST in sequence. Line 92 calls procedure PREORDER to traverse the BST in preorder. Line 93 calls procedure FIND to determine whether character '9' is in the BST. Line 94 prints out the message if '9' is not found in the BST. Lines 96 to 97 are similar to FIND character 'A'. Line 98 deallocates the BST pointed by the root HEAD. The following is the message generated from the actual simulation. Note the increasing sequence when the INORDER is performed.

180 NS

Assertion NOTE at 180 NS in design unit BSTREE(BEH) from process /BSTREE/BST: "Visit node 0"

Assertion NOTE at 180 NS in design unit BSTREE(BEH) from process /BSTREE/BST: "Visit node 1"

Assertion NOTE at 180 NS in design unit BSTREE(BEH) from process /BSTREE/BST: "Visit node 2"

Assertion NOTE at 180 NS in design unit BSTREE(BEH) from process /BSTREE/BST: "Visit node 3"

Assertion NOTE at 180 NS in design unit BSTREE(BEH) from process /BSTREE/BST: "Visit node 4"

Assertion NOTE at 180 NS in design unit BSTREE(BEH) from process /BSTREE/BST: "Visit node 5"

Assertion NOTE at 180 NS in design unit BSTREE(BEH) from process /BSTREE/BST: "Visit node 6"

Assertion NOTE at 180 NS in design unit BSTREE(BEH) from process /BSTREE/BST: "Visit node 7"

Assertion NOTE at 180 NS in design unit BSTREE(BEH) from process /BSTREE/BST: "Visit node 8"

Assertion NOTE at 180 NS in design unit BSTREE(BEH) from process /BSTREE/BST: "Visit node 9"

200 NS

Assertion NOTE at 200 NS in design unit BSTREE(BEH) from process /BSTREE/BST: "Visit node 0"

Assertion NOTE at 200 NS in design unit BSTREE(BEH) from process /BSTREE/BST: "Visit node 3"

Assertion NOTE at 200 NS in design unit BSTREE(BEH) from process /BSTREE/BST: "Visit node 2"

Assertion NOTE at 200 NS in design unit BSTREE(BEH) from process /BSTREE/BST: "Visit node 1"

Assertion NOTE at 200 NS in design unit BSTREE(BEH) from process /BSTREE/BST: "Visit node 6"

Assertion NOTE at 200 NS in design unit BSTREE(BEH) from process /BSTREE/BST: "Visit node 4"

Assertion NOTE at 200 NS in design unit BSTREE(BEH) from process /BSTREE/BST: "Visit node 5"

Assertion NOTE at 200 NS in design unit BSTREE(BEH) from process /BSTREE/BST: "Visit node 8"

Assertion NOTE at 200 NS in design unit BSTREE(BEH) from process /BSTREE/BST: "Visit node 7"

Assertion NOTE at 200 NS in design unit BSTREE(BEH) from process /BSTREE/BST: "Visit node 9"

220 NS

Assertion NOTE at 220 NS in design unit BSTREE(BEH) from process /BSTREE/BST: "Found 9"

240 NS

Assertion NOTE at 240 NS in design unit BSTREE(BEH) from process /BSTREE/BST: "A not found"

The access type is provided in VHDL for simulation. However, it is rarely used in the actual hardware design.

10.6 GUARDED BLOCK

A guarded block is a block statement with guarded expression (of type BOOLEAN) specified. An implicit signal (of type BOOLEAN) GUARD is declared as the value of the guarded expression. It can be used to control the operation of certain statements inside the block, and it cannot be the target of a signal assignment statement. The following VHDL code shows a simple guarded block example. Line 6 declares signals. Lines 8 to 15 depict a guarded block with guarded expression EN = '1'. Lines 16 to 18 assign values to signal A, B, and EN for the simulation. Lines 10 to 14 are concurrent signal assignment statements. Lines 11 and 13 use the reserved word **guarded** in concurrent signal assignment statements which cannot appear in the sequential signal assignment statement. Lines 12 and 14 use the implicit BOOLEAN signal GUARD.

```
1    library IEEE;
2    use IEEE.std_logic_1164.all;
3    entity GARD is
4    end GARD;
5    architecture BEH of GARD is
6       signal A, B, C, D, E, EN, G : std_logic := '0';
7    begin
8       gblock : block (EN = '1')
9       begin
10         C <=          A nand B;
11         D <= guarded A nand B;
12         E <= '0' when GUARD else 'Z';
13         E <= guarded C;
14         G <= B when GUARD else '0';
15      end block;
16      A  <= not A after 20 ns;
17      B  <= not B after 10 ns;
18      EN <= '1', '0' after 25 ns, '1' after 55 ns, '0' after 80 ns;
19   end BEH;
```

Figure 10.5 shows the simulation waveform of GARD VHDL code. Signal C is assigned as A nand B which has nothing to do with the value of EN = '1' (GUARD) value. Signal D is guarded and it is assigned A nand B only when the EN = '1'. Signal G is assigned to the signal B when signal GUARD is TRUE, otherwise '0'. Signal E has two drivers from lines 12 and 13. It has unknown value 'X' when one driver has value '1' and the other driver has value '0'. The implicit signal GUARD is also shown.

10.7 GUARDED SIGNAL AND NULL WAVEFORM

A guarded signal is a signal with signal kind (reserved word **register** or **bus**) specified in the signal declaration. A **register** guarded signal retains its value when the signal is not driven. This can be used to model a signal with storage (memory) behavior. A **bus** guarded signal does not retain its value when the signal is not driven. The value is determined by the associated resolution function. The following VHDL code shows the behaviors of different types of signals. Note that a guarded signal must be a resolved type. The following GARDPACK VHDL code declares subtype SMALL_INT2 and SMALL_INT in lines 2 and 3. Line 4 declares an array type of SMALL_INT to be used in the resolution function RESOLVE_INT_FUNC in line 5. Line 6 declares a resolved type RESOLVE_INT for signals. Lines 7 to 9 declare three constants.

```
1    package GARDPACK is
2       subtype SMALL_INT2 is integer    range -500 to 700;
3       subtype SMALL_INT  is SMALL_INT2 range -20  to 30;
```

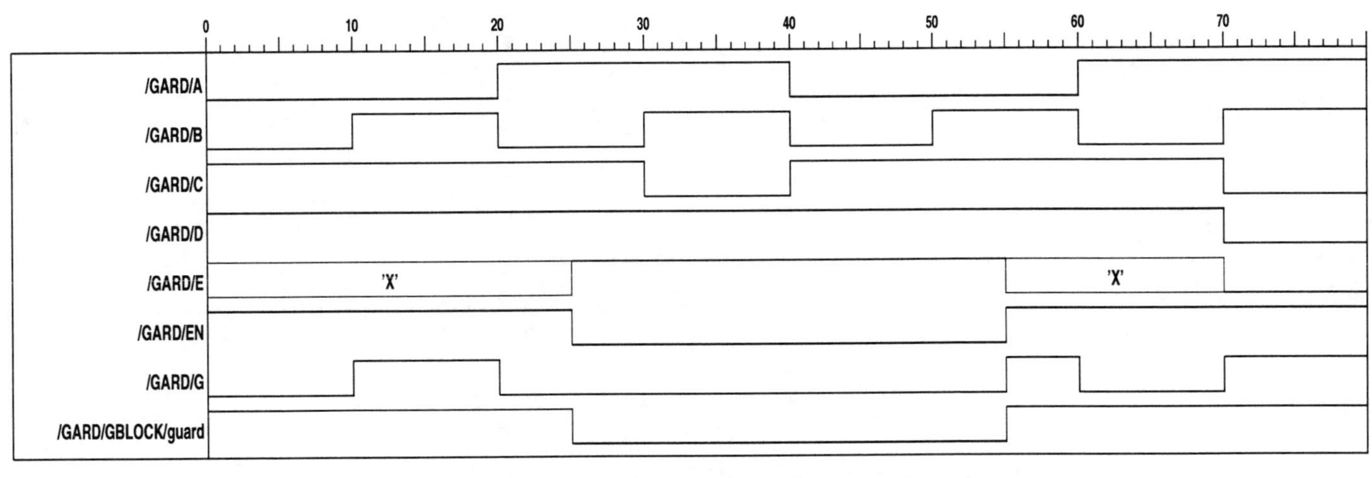

FIGURE 10.5 Simulation waveform of GARD VHDL code.

```
4        type INT_ARY is array (integer range <>) of SMALL_INT;
5        function RESOLVE_INT_FUNC (D : in INT_ARY) return SMALL_INT2;
6        subtype  RESOLVE_INT is RESOLVE_INT_FUNC SMALL_INT;
7        constant CYCLES    : integer := 3;
8        constant PERIOD    : time    := 30 ns;
9        constant TIME_SPAN : time    := CYCLES * PERIOD;
10    end GARDPACK;
11    package body GARDPACK is
12       function RESOLVE_INT_FUNC (D : in INT_ARY) return SMALL_INT2 is
13          variable result : SMALL_INT2;
14       begin
15          if (D'length = 0) then
16             return -10;
17          else
18             result := 0;
19             for i in D'range loop
20                result := result + D(i);
21             end loop;
22             return result;
23          end if;
24       end RESOLVE_INT_FUNC;
25    end GARDPACK;
```

The RESOLVE_INT_FUNC function body is specified in the GARDPACK package body from lines 12 to 24. It takes an array of SMALL_INT and returns a value of SMALL_INT2 type. The returned value is simply the sum of all its drivers. The function returns value –10 (lines 15 and 16) when the associated signal is not driven at all (the array length is 0).

The following DIFFGARD VHDL code shows entity DIFFGARD, which references GARDPACK package in line 28. Lines 32, 33, and 34 declare guarded signals of **register** and **bus** kind and normal nonguarded signals, respectively. Line 35 declares signal EN of INTEGER type which is assigned with values 0, 1, 2, and 3 in lines 37 and 38 to be used in the guarded expressions of guarded blocks. Each block statement has guarded expression EN = 0, EN = 1, EN = 2, respectively. When EN = 3, all guarded expressions are FALSE. Each block has a process which assigns an INTEGER value. For example, block b0 (lines 39 to 55) assigns values 1 to CYCLES (3), block b1 assigns values 11 to 13, block b2 assigns values from 21 to 23. Each block has a local signal (REGSIG0, REGSIG1, REGSIG2) declared (lines 40, 57, and 76) of normal, **register**, and **bus** kind. Lines 49 to 54 are concurrent (and guarded) signal assignments in block b0. The similar statements appear also for block b1 and b2.

```
26    library IEEE;
27    use IEEE.std_logic_1164.all;
28    use work.GARDPACK.all;
29    entity DIFFGARD is
30    end DIFFGARD;
31    architecture BEH of DIFFGARD is
32       signal REG_SIG,REG_SIG0,REG_SIG1,REG_SIG2:RESOLVE_INT register;
```

```
33     signal BUS_SIG,BUS_SIG0,BUS_SIG1,BUS_SIG2:RESOLVE_INT bus      ;
34     signal NOR_SIG,NOR_SIG0,NOR_SIG1,NOR_SIG2:RESOLVE_INT          ;
35     signal EN                                          : integer;
36   begin
37     EN <= 0, 1 after TIME_SPAN, 2 after 2 * TIME_SPAN,
38         3 after 3 * TIME_SPAN;
39     b0 : block (EN = 0)
40        signal REGSIG0 : SMALL_INT;
41     begin
42        phase0 : process
43        begin
44           for i in 1 to CYCLES loop
45              REGSIG0 <= i;
46              wait for PERIOD;
47           end loop;
48        end process;
49        REG_SIG0 <= guarded REGSIG0;
50        BUS_SIG0 <= guarded REGSIG0;
51        NOR_SIG0 <= guarded REGSIG0;
52        REG_SIG  <= guarded REG_SIG0;
53        BUS_SIG  <= guarded BUS_SIG0;
54        NOR_SIG  <= guarded NOR_SIG0;
55     end block;
56     b1 : block (EN = 1)
57        signal REGSIG1 : RESOLVE_INT register;
58     begin
59        phase1 : process
60        begin
61           REGSIG1 <= null;
62           wait for TIME_SPAN;
63           for i in 1 to CYCLES loop
64              REGSIG1 <= i + 10;
65              wait for PERIOD;
66           end loop;
67        end process;
68        REG_SIG1 <= guarded REGSIG1;
69        BUS_SIG1 <= guarded REGSIG1;
70        NOR_SIG1 <= guarded REGSIG1;
71        REG_SIG  <= guarded REG_SIG1;
72        BUS_SIG  <= guarded BUS_SIG1;
73        NOR_SIG  <= guarded NOR_SIG1;
74     end block;
75     b2 : block (EN = 2)
76        signal REGSIG2 : RESOLVE_INT bus;
77     begin
78        phase1 : process
79        begin
80           REGSIG2 <= null;
81           wait for 2 * TIME_SPAN;
82           for i in 1 to CYCLES loop
```

```
83                REGSIG2 <= i + 20;
84                wait for PERIOD;
85             end loop;
86          end process;
87          REG_SIG2 <= guarded REGSIG2;
88          BUS_SIG2 <= guarded REGSIG2;
89          NOR_SIG2 <= guarded REGSIG2;
90          REG_SIG  <= guarded REG_SIG2;
91          BUS_SIG  <= guarded BUS_SIG2;
92          NOR_SIG  <= guarded NOR_SIG2;
93       end block;
94    end BEH;
```

Figure 10.6 shows the simulation waveform of DIFFGARD VHDL code. For the first TIME_SPAN, EN = 0, block b0 guarded expression is TRUE. The other two block statement guarded expressions are FALSE. Signals REG_SIG0 has value 1, 2, and 3. Signals REG_SIG1 and REG_SIG2 have values –20. They are not driven (line 61 and 80 null waveform) and they are **register** kind to retain values which are initialized as the left-most value of their type (–20). Signal BUS_SIG0 has value of 1. Both signals BUS_SIG1 and BUS_SIG2 have value –10. This is because they are **bus** kind signal. When they are not driven, they would not retain old values but take the result of the bus resolution function (which returns –10 when there is no driver). Normal signals NOR_SIG0, NOR_SIG1, and NOR_SIG2 have the same values as signals REG_SIG0, REG_SIG1, and REG_SIG2. Signals REG_SIG and BUS_SIG have values 1, 2, 3 since they are only driven by REG_SIG0 and BUS_SIG0. Normal signals are driven by NOR_SIG0, NOR_SIG1, and NOR_SIG2 with value –39 (1, –20, –20), –38 (2, –20, –20), –37 (3, –20, –20).

For the second TIME_SPAN, EN = 1, block b1 guarded expression is TRUE, while the other two block guarded expressions are FALSE. REG_SIG0 retains its value 3, while BUS_SIG0 has value –10. REG_SIG and BUS_SIG have the same values of 11, 12, 13 driven only by REG_SIG1 and BUS_SIG1, respectively. Normal signal NOR_SIG has value –6 (3, 11, –20), –5 (3, 12, –20), –4 (3, 13, –20).

For the third TIME_SPAN, EN = 2, block b2 guarded expression is TRUE, while the other two block guarded expressions are FALSE. REG_SIG0 retains its value 3, while BUS_SIG0 has value –10. REG_SIG1 retains its value 13 while BUS_SIG1 has value –10. REG_SIG and BUS_SIG have the same values of 21, 22, 23 driven only by REG_SIG2 and BUS_SIG2, respectively. Normal signal NOR_SIG has value 37 (3, 13, 21), 38 (3, 13, 22), 39 (3, 13, 23).

For the fourth TIME_SPAN, EN = 3, all three block guarded expressions are FALSE. REG_SIG0 retains its value 3, while BUS_SIG0 has value –10. REG_SIG1 retains its value 13, while BUS_SIG1 has value –10. REG_SIG2 retains its value 23 while BUS_SIG1 has value –10. REG_SIG retains its value of 23, while BUS_SIG has value –10 (neither is driven). Normal signal NOR_SIG retains its value 39 (3, 13, 23), which is continuously driven by three signals.

10.8 DISCONNECTION SPECIFICATION

A disconnection specification defines the time delay for the implicit disconnection of drivers of a guarded signal in a guarded signal assignment.

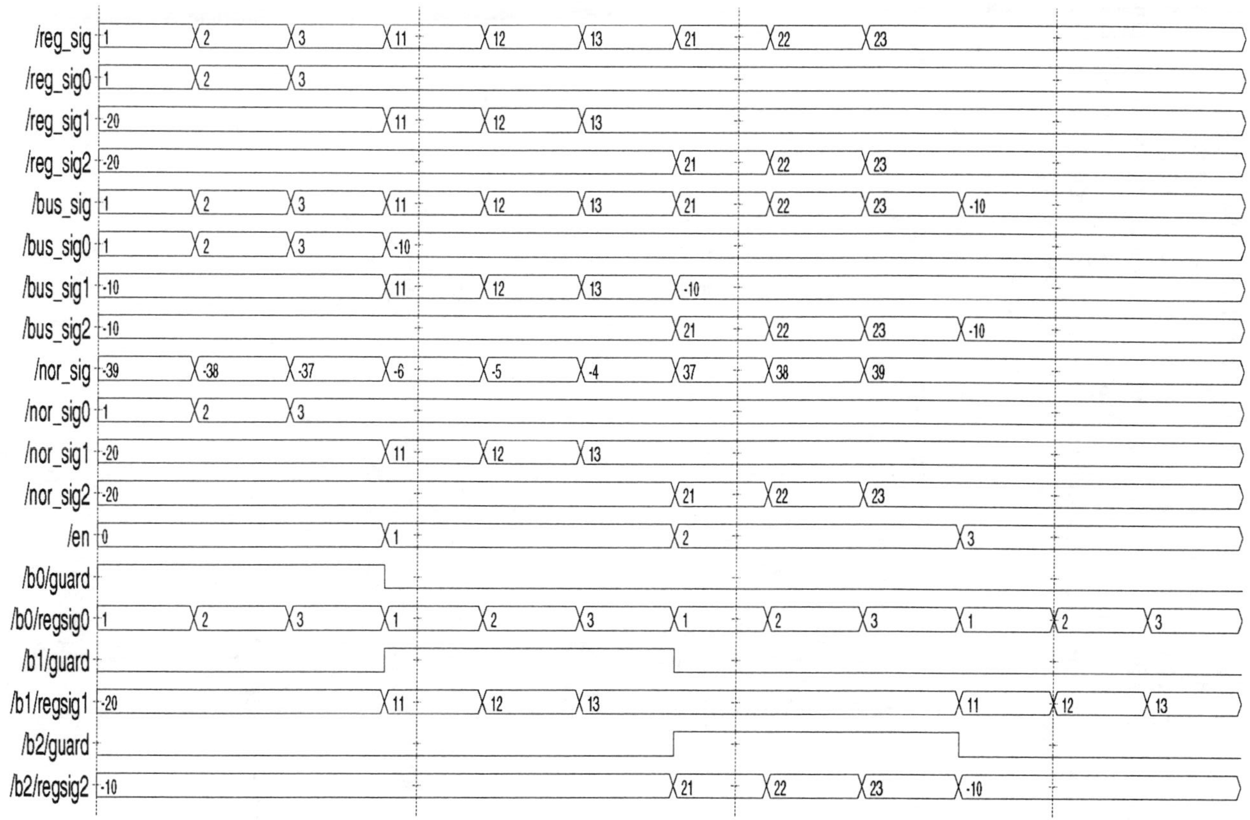

/reg_sig 1 2 3 11 12 13 21 22 23
/reg_sig0 1 2 3
/reg_sig1 -20 11 12 13
/reg_sig2 -20 21 22 23
/bus_sig 1 2 3 11 12 13 21 22 23 -10
/bus_sig0 1 2 3 -10
/bus_sig1 -10 11 12 13 -10
/bus_sig2 -10 21 22 23 -10
/nor_sig -39 -38 -37 -6 -5 -4 37 38 39
/nor_sig0 1 2 3
/nor_sig1 -20 11 12 13
/nor_sig2 -20 21 22 23
/en 0 1 2 3
/b0/guard
/b0/regsig0 1 2 3 1 2 3 1 2 3 1 2 3
/b1/guard
/b1/regsig1 -20 11 12 13 11 12 13
/b2/guard
/b2/regsig2 -10 21 22 23 -10

FIGURE 10.6 Simulation waveform of DIFFGARD VHDL code.

disconnection_specification ::=
  disconnect guarded_signal_specification **after** time_expression ;
guarded_signal_specification ::= signal_name { , signal_name } | **others** |
  all : type_mark

The signals refer to the immediately enclosing declaration region of the specified type. The time expression must be static and not negative. Note that reserved words **others** (**all**) can only appear as the last disconnection specification when it is used. Each guarded signal GS of type T has an implicit disconnection specification of **disconnect** GS : T after 0 ns so that implicit disconnection delay is defined for all guarded signals. For example, there are no explicit disconnection specifications for the guarded signals in DIFFGARD VHDL code. When the guarded expression is FALSE, the driver is immediately (after 0 ns) disconnected.

The following VHDL code illustrates the effect of disconnect specifications. The VHDL code is very similar to DIFFGARD VHDL code. Line 5 declares a port signal of **bus** kind (a port signal cannot be a **register** kind). Lines 6, 12, 19, 41, and 42 are various disconnection specifications.

```
1     library IEEE;
2     use IEEE.std_logic_1164.all;
3     use work.GARDPACK.all;
4     entity DISCON is
5        port (BUS_SIG   : out RESOLVE_INT bus);
6        disconnect BUS_SIG : RESOLVE_INT after 20 ns;
7     end DISCON;
8     architecture BEH of DISCON is
9        signal   REG_SIG    : RESOLVE_INT register ;
10       signal   NOR_SIG    : RESOLVE_INT          ;
11       signal   EN         : integer;
12       disconnect REG_SIG : RESOLVE_INT after 20 ns;
13    begin
14       EN <= 0, 1 after TIME_SPAN, 2 after 2 * TIME_SPAN;
15       b0 : block (EN = 0)
16          signal REG_SIG0, REGSIG0 : RESOLVE_INT register ;
17          signal BUS_SIG0          : RESOLVE_INT bus ;
18          signal NOR_SIG0          : RESOLVE_INT ;
19          disconnect all           : RESOLVE_INT after 20 ns;
20       begin
21          phase0 : process
22          begin
23             for i in 1 to CYCLES loop
24                REGSIG0 <= i after 10 ns;
25                wait for PERIOD;
26             end loop;
27             REGSIG0 <= null after 10 ns;
28             wait;
29          end process;
30          REG_SIG0 <= guarded REGSIG0;
31          BUS_SIG0 <= guarded REGSIG0;
32          NOR_SIG0 <= guarded REGSIG0;
```

```
33          REG_SIG  <= guarded REG_SIG0;
34          BUS_SIG  <= guarded REG_SIG0;
35          NOR_SIG  <= guarded REG_SIG0;
36       end block;
37       b1 : block (EN = 1)
38          signal REG_SIG1          : RESOLVE_INT register ;
39          signal BUS_SIG1, REGSIG1 : RESOLVE_INT bus ;
40          signal NOR_SIG1          : RESOLVE_INT ;
41          disconnect REGSIG1, REG_SIG1 : RESOLVE_INT after 20 ns;
42          disconnect others            : RESOLVE_INT after 20 ns;
43       begin
44          phase1 : process
45          begin
46             REGSIG1 <= null after 10 ns;
47             wait for TIME_SPAN;
48             for i in 1 to CYCLES loop
49                REGSIG1 <= i + 10 after 10 ns;
50                wait for PERIOD;
51             end loop;
52             REGSIG1 <= null after 10 ns;
53             wait;
54          end process;
55          REG_SIG1 <= guarded REGSIG1;
56          BUS_SIG1 <= guarded REGSIG1;
57          NOR_SIG1 <= guarded REGSIG1;
58          REG_SIG  <= guarded REG_SIG1;
59          BUS_SIG  <= guarded REG_SIG1;
60          NOR_SIG  <= guarded REG_SIG1;
61       end block;
62    end BEH;
```

Figure 10.7 shows the simulation waveform for the DISCON VHDL code. Note that BUS_SIG0 is disconnected at time 110 ns (20 ns after block b0 guarded expression is FALSE at time 90 ns). You are encouraged to study the simulation waveform in Figure 10.7 and compare your findings with the signal values provided.

10.9 WRITING EFFICIENT VHDL CODE

As with any other high-level programming language, VHDL has different programs that produce the same results. Sometimes, one is more efficient than others. The efficiency can be measured as the run time, memory space used, or the program size. For example, a common term inside a loop statement can be done outside of the loop if the term is independent of the looping variable. In this section, a few examples are illustrated showing how to improve efficiency. These examples fall into the following categories: splitting, combining, concurrency, hiding, and event reduction.

Splitting. Split a process statement into smaller unrelated processes statements. The following VHDL code infers two latches for A_OUT and B_OUT output ports. Latches A_OUT and B_OUT have unrelated data inputs A_DATA, B_DATA and enable signals A_EN and B_EN. In architecture RTL_OK VHDL code, these two

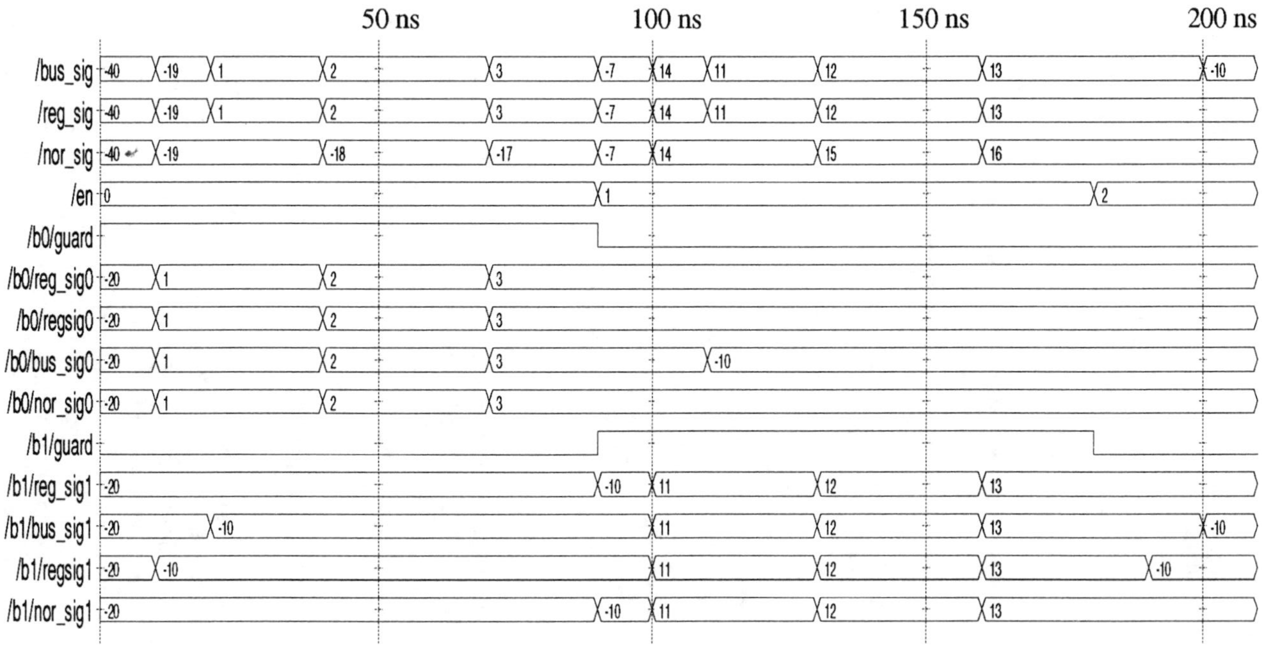

FIGURE 10.7 Simulation waveform for DISCON VHDL code.

latches are inferred with a process statement from lines 11 to 19. Whenever there is an event on any signal A_DATA, A_EN, B_DATA, B_EN, the process is executed. For example, when only signal A_EN has an event, lines 16 to 18 for latch B_OUT are also evaluated. This is not necessary.

Architecture RTL_BETTER split the latch inference into two separate unrelated process statements (lines 24 to 35) with only two signals for each process sensitivity list. When only one signal has an event, only one process statement is executed. This improves run time efficiency, especially when the process statement has more sequential statements inside. Note that architectures RTL_OK and RTL_BETTER would have the same simulation and synthesis results.

```
1   library IEEE;
2   use IEEE.std_logic_1164.all;
3   entity SPLIT is
4      port (
5         A_DATA, A_EN, B_DATA, B_EN : in  std_logic;
6         A_OUT, B_OUT               : out std_logic);
7   end SPLIT;
8
9   architecture RTL_OK of SPLIT is
10  begin
11     p0 : process (A_DATA, A_EN, B_DATA, B_EN)
12     begin
13        if (A_EN = '1') then
14           A_OUT <= A_DATA;
15        end if;
16        if (B_EN = '1') then
17           B_OUT <= B_DATA;
18        end if;
19     end process;
20  end RTL_OK;
21
22  architecture RTL_BETTER of SPLIT is
23  begin
24     p0 : process (A_DATA, A_EN)
25     begin
26        if (A_EN = '1') then
27           A_OUT <= A_DATA;
28        end if;
29     end process;
30     p1 : process (B_DATA, B_EN)
31     begin
32        if (B_EN = '1') then
33           B_OUT <= B_DATA;
34        end if;
35     end process;
36  end RTL_BETTER;
```

Combining. Combine process statements with the same sensitivity list into one process statement. The following VHDL code shows the inference of two flipflops A_OUT and B_OUT with the same clock signal CLK and reset signal RSTn. Architecture RTL_OK uses two separate process statements with the same sensitivity list (lines 11 and 19). Inside the process statement, the if statement (lines 13 and 21, lines 15 and 23) conditions are the same. They can be combined as one process statement as shown in architecture RTL_BETTER (lines 29 to 41) so that the if statement condition (RSTn = '0', CLK'event and CLK = '1') is evaluated only once. Again, these two architectures would have the same simulation and synthesis. However, using architecture RTL_OK with a separate process statement is often used for readability and documentation. For example, you may use one process statement for each 32-bit register. It would be easier to maintain the VHDL code when changes are necessary.

```
1     library IEEE;
2     use IEEE.std_logic_1164.all;
3     entity COMBINE is
4        port (
5           RSTn, CLK, A_DATA, B_DATA : in  std_logic;
6           A_OUT, B_OUT              : out std_logic);
7     end COMBINE;
8
9     architecture RTL_OK of COMBINE is
10    begin
11       p0 : process (RSTn, CLK)
12       begin
13          if (RSTn = '0') then
14             A_OUT <= '0';
15          elsif (CLK'event and CLK = '1') then
16             A_OUT <= A_DATA;
17          end if;
18       end process;
19       p1 : process (RSTn, CLK)
20       begin
21          if (RSTn = '0') then
22             B_OUT <= '0';
23          elsif (CLK'event and CLK = '1') then
24             B_OUT <= B_DATA;
25          end if;
26       end process;
27    end RTL_OK;
28
29    architecture RTL_BETTER of COMBINE is
30    begin
31       p0 : process (RSTn, CLK)
32       begin
33          if (RSTn = '0') then
34             A_OUT <= '0';
35             B_OUT <= '0';
36          elsif (CLK'event and CLK = '1') then
37             A_OUT <= A_DATA;
38             B_OUT <= B_DATA;
39          end if;
```

```
40        end process;
41     end RTL_BETTER;
```

Concurrency. Use concurrent statements to reduce the amount of VHDL coding and to avoid the process sensitivity list. In the architecture RTL_OK of the following VHDL code, process statement p0 (lines 11 to 14) has a signal assignment statement (line 13), and process statement p1 (lines 15 to 22) infers a tristate buffer BUF3_OUT. These two process statements (12 lines) can be replaced with two concurrent signal assignment statements (lines 27 and 28) as shown in architecture RTL_BETTER. The amount of VHDL coding is reduced. The burden of having the right process sensitivity list is gone.

```
1      library IEEE;
2      use IEEE.std_logic_1164.all;
3      entity CONCURR is
4         port (
5            A, B, C, D, E, F, G, H, K, EN : in  std_logic;
6            COMB_OUT, BUF3_OUT            : out std_logic);
7      end CONCURR;
8
9      architecture RTL_OK of CONCURR is
10     begin
11        p0 : process (A, B, C, D, E, F, G, H, K)
12        begin
13           COMB_OUT <= A or B or C or D or E or F or G or H or K;
14        end process;
15        p1 : process (A, EN)
16        begin
17           if (EN = '1') then
18              BUF3_OUT <= A;
19           else
20              BUF3_OUT <= 'Z';
21           end if;
22        end process;
23     end RTL_OK;
24
25     architecture RTL_SHORT of CONCURR is
26     begin
27        COMB_OUT <= A or B or C or D or E or F or G or H or K;
28        BUF3_OUT <= A when EN = '1' else 'Z';
29     end RTL_SHORT;
```

Hiding. Hide the combination circuits with the sequential circuits in one process statement. The following VHDL code models a parity generation circuit (of input 32-bit DATA) with flipflop output parity (PAR). In architecture RTL_OK, process statement p0 (lines 13 to 21) generates a signal PAR_comb with combinational circuits. This signal then goes to the data input of the flipflop as inferred in process statement p1 (lines 22 to 29). Process statement p0 would be executed whenever any 1 bit of DATA has an event. Signal PAR_comb would not be clocked into flipflop PAR until the rising edge of the clock CLK. In practical use, it is better to hide the parity generation inside the process statement of inferring the flip-flop, as shown in architecture RTL_BETTER (lines 34 to 46). The parity generation is only evaluated at each rising edge of the clock. The run-time efficiency is improved. Both architectures would synthesize to the same schematic. However, architecture RTL_OK has the advantage of monitoring signal PAR_comb to check whether a spike is generated, caused by a spike of the input signal DATA. In practice, the goal of VHDL code is to obtain the correct functionality (which would be implemented with the correct schematic through synthesis). The spike and detailed timing would be performed at the structural gate level simulation.

```
1     library IEEE;
2     use IEEE.std_logic_1164.all;
3     entity COMBSEQ is
4        port (
5           RSTn, CLK : in  std_logic;
6           DATA      : in  std_logic_vector(31 downto 0);
7           PAR       : out std_logic);
8     end COMBSEQ;
9
10    architecture RTL_OK of COMBSEQ is
11       signal PAR_comb : std_logic;
12    begin
13       p0 : process (DATA)
14          variable TEMP : std_logic;
15       begin
16          TEMP := DATA(DATA'high);
17          for i in DATA'high-1 downto DATA'low loop
18             TEMP := TEMP xor DATA(i);
19          end loop;
20          PAR_comb <= TEMP;
21       end process;
22       p1 : process (RSTn, CLK)
23       begin
24          if (RSTn = '0') then
25             PAR <= '0';
26          elsif (CLK'event and CLK = '1') then
27             PAR <= PAR_comb;
28          end if;
29       end process;
30    end RTL_OK;
31
32    architecture RTL_BETTER of COMBSEQ is
33    begin
34       p0 : process (RSTn, CLK)
35          variable TEMP : std_logic;
36       begin
37          if (RSTn = '0') then
38             PAR <= '0';
39          elsif (CLK'event and CLK = '1') then
40             TEMP := DATA(DATA'high);
41             for i in DATA'high-1 downto DATA'low loop
```

```
42                TEMP := TEMP xor DATA(i);
43             end loop;
44             PAR <= TEMP;
45          end if;
46       end process;
47    end RTL_BETTER;
```

Event reduction. Reduce the number of events. Signal events are the causes of processes to be activated and evaluated. Therefore, the simulation speed is closely related to the number of events. Reducing the number of events would improve simulation run time. The following VHDL code is to generate a parity output PAR of combination circuits. Architecture RTL_BAD (lines 9 to 20) declares a 32-bit signal TEMP in line 10. The process statement p0 has DATA and TEMP as the sensitivity list. Even though line 14 can be taken out of the process statement, the process statement assigns signal TEMP (in line 16), which causes an event on signal TEMP (one bit at a time). The process statement would be evaluated 32 times (caused by the event on signal TEMP) after DATA has the event. This is not efficient and should be avoided.

Architecture RTL_OK (lines 22 to 33) declares a 32-bit variable instead of a 32-bit signal. The process statement has only signal DATA as the sensitivity list (line 24). There is no event generated by variable TEMP. The simulation run time is improved. Architecture RTL_BETTER (lines 35 to 46) uses only 1-bit variable. These three architectures have the same simulation and synthesis results with various simulation efficiencies.

```
1     library IEEE;
2     use IEEE.std_logic_1164.all;
3     entity LESSEVENT is
4        port (
5           DATA      : in  std_logic_vector(31 downto 0);
6           PAR       : out std_logic);
7     end LESSEVENT;
8
9     architecture RTL_BAD of LESSEVENT is
10       signal TEMP : std_logic_vector(31 downto 0);
11    begin
12       p0 : process (DATA, TEMP)
13       begin
14          TEMP(31) <= DATA(31);
15          for i in 30 downto 0 loop
16             TEMP(i) <= TEMP(i+1) xor DATA(i);
17          end loop;
18          PAR <= TEMP(0);
19       end process;
20    end RTL_BAD;
21
22    architecture RTL_OK of LESSEVENT is
23    begin
24       p0 : process (DATA)
25          variable TEMP : std_logic_vector(31 downto 0);
```

```
26       begin
27          TEMP(31) := DATA(31);
28          for i in 30 downto 0 loop
29             TEMP(i) := TEMP(i+1) xor DATA(i);
30          end loop;
31          PAR <= TEMP(0);
32       end process;
33    end RTL_OK;
34
35    architecture RTL_BETTER of LESSEVENT is
36    begin
37       p0 : process (DATA)
38          variable TEMP : std_logic;
39       begin
40          TEMP := DATA(31);
41          for i in 30 downto 0 loop
42             TEMP := TEMP xor DATA(i);
43          end loop;
44          PAR <= TEMP;
45       end process;
46    end RTL_BETTER;
```

10.10 EXERCISES

10.1 Determine the TEXTIO package of library STD to see what other READ and WRITE procedures are available.

10.2 What would happen if multiple file objects of mode **in** have the same logical file name? Write a VHDL code to check how your VHDL simulator would handle this.

10.3 What would happen if multiple file objects of mode **out** have the same logical file name? Write a VHDL code to check how your VHDL simulator would handle this.

10.4 Try to find whether the following functions used in the ROMRF exist in your system. If not, implement them. TO_INTEGER, TO_STDLOGICVECTOR, TO_HEX_STRING (should convert any number of bits), IS_X

10.5 Add an input signal OEn to the ROMRF VHDL code to indicate output enable from the ROM. When OEn = '0', normal DATA is output, otherwise, DATA is tristated. Make some timing checks such as the address ADDR width, setup, hold time relative to OEn signal and the minimum width of OEn signal.

10.6 Random-access memory (RAM) is also used in the design. Write a RAM model with generics and ports CSn (low active chip select, mode **in**), ADDR (address, mode **in**), DATA (data bus of mode **inout**), WEn (low active write enable pulse, mode **in**), OEn (low active output enable, mode **in**). Draw a timing diagram with all timing (such as address setup time, and so on.) passed in as generics.

10.7 Write a behavioral (not to be synthesized) VHDL code to model a dual port RAM.

10.8 Using the dual port RAM model in the last exercise, write a first-in first-out (FIFO) VHDL model so that one port is only pushing (writing), the other port is doing pulling (reading). The FIFO VHDL code should instantiate the dual port RAM. The rest of the circuits are to be synthesizable. Verify your VHDL code through simulation and synthesis.

10.9 Large arrays are the popular choice for modeling a ROM or RAM, but they are not without their problems. This causes the difficulty of allocating large block of consecutive memory during VHDL simulation. This can be overcome, however, by modifying the ROM model so that the large array is broken down to smaller arrays with access types. This allows the smaller arrays to be dynamically separately allocated. Modify the ROM VHDL code to use the access type with a specified maximum array size.

10.10 In the BIPAD VHDL model, the type of the of ports are unconstrained std_logic_vector with no range specified. What are the advantages? Are there any situations when the model would not function correctly?

10.11 In the CONVERT package VHDL code, function CONVERT parameter and the function return value are also unconstrained. What are the advantages? Are there any situations when the model would not function correctly?

10.12 In the BSTREE VHDL code, what would happen if a value to be inserted has been in the BST already? What would you change so that the FIND procedure could handle the following? (a) a value is already in the BST, MATCH is TRUE, and the value is not inserted. (b) the value is not in the BST, MATCH is FALSE, and the value is inserted in the BST.

10.13 In the BSTREE VHDL code, the memory for the BST is deallocated altogether by deallocate(HEAD). Write a procedure DELETE that would delete (and deallocate) a node if the value is found in the BST.

10.14 With the sequence of INORDER and PREORDER for a binary tree, the binary tree can be uniquely defined. Draw the BST from the message generated in the simulation of BSTREE VHDL code. Determine whether the BSTREE VHDL code can correctly generate the BST.

10.15 In the DISCON VHDL code and its simulation waveform, explain the reasons for each signal value and for each value change.

10.16 Describe a situation where the guarded signals and disconnection specifications can be used to take advantages of their unique features.

Chapter 11

A Design Case and Test Bench

11.1 DESIGN DESCRIPTION

A design problem was given in a design contest (see *Integrated System Design*, July 1995, pp. 56–60) for designers who use either VHDL or Verilog. The design is to have a 9-bit up-by-3 and down-by-5 counter with even parity (PARITY), carry (CARRY_OUT), borrow (BORROW_OUT), and counter value (COUNT) as outputs. The input signals are a clock, 9-bit data to be parallel loaded into the counter, a bit UP and a bit DOWN to control the operation of the counter. The VHDL entity is shown as follows:

```
1     library IEEE;
2     use IEEE.std_logic_1164.all;
3     use IEEE.std_logic_arith.all;
4     entity COUNT9 is
5        port (
6           RSTn, CLK, UP, DOWN : in  std_logic;
7           DATA                : in  std_logic_vector(8 downto 0);
8           COUNT               : out std_logic_vector(8 downto 0);
9           CARRY_OUT           : out std_logic;
10          BORROW_OUT          : out std_logic;
11          PARITY              : out std_logic);
12    end COUNT9;
```

All output signals are flipflop outputs. When UP & DOWN = “00”, the counter is parallel loaded with input DATA. When UP & DOWN = “01”, the counter is decreased by 5. When UP & DOWN = “10”, the counter is increased by 3. When UP & DOWN = “11”, the counter is not changed. RSTn is the asynchronous reset signal to be connected to all asynchronous reset inputs of flipflops. Lines 1 to 3 reference library IEEE and use std_logic_1164 and std_logic_arith packages. CLK is the clock signal. We call the design COUNT9 as the entity name.

11.2 WRITING VHDL MODEL

Before we start to implement the design with VHDL, conceptually, we have the following rough schematic in mind as shown in Figure 11.1. Input signals are shown on the left. Output signals are shown in the right. Combinational circuits are shown in round-cornered blocks. Flipflops are shown with rectangle blocks. Internal signals are also shown to be used to communicate between VHDL concurrent statements. Note that CLK and RSTn are directly connected to flipflops, but not drawn in the figure.

From the conceptual schematic, we can use a process statement to generate new counter value NEW_COUNT with combinational circuits. Use another process statement to generate even parity PAR_EVEN with combinational circuits and a process to generate all flipflops. The following VHDL shows the architecture RTL of COUNT9. Lines 14 to 16 declare signals. Process ffs (lines 19 to 32) infer flipflops with asynchronous reset for COUNT_FF, BORROW_OUT, CARRY_OUT, and PARITY. Note that output signal COUNT is directly connected to COUNT_FF in line 18.

Signal COUNT_FF and input signals DATA, UP, DOWN are included in the sensitivity list of process new_cnt in line 33. Inside the process statement (lines 33 to 54), a local variable MODE is declared in line 34 which is the concatenation of UP and DOWN in line 39. It is then used as the case statement expression in line 40. Another variable temp of 10 bits is declared in line 35 which is used as the sum or the difference in lines 44 and 48. These operators "+" and "–" are defined in the std_logic_arith which is referenced in line 3. The process statement first assigns default values to NEW_CARRY and NEW_BORROW to be '0' (lines 37 and 38). If MODE is "00" or "11", they should be '0' since no increment of 3 or decrement of 5 is performed. The case statement based on the value of MODE performs parallel load ("00"), decreasing by 5 ("01"), increasing by 3 ("10"), or no change ("11"). Note that NEW_CARRY and NEW_BORROW get the left-most bit of temp in lines 46 and 50 which will override the default value in lines 37 and 38.

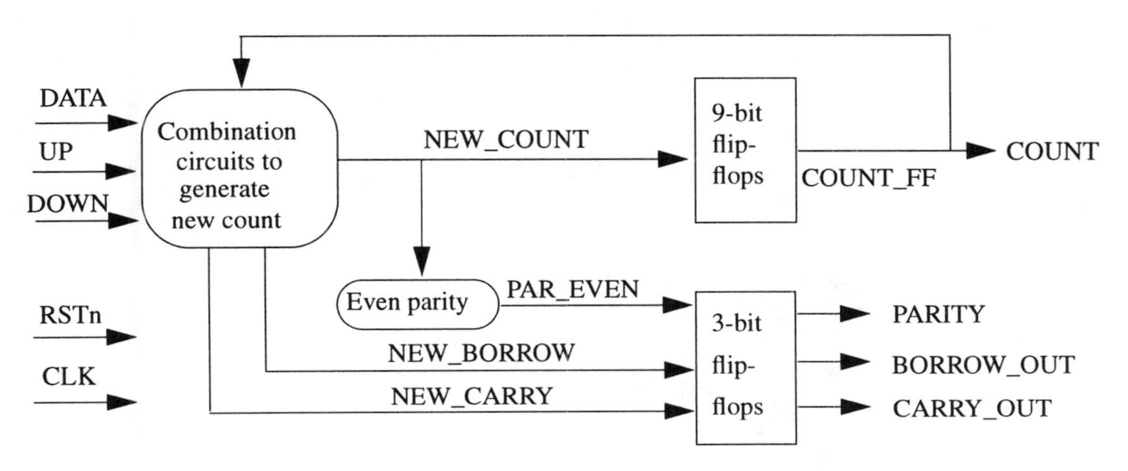

FIGURE 11.1 Conceptual schematic for COUNT9.

```
13   architecture RTL of COUNT9 is
14      signal COUNT_FF   : std_logic_vector(8 downto 0);
15      signal NEW_COUNT : std_logic_vector(8 downto 0);
16      signal PAR_EVEN, NEW_CARRY, NEW_BORROW : std_logic;
17   begin
18      COUNT <= COUNT_FF;
19      ffs : process (RSTn, CLK)
20      begin
21         if (RSTn = '0') then
22            COUNT_FF    <= (others => '0');
23            PARITY      <= '0';
24            CARRY_OUT   <= '0';
25            BORROW_OUT <= '0';
26         elsif (CLK'event and CLK = '1') then
27            COUNT_FF    <= NEW_COUNT;
28            PARITY      <= PAR_EVEN;
29            CARRY_OUT   <= NEW_CARRY;
30            BORROW_OUT <= NEW_BORROW;
31         end if;
32      end process;
33      new_cnt : process (DATA, COUNT_FF, UP, DOWN)
34         variable MODE : std_logic_vector(1 downto 0);
35         variable temp : std_logic_vector(9 downto 0);
36      begin
37         NEW_CARRY    <= '0';
38         NEW_BORROW   <= '0';
39         MODE := UP & DOWN;
40         case MODE is
41            when "00"    =>
42               NEW_COUNT   <= DATA;
43            when "01"    =>
44               temp        := unsigned('0' & COUNT_FF) - 5;
45               NEW_COUNT   <= temp(8 downto 0);
46               NEW_BORROW <= temp(9);
47            when "10"    =>
48               temp        := unsigned('0' & COUNT_FF) + 3;
49               NEW_COUNT   <= temp(8 downto 0);
50               NEW_CARRY   <= temp(9);
51            when others =>
52               NEW_COUNT   <= COUNT_FF;
53         end case;
54      end process;
55      par_gen : process (NEW_COUNT)
56         variable PAR : std_logic;
57      begin
58         PAR := NEW_COUNT(0);
59         for i in 1 to 8 loop
60            PAR := PAR xor NEW_COUNT(i);
61         end loop;
62         PAR_EVEN <= PAR;
63      end process;
64   end RTL;
```

Lines 55 to 63 describe the generation of even parity for the NEW_COUNT which is used as the sensitivity list of the process (line 55). A local variable PAR is declared in line 56. PAR is first assigned as the NEW_COUNT(0) in line 58. A loop statement (lines 59 to 61) goes through the index from 1 to 8. The variable PAR is then assigned to the signal PAR_EVEN in line 62. PAR_EVEN, NEW_COUNT, NEW_CARRY, and NEW_BORROW are clocked into flipflops in process ffs (lines 27 to 30).

11.3 ANOTHER ARCHITECTURE

Architecture RTL uses std_logic_arith package to perform the arithmetic operations with operators "+" and "–" (lines 44 and 48). This package may not be available to you. Also, increment by 3 and decrement by 5 have something in common. The following architecture RTL1 shows another implementation of the same design COUNT9. Processes ffs and par_gen are the same as in architecture RTL. The only difference is in the process new_cnt. When the MODE is "00" or "11", they are the same. When MODE is "01", the decrement by 5 has the least significant bit (bit 0, LSB) reversed as in line 96. Bit 1 of NEW_COUNT is specified in line 97. A local variable CARRY is declared in line 87 to be used as the carry from the previous bit. Lines 100 and 101 set up the first carry for the loop statement in lines 102 to 105 to generate bits 3 to 8. After the loop, the CARRY is assigned to the signal NEW_BORROW. The same idea is used for MODE = "10" to increase by 3 in lines 108 to 115.

```
65    architecture RTL1 of COUNT9 is
66       signal COUNT_FF  : std_logic_vector(8 downto 0);
67       signal NEW_COUNT : std_logic_vector(8 downto 0);
68       signal PAR_EVEN, NEW_CARRY, NEW_BORROW : std_logic;
69    begin
70       COUNT <= COUNT_FF;
71       ffs : process (RSTn, CLK)
72       begin
73          if (RSTn = '0') then
74             COUNT_FF   <= (others => '0');
75             PARITY     <= '0';
76             CARRY_OUT  <= '0';
77             BORROW_OUT <= '0';
78          elsif (CLK'event and CLK = '1') then
79             COUNT_FF   <= NEW_COUNT;
80             PARITY     <= PAR_EVEN;
81             CARRY_OUT  <= NEW_CARRY;
82             BORROW_OUT <= NEW_BORROW;
83          end if;
84       end process;
85       new_cnt : process (DATA, COUNT_FF, UP, DOWN)
86          variable MODE  : std_logic_vector(1 downto 0);
87          variable CARRY : std_logic;
88       begin
89          NEW_CARRY   <= '0';
```

```
90          NEW_BORROW  <= '0';
91          MODE := UP & DOWN;
92          case MODE is
93             when "00"    =>
94                NEW_COUNT <= DATA;
95             when "01"    =>
96                NEW_COUNT(0) <= not COUNT_FF(0);
97                NEW_COUNT(1) <= not (COUNT_FF(0) xor COUNT_FF(1));
98                NEW_COUNT(2) <= (COUNT_FF(0) or COUNT_FF(1)) xor
99                                COUNT_FF(2);
100               CARRY        := (COUNT_FF(0) nor COUNT_FF(1)) or
101                               not COUNT_FF(2);
102               for i in 3 to 8 loop
103                  NEW_COUNT(i) <= CARRY xor COUNT_FF(i);
104                  CARRY        := CARRY and not COUNT_FF(i);
105               end loop;
106               NEW_BORROW <= CARRY;
107            when "10"    =>
108               NEW_COUNT(0) <= not COUNT_FF(0);
109               NEW_COUNT(1) <= not (COUNT_FF(0) xor COUNT_FF(1));
110               CARRY := COUNT_FF(0) or COUNT_FF(1);
111               for i in 2 to 8 loop
112                  NEW_COUNT(i) <= CARRY xor COUNT_FF(i);
113                  CARRY        := CARRY and COUNT_FF(i);
114               end loop;
115               NEW_CARRY <= CARRY;
116            when others =>
117               NEW_COUNT <= COUNT_FF;
118         end case;
119      end process;
120      par_gen : process (NEW_COUNT)
121         variable PAR : std_logic;
122      begin
123         PAR := NEW_COUNT(0);
124         for i in 1 to 8 loop
125            PAR := PAR xor NEW_COUNT(i);
126         end loop;
127         PAR_EVEN <= PAR;
128      end process;
129   end RTL1;
```

11.4 A TEST BENCH

Now, let's verify the VHDL code to make sure it works as required. The idea of a test bench is to set up an easy way to verify the VHDL code without typing in simulation commands in the VHDL simulator every time we make a change in the VHDL code. Also, we want to verify with some self-checking capability so that the test will give us an answer "pass" or "fail" without staring at the waveforms on the screen while zooming in or out, panning left or right. After the test bench is set up, the simulation can be run overnight with a pass or fail result available the next morning.

One way to arrange a test bench is to have a known test vector set with expected output vectors in a text file. The text file can be read in, translated into input signals to the Design Under Test (DUT). The outputs from the DUT are compared with the translation of the expected signals from the text file. The number of mismatches that indicate design error can be counted (may be with the simulation time) and reported. If there is no mismatch, the test passes. Figure 11.2 shows this concept of the test bench.

The author would like to thank John Cooley for providing the test vector file used in the design contest shown here:

```
1     00111111101111111101000
2     10000000000000000000100
3     00111111110111111110000
4     10XXXXXXXXX000000001101
5     00111111111111111111001
6     10000000000000000010101
7     00000000101000000101000
8     01111111111000000000000
9     00000000100000000100001
10    01010101010111111111011
11    00000000011000000011000
12    01101010101111111110010
13    00000000010000000010001
14    01000000010111111101010
15    00000000001000000001001
16    01000000000111111100011
17    00000000000000000000000
18    01XXXXXXXXX111111011010
```

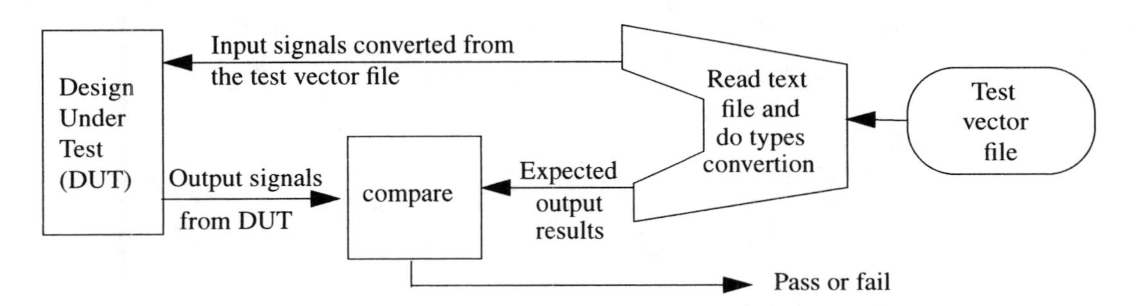

FIGURE 11.2 A test bench organization.

Each vector has 23 bits (22 downto 0). Bit 22 is for input UP. Bit 21 is for input DOWN. Bits 20 to 12 (9 bits) are for input DATA. The expected output COUNT value is in bits 11 to 3. Bits 2, 1, and 0 are for expected CARRY_OUT, BORROW_OUT, and PARITY, respectively. We want to supply the input vectors at the falling edge of the clock. The counter value would be updated after the rising edge of the clock. At the next falling edge of the clock, the expected output results are compared with the DUT output signals. At the same time, a new input vector is supplied. The following VHDL code shows the test bench.

Lines 1 to 3 reference library IEEE and use packages std_logic_1164, std_logic_textio. The library STD is visible always without a library clause. The package textio in STD library is referenced in line 4. The test bench TBCOUNT9A entity does not have any ports as described in lines 5 to 6.

Lines 9 to 17 declare the DUT component COUNT9. Lines 18 to 20 declare signals for interface to the DUT COUNT9 and expected outputs. Line 21 defines a constant HALF_PERIOD to be used as the half period of the clock with a value of 20 ns. Line 22 uses the hard binding to bind the component COUNT9 to the architecture RTL of COUNT9 in the library WORK. Line 24 assigns signal RSTn for asynchronous reset. Lines 25 to 29 use a process statement to assign the clock signal CLK.

```
1     library IEEE;
2     use IEEE.std_logic_1164.all;
3     use IEEE.std_logic_textio.all;
4     use std.textio.all;
5     entity TBCOUNT9A is
6     end TBCOUNT9A;
7
8     architecture BEH of TBCOUNT9A is
9        component COUNT9
10       port (
11          RSTn, CLK, UP, DOWN : in  std_logic;
12          DATA                : in  std_logic_vector(8 downto 0);
13          COUNT               : out std_logic_vector(8 downto 0);
14          CARRY_OUT           : out std_logic;
15          BORROW_OUT          : out std_logic;
16          PARITY              : out std_logic);
17       end component;
18       signal RSTn, CLK, UP, DOWN, PARITY0 : std_logic;
19       signal CARRY_OUT0, BORROW_OUT0      : std_logic;
20       signal DATA, COUNT0  : std_logic_vector(8 downto 0);
21       constant HALF_PERIOD : time := 20 ns;
22       for count9_inst0 : COUNT9 use entity work.COUNT9(RTL);
23    begin
24       RSTn <= '0', '1' after HALF_PERIOD;
25       clk_gen : process
26       begin
27          CLK <= '1'; wait for HALF_PERIOD;
28          CLK <= '0'; wait for HALF_PERIOD;
29       end process;
```

The next process is the major process that reads the vector file and compares the results. Inside the process, a file INFILE is declared as type text with file name "vector.in" in line 31. Line 32 declares a variable LINE_BUFFER of type line to be used in the read procedure in line 32. Line 33 declares variable VECTOR as the vector translated from the input text file. Line 34 declares a variable NUMBER_ERR to count the number of mismatches. Line 35 declares variable EXPECT_RESULT for the expected output value. Line 36 declares variable COUNT9_OUTPUT to indicate the output value from COUNT9.

The process statement does not have a sensitivity list in line 30 since a couple of *wait statements* are inside the process. The process waits until the first falling edge of the clock in line 38. Lines 39 to 52 define a *while loop statement* until the end of file is reached. Function *endfile* returns a BOOLEAN value TRUE when the file reaches to the end. Line 40 reads a line from the file to the LINE_BUFFER. Line 41 reads the 23 bits from the buffer LINE_BUFFER. Input signals UP, DOWN, and DATA are assigned in lines 42 to 44. Line 45 waits for the falling edge of the clock. It is important to know that entity COUNT9 processes would be executed following the rising edge of the clock. They will be done before line 45 is completed and moving to line 46. Line 46 gets the expected results. Line 47 gets the output value from COUNT9. They are compared in line 48. The number of mismatches is increased and a message is generated if results are not the same. Line 52 ends the loop. The execution goes back to line 40 to 44 and waits in line 45. The clock signal CLK rising edge wakes up COUNT9 processes to be executed. After COUNT9 processes are completed, the clock signal CLK falling edge allows the process read_vector to be continued at line 46. Note that the process clk_gen in lines 26 to 29 generates clock signal CLK which wakes up process read_vector at the falling edge of CLK and wakes up COUNT9 processes at the rising edge of the clock CLK.

After the while loop is completed, the execution continues at line 53. If the number of mismatches is 0, the test is passed. Otherwise, a message to indicate the test failed is generated with *assert statements.*

```
30      read_vector : process
31         file      INFILE              : text is in "vector.in";
32         variable LINE_BUFFER    : line;
33         variable VECTOR             : std_logic_vector(22 downto 0);
34         variable NUMBER_ERR      : integer := 0;
35         variable EXPECT_RESULT : std_logic_vector(11 downto 0);
36         variable COUNT9_OUTPUT : std_logic_vector(11 downto 0);
37      begin
38         wait until CLK = '0' and CLK'event;
39         while (not endfile(INFILE)) loop
40            readline (INFILE, LINE_BUFFER);
41            read (LINE_BUFFER, VECTOR) ;
42            UP          <= VECTOR(22);
43            DOWN        <= VECTOR(21);
44            DATA        <= VECTOR(20 downto 12);
45            wait until CLK = '0' and CLK'event;
46            EXPECT_RESULT := VECTOR(11 downto 0);
47            COUNT9_OUTPUT := COUNT0 & CARRY_OUT0 & BORROW_OUT0 &
                               PARITY0;
```

```
48              if (EXPECT_RESULT /= COUNT9_OUTPUT) then
49                 NUMBER_ERR := NUMBER_ERR + 1;
50                 assert FALSE report "Results mismatch" severity NOTE;
51              end if;
52           end loop;
53           if (NUMBER_ERR = 0) then
54              assert FALSE report "TEST PASSED!" severity NOTE;
55           else
56              assert FALSE report "TEST FAILED! However, the road to " &
57                 "success is always under construction." severity NOTE;
58           end if;
59           wait;
60        end process;
61        count9_inst0 :  COUNT9
62        port map (
63           RSTn      => RSTn,       CLK        => CLK,
64           UP        => UP,         DOWN       => DOWN,
65           DATA      => DATA,       COUNT      => COUNT0,
66           CARRY_OUT => CARRY_OUT0, BORROW_OUT => BORROW_OUT0,
67           PARITY    => PARITY0);
68     end BEH;
```

Lines 61 to 67 are for the COUNT9 component instantiation with ports mapped to signals. The COUNT9 component is hard bound in line 22 so that the simulation can be run without a separate configuration (soft binding).

The test bench TBCOUNT9A is simulated. Figures 11.3 and 11.4 show the simulation waveforms. Note that the inputs are changed at the falling edge of the clock. The output values are changed at the rising edge of the clock. The following message is generated from the VHDL simulator.

Assertion NOTE at 740 NS in design unit TBCOUNT9A(BEH) from process /TBCOUNT9A/READ_VECTOR: "TEST PASSED!"

11.5 ANOTHER TEST BENCH

We can use test bench TBCOUNT9A to verify architecture RTL1 of COUNT9 by changing the hard binding to architecture RTL1 instead of RTL in line 22. Another approach is to delete line 22 and write a separate configuration to configure entity COUNT9. This way, the recompiling of the VHDL code for TBCOUNT9A is not needed. The configuration can be changed using RTL or RTL1. This configuration file is a lot shorter and more easily recompiled. Of course, we can have two separate configurations, one configuration uses architecture RTL of COUNT9. The other uses architecture RTL1 of COUNT9. This is left as an exercise. In this section, we discuss another way of doing the test bench.

Test bench TBCOUNT9A verifies a COUNT9 with a known test vector set. Sometimes, the known test vector set is not available. The input test vectors may be generated randomly as in the case of verifying a 16 by 16 Booth Wallace tree multiplier BWMULT with inputs BWA (16 bits), BWB (16 bits) and output BWC (32 bits). The input INTEGER vectors A and B can be generated randomly. The multiplier can be described with a simple VHDL architecture, simply with a signal assignment statement

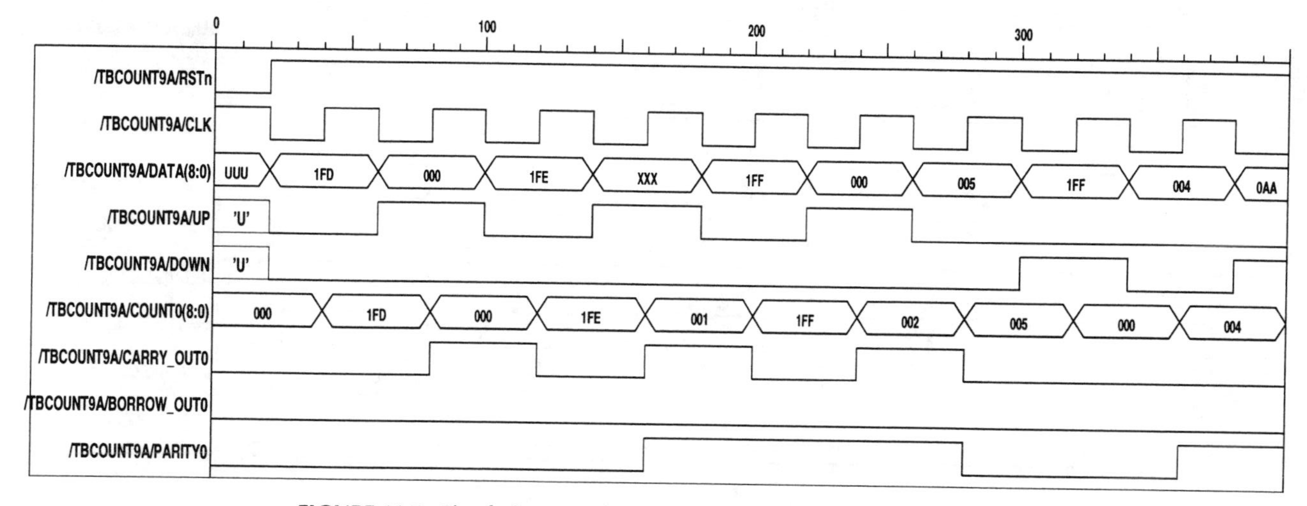

FIGURE 11.3 Simulation waveform for test bench TBCOUNT9A.

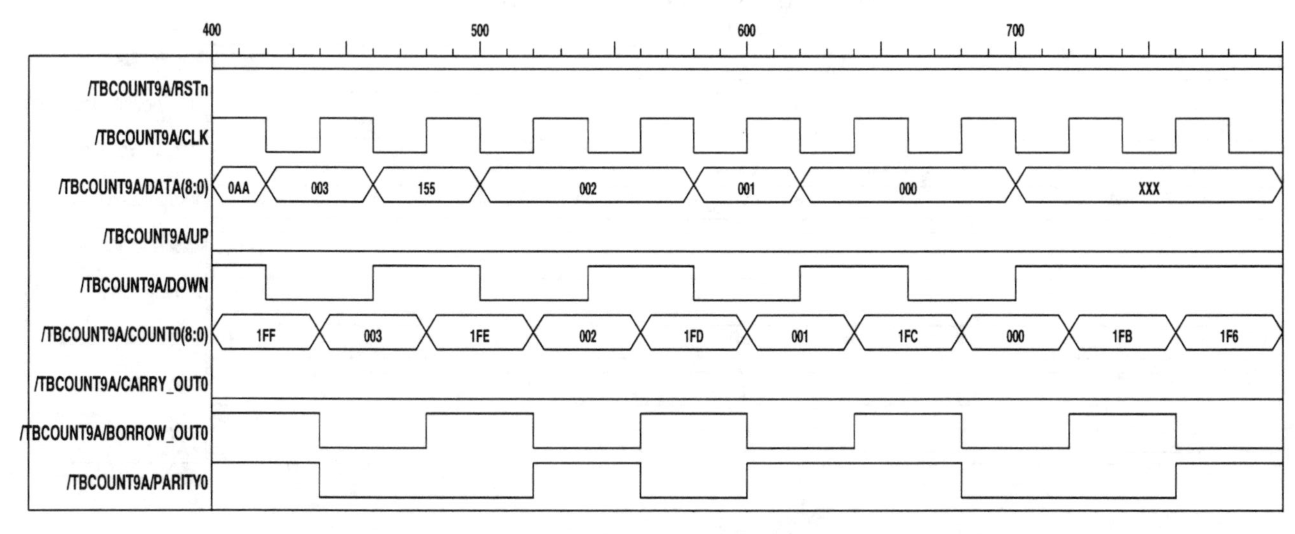

FIGURE 11.4 Simulation waveform for test bench TBCOUNT9A. - (continue)

as C <= A * B;. It is simple enough to ensure it functions correctly. Signals A and B can be converted to 16 bits std_logic_vector (15 downto 0) types as inputs to BWA and BWB, respectively. The output signal BWC from BWMULT is converted to an integer and compared with C. The test bench can run for a long time without human interaction. The idea is to use a benchmark architecture which is appropriate calibrating another architecture. For a typical application example, the functional RTL VHDL code can be used for comparison with the structural VHDL code after the synthesis and layout. The comparison is done by strobing at a particular time based on the clock. Figure 11.5 illustrates this test bench concept. Note that the input test vector can be obtained by reading a file as the previous test bench or by VHDL code.

To illustrate this, we will use architecture RTL to calibrate architecture RTL1 since architecture RTL has been verified with TBCOUNT9A. The following VHDL code shows the test bench TBCOUNT9B.

Library and package references are about the same as TBCOUNT9A. No file IO is used, so the textio package is not needed. Component declaration (lines 10 to 16) is the same. More signals are declared in lines 17 to 20. Constants, PERIOD and DELAY, are declared in lines 21 and 22. Lines 23 and 24 bind the component COUNT9 with architectures RTL and RTL1. The component COUNT9 is instantiated twice in lines 52 to 66. The RSTn and CLK signals are generated as before. Signals DATA, UP, and DOWN are generated with two processes. These two processes are used to illustrate different ways that the values of signals can be generated (may not be the best way for this design COUNT9). Note that the constant DELAY is used to assign values after the rising edge of the clock. The constants PERIOD and DELAY can be generics so that the test bench can be used by delaying the input signals relative to the rising edge of the clock. This allows the designer to check whether the design still works with different delays for the real structural netlist (may be post-layout).

```
1     library IEEE;
2     use IEEE.std_logic_1164.all;
3     use IEEE.std_logic_arith.all;
4     entity TBCOUNT9B is
5     end TBCOUNT9B;
6
```

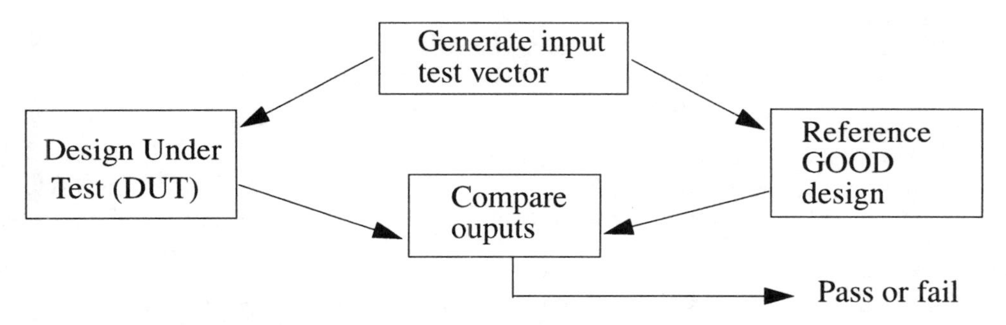

FIGURE 11.5 A test bench with a reference known GOOD design.

```
7     architecture BEH of TBCOUNT9B is
8        component COUNT9
9        port (
10          RSTn, CLK, UP, DOWN : in  std_logic;
11          DATA                : in  std_logic_vector(8 downto 0);
12          COUNT               : out std_logic_vector(8 downto 0);
13          CARRY_OUT           : out std_logic;
14          BORROW_OUT          : out std_logic;
15          PARITY              : out std_logic);
16       end component;
17       signal RSTn, CLK, UP, DOWN, PARITY0, PARITY1 : std_logic;
18       signal CARRY_OUT0, CARRY_OUT1   : std_logic;
19       signal BORROW_OUT0, BORROW_OUT1 : std_logic;
20       signal DATA, COUNT0, COUNT1  : std_logic_vector(8 downto 0);
21       constant PERIOD  : time := 30 ns;
22       constant DELAY   : time :=  2 ns;
23       for count9_inst0 : COUNT9 use entity work.COUNT9(RTL);
24       for count9_inst1 : COUNT9 use entity work.COUNT9(RTL1);
25    begin
26       RSTn <= '0', '1' after 2.5 * PERIOD;
27       data_gen : process
28       begin
29          DATA <= "111111111",
30                  "000000000" after 12 * PERIOD + DELAY,
31                  "010101010" after 24 * PERIOD + DELAY,
32                  "101010101" after 36 * PERIOD + DELAY,
33                  "000000001" after 48 * PERIOD + DELAY,
34                  "111111110" after 60 * PERIOD + DELAY;
35          wait for 72 * PERIOD;
36       end process;
37       up_down : process
38          variable temp : std_logic_vector(2 downto 0) := "000";
39       begin
40          wait until CLK'event and CLK = '1';
41          DOWN <= temp(2) after DELAY;
42          UP   <= temp(1) after DELAY;
43          temp := unsigned(temp) + 1;
44       end process;
45
46       clk_gen : process
47       begin
48          CLK <= '1'; wait for 0.5 * PERIOD;
49          CLK <= '0'; wait for 0.5 * PERIOD;
50       end process;
51
52       count9_inst0 :  COUNT9
53       port map (
54          RSTn      => RSTn,       CLK        => CLK,
55          UP        => UP,         DOWN       => DOWN,
56          DATA      => DATA,       COUNT      => COUNT0,
57          CARRY_OUT => CARRY_OUT0, BORROW_OUT => BORROW_OUT0,
58          PARITY    => PARITY0);
```

```
59
60        count9_inst1  :  COUNT9
61        port map (
62           RSTn       => RSTn,          CLK          => CLK,
63           UP         => UP,            DOWN         => DOWN,
64           DATA       => DATA,          COUNT        => COUNT1,
65           CARRY_OUT  => CARRY_OUT1,    BORROW_OUT   => BORROW_OUT1,
66           PARITY     => PARITY1);
67
68        check : process
69        begin
70           wait until CLK'event and CLK = '1';
71           wait for PERIOD - DELAY;
72           assert COUNT0 = COUNT1
73              report "COUNT0 and COUNT1 are not the same"
74              severity NOTE;
75           assert PARITY0 = PARITY1
76              report "PARITY0 and PARITY1 are not the same"
77              severity NOTE;
78           assert (CARRY_OUT0 = CARRY_OUT1)
79              report "CARRY_OUT0 and CARRY_OUT1 are not the same"
80              severity NOTE;
81           assert (BORROW_OUT0 = BORROW_OUT1)
82              report "BORROW_OUT0 and BORROW_OUT1 are not the same"
83              severity NOTE;
84        end process;
85     end BEH;
```

The verification of both architectures is in lines 68 to 84. Note that line 70 waits until the rising edge of the clock. Line 71 waits to the DELAY before the next rising edge of the clock. This is the time when the values are strobed and compared. Figures 11.6, 11.7, and 11.8 show the simulation waveform of TBCOUNT9B. Note that the signals from both architectures are being compared. No assert string is generated.

11.6 SYNTHESIZING THE DESIGN

Both architectures RTL and RTL can be synthesized. Figures 11.9 and 11.10 show the synthesized schematics. They are synthesized with the same operating condition (WCCOM), library (LSI lca300k), wire load model (B5X5), timing constraints, and driving strength of input signals (ND2). Architecture RTL has a total area of 1161 and a timing delay of 7.33 ns. Architecture RTL1 has a total area of 963 and timing delay of 6.91 ns.

Note that the adder and subtracter in the Figure 11.9 schematic are shown as blocks. Their individual schematics are not shown here. In the Figure 11.10 schematic, 9-bit flipflops of the counter are grouped into one block to reduce the size of the schematic.

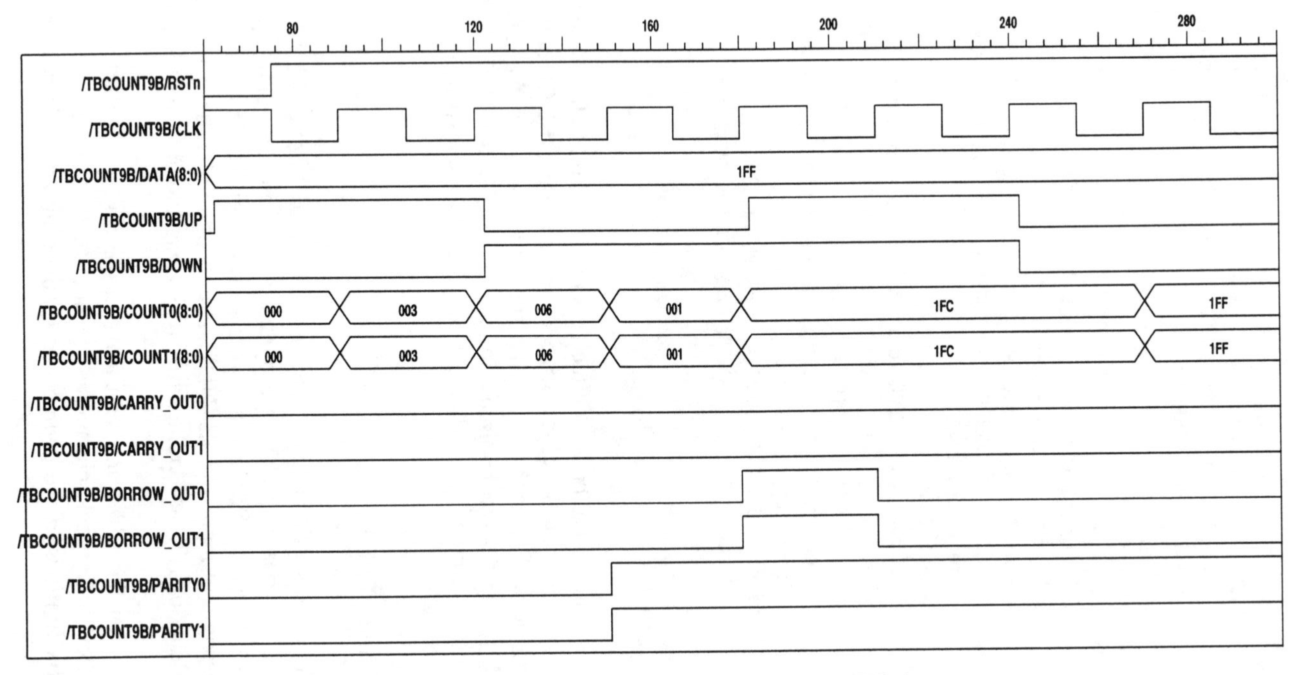

FIGURE 11.6 Simulation waveform for TBCOUNT9B-1.

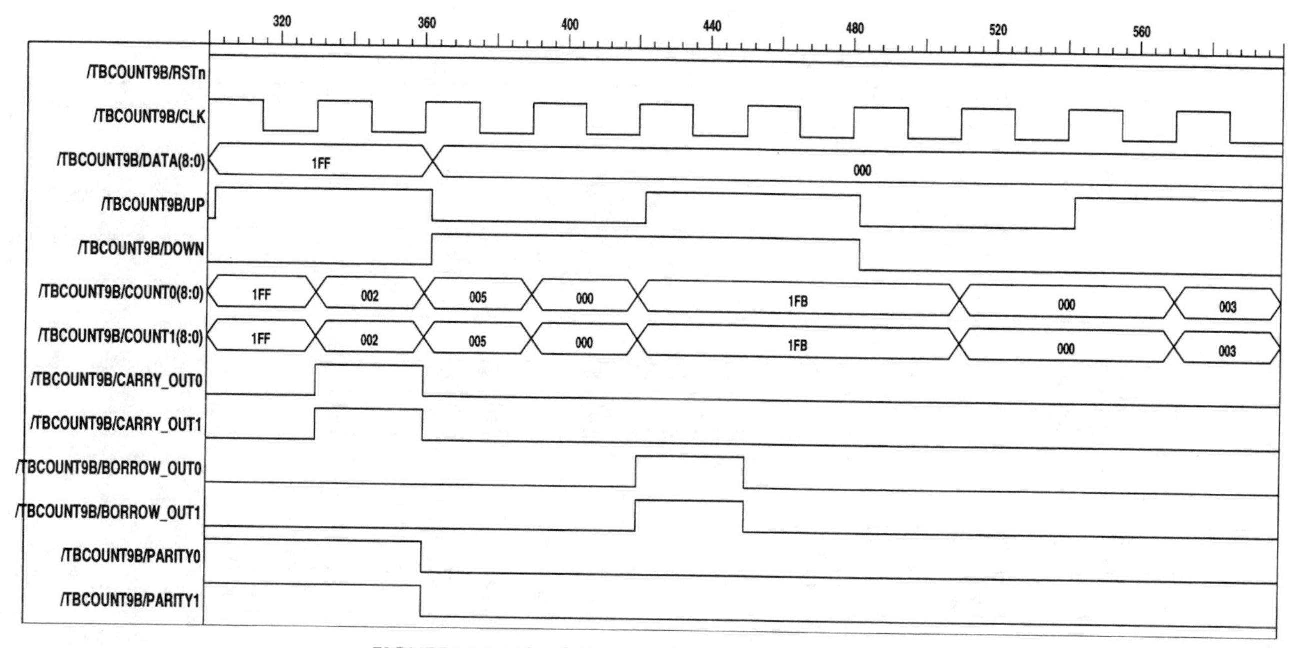

FIGURE 11.7 Simulation waveform for TBCOUNT9B-2.

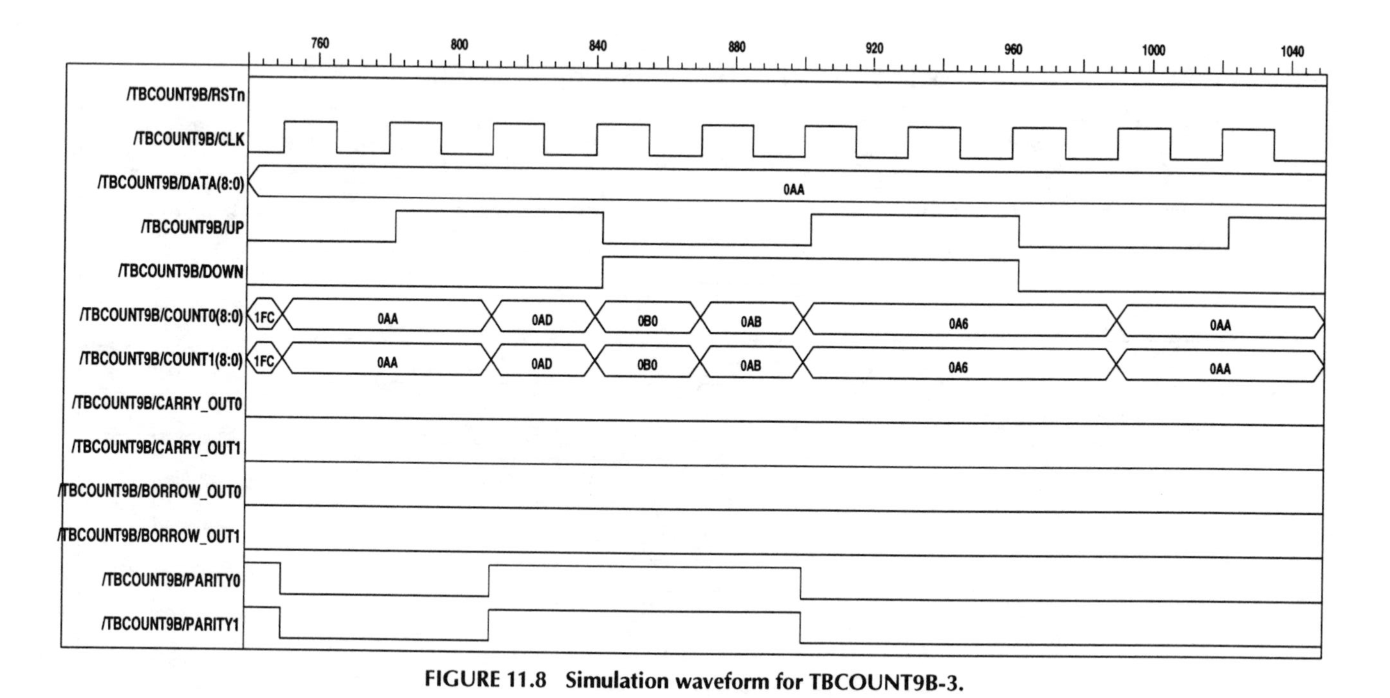

FIGURE 11.8 Simulation waveform for TBCOUNT9B-3.

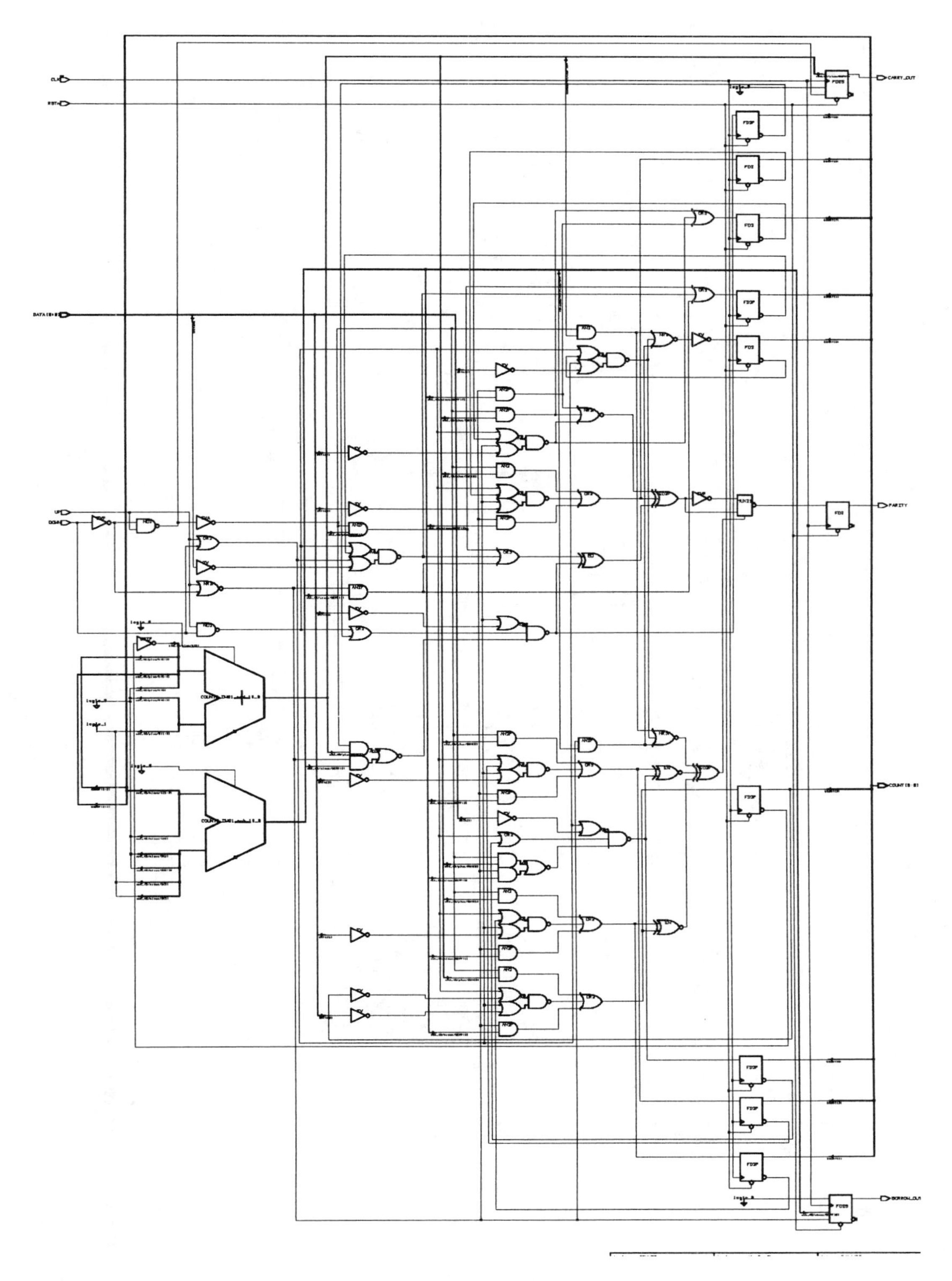

FIGURE 11.9 Synthesized schematic of COUNT9 architecture RTL.

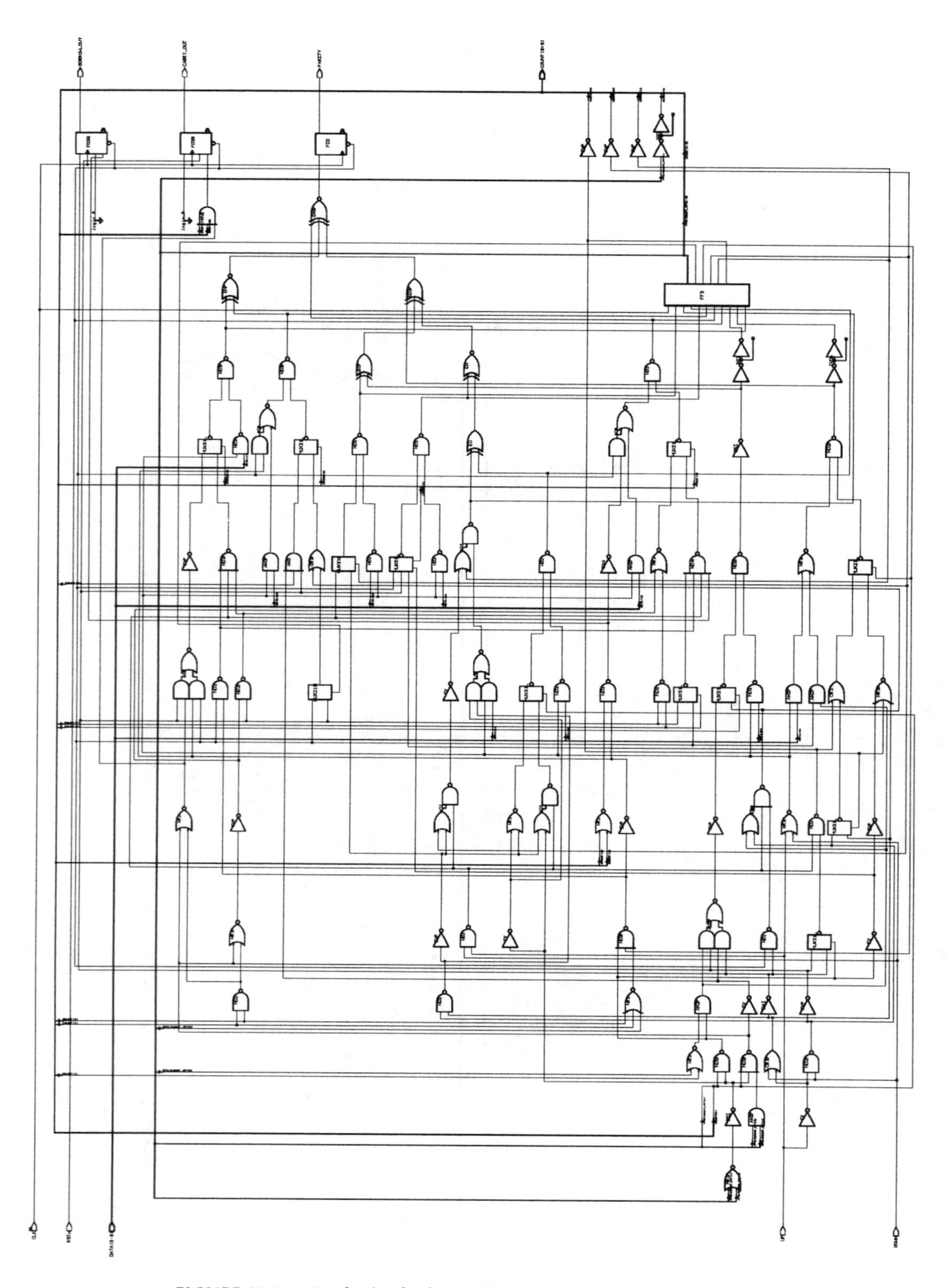

FIGURE 11.10 Synthesized schematic of COUNT9 architecture RTL1.

Another model for a test bench is shown in Figure 11.11. We will call it a handshake test bench model. This model is useful in bus protocols, where a bus master is initiating a transaction and a bus target agent responds. A bus monitor is used to record the address, data, and other violations into a text file. The bus agent can be a microprocessor or a memory controller that is connected to a bus. The bus agent can read in a command file to initiate a bus transaction by converting the bus command into signal values with appropriate timing and handshaking, defined by the bus specification. In Chapter 13, this test bench model will be used to verify the design.

11.7 EXERCISES

11.1 Write a configuration for TBCOUNT9A so that soft binding is used. What needs to be changed in TBCOUNT9A?

11.2 Write a configuration for TBCOUNT9B so that soft binding is used. What needs to be changed in TBCOUNT9B?

11.3 Change TBCOUNT9B to have DELAY and PERIOD as generics. Use the configuration to pass in the generic values from the configuration. Simulate to make sure the generics are passed in correctly.

11.4 Discuss whether the values for DATA, UP, and DOWN are sufficient to verify COUNT9. Change the processes to generate DATA, UP, and DOWN to cover more test combinations. Simulate your test bench.

11.5 After synthesizing COUNT9, a structural netlist can be generated by the synthesis tool. Use that netlist as another architecture. Configure your test bench to verify the correctness of the netlist and its timing delay.

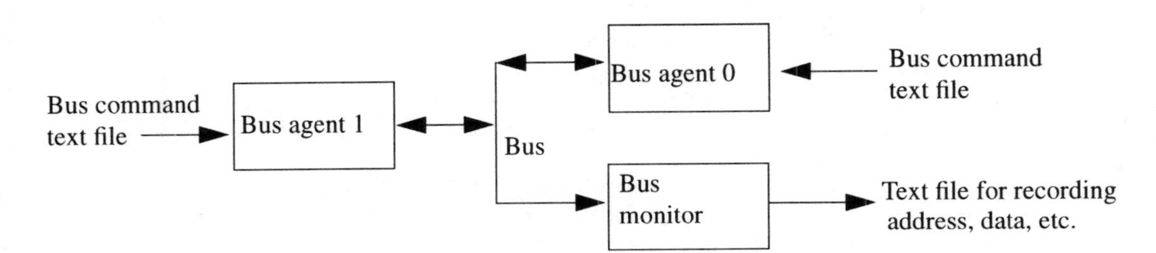

FIGURE 11.11 Handshake test bench model.

Chapter 12

ALU Design

Arithmetic logic units (ALU) are essential elements for a microprocessor. The design requirements for an ALU are first introduced with a truth table. It is implemented with two VHDL architectures, both of which are discussed and synthesized. The results are compared to illustrate ways to improve VHDL coding that will yield more efficient hardware implementation. A test bench is written to test both architectures, with simulation waveforms shown.

12.1 ALU DESIGN REQUIREMENTS

ALUs perform basic arithmetic and logical functions. The following VHDL shows the entity for an ALU. It has two N-bit wide input data A and B, as shown in line 9. The generic N is used to specify the width of the data, with 8 as the default value. Input signal MODE is used to indicate an arithmetic operation (when MODE = '1') or a logical operation (when MODE = '0'). Input signal CI indicates carry input for the arithmetic operations. Input signal OP selects the function according to the truth table in Figure 12.1. Output signals COUT and OVF indicate carry and overflow conditions, respectively. Output signal Z is the result of the operation.

```
1     --------------------- file alu.vhd
2     library IEEE;
3     use IEEE.std_logic_1164.all;
4     use IEEE.std_logic_arith.all;
5
6     entity ALU is
7        generic ( N  : in  integer := 8);
8        port (
9           A, B      : in  std_logic_vector(N-1 downto 0);
10          OP        : in  std_logic_vector(3 downto 0);
11          CI, MODE  : in  std_logic;
12          COUT, OVF : out std_logic;
13          Z         : out std_logic_vector(N-1 downto 0));
14    end ALU;
```

OP	Logical operations (MODE = '0')	Arithmetic operations (MODE = '1', CI = '0')	Arithmetic operations (MODE = '1', CI = '1')
"0000"	0	A	A – 1
"0001"	A	A – B	A – B – 1
"0010"	B	(A and B)	(A and B) – 1
"0011"	not A	(A and (not B))	(A and (not B)) – 1
"0100"	not B	A	A + 1
"0101"	A and B	A + (A or (not B))	A + (A or (not B)) + 1
"0110"	A nand B	(A and B) + (A or (not B))	(A and B) + (A or (not B))+ 1
"0111"	A or B	A or (not B)	(A or (not B)) + 1
"1000"	A nor B	A + (A or B)	A + (A or B) + 1
"1001"	A xor B	(A and (not B)) + (A or B)	(A and (not B)) + (A or B) + 1
"1010"	(not A) or B	A or B	(A or B) + 1
"1011"	not (A xor B)	A + A	A + A + 1
"1100"	A or (not B)	A + B	A + B + 1
"1101"	(not A) and B	A + (A and B)	A + (A and B) + 1
"1110"	A and (not B)	A + (A and (not B))	A + (A and (not B)) + 1
"1111"	1	0	–1

FIGURE 12.1 ALU functions truth table.

12.2 DESCRIBING ALU WITH VHDL

ALU functions can be described in a process statement with all input signals in the process sensitivity list (line 17). Variables are declared in line 18 as N+1 bits of unsigned type to be used in the operators "+" and "–". Variable MOP is declared in line 19 as the concatenation of MODE and OP in line 28. Lines 21 to 26 set up the default values for variables and outputs so that latches would not be inferred.

Inside the process statement, a case statement is used (starting at line 29). Based on the value of MOP, each operation is defined. Outputs OVF and COUT remain '0' for the logical operations. For the arithmetic operations, COUT will be the result of the addition or subtraction carry output. This is done by extending the addition and subtraction operands by 1 bit as shown in lines 58 and 59. COUT is then the left-most bit of the arithmetic operation result. Output OVF depends on the leftmost bit of the operand (sign bit for the 2's complement data), and the result sign bit is shown in lines 65 to 68. Lines 220 to 223 assign output Z and COUT for the arithmetic operations.

```
15    architecture RTL of ALU is
16    begin
```

```
17      proc : process (A, B, CI, MODE, OP)
18         variable temp1, temp2, result : unsigned(N downto 0);
19         variable MOP : std_logic_vector(4 downto 0);
20      begin
21         Z      <= (Z'range => '0');
22         COUT   <= '0';
23         OVF    <= '0';
24         result := (result'range => '0');
25         temp1  := (temp1'range => '0');
26         temp2  := (temp2'range => '0');
27
28         MOP    := MODE & OP;
29         case MOP is
30            when "00000" => Z <= (Z'range => '0');
31            when "00001" => Z <= A;
32            when "00010" => Z <= B;
33            when "00011" => Z <= not A;
34            when "00100" => Z <= not B;
35            when "00101" => Z <= A and B;
36            when "00110" => Z <= A nand B;
37            when "00111" => Z <= A or B;
38            when "01000" => Z <= A nor B;
39            when "01001" => Z <= A xor B;
40            when "01010" => Z <= (not A) or B;
41            when "01011" => Z <= not(A xor B);
42            when "01100" => Z <= A or (not B);
43            when "01101" => Z <= (not A) and B;
44            when "01110" => Z <= A and (not B);
45            when "01111" => Z(0) <= '1';
46               -- starting arithematic operation
47            when "10000" => -- A   - 0         - CI
48               temp1  := '0' & unsigned(A);
49               if (CI = '1') then
50                  result := temp1 - 1;
51               else
52                  result := temp1;
53               end if;
54               if (('0' = result(N - 1)) and (A(N - 1) = '1')) then
55                  OVF <= '1';
56               end if;
57            when "10001" => -- A   - B         - CI
58               temp1  := '0' & unsigned(A);
59               temp2  := '0' & unsigned(B);
60               if (CI = '1') then
61                  result := temp1 - temp2 - 1;
62               else
63                  result := temp1 - temp2;
64               end if;
65               if (((A(N - 1) xor B(N - 1)) = '1') and
66                    (A(N - 1) /= result(N - 1))) then
67                  OVF <= '1';
68               end if;
```

```
69              when "10010" => -- AB  - 0          - CI
70                 temp1 := '0' & unsigned(A and B);
71                 if (CI = '1') then
72                    result := temp1 - 1;
73                 else
74                    result := temp1;
75                 end if;
76                if (('0' = result(N - 1)) and (temp1(N - 1) = '1')) then
77                    OVF <= '1';
78                 end if;
79              when "10011" => -- ABn - 0          - CI
80                 temp1  := '0' & unsigned(A and (not B));
81                 if (CI = '1') then
82                    result := temp1 - 1;
83                 else
84                    result := temp1;
85                 end if;
86                if (('0' = result(N - 1)) and (temp1(N - 1) = '1')) then
87                    OVF <= '1';
88                 end if;
89              when "10100" => -- A   + 0          + CI
90                 temp1  := '0' & unsigned(A);
91                 if (CI = '1') then
92                    result := temp1 + 1;
93                 else
94                    result := temp1;
95                 end if;
96                 if ((temp1((N - 1)) = '0') and
97                      (result((N - 1)) = '1')) then
98                    OVF <= '1';
99                 end if;
100             when "10101" => -- A   + (A or Bn) + CI
101                temp1  := '0' & unsigned(A);
102                temp2  := '0' & unsigned(A or ( not B));
103                if (CI = '1') then
104                   result := temp1 + temp2 + 1;
105                else
106                   result := temp1 + temp2;
107                end if;
108                if (('0' = (temp1((N - 1)) xor temp2((N - 1)))) and
109                   (temp1((N - 1)) /= result((N - 1)))) then
110                   OVF <= '1';
111                end if;
112             when "10110" => -- AB  + (A or Bn) + CI
113                temp1  := '0' & unsigned(A and B);
114                temp2  := '0' & unsigned(A or (not B));
115                if (CI = '1') then
116                   result := temp1 + temp2 + 1;
117                else
118                   result := temp1 + temp2;
119                end if;
```

```
120             if (('0' = (temp1((N - 1)) xor temp2((N - 1)))) and
121                (temp1((N - 1)) /= result((N - 1)))) then
122                OVF <= '1';
123             end if;
124          when "10111" => -- 0   + (A or Bn) + CI
125             temp2  := '0' & unsigned(A or (not B));
126             if (CI = '1') then
127                result := temp2 + 1;
128             else
129                result := temp2;
130             end if;
131             if ((temp2(N - 1) = '0') and
132                 (result(N - 1) = '1')) then
133                OVF <= '1';
134             end if;
135          when "11000" => -- A   + (A or B)  + CI
136             temp1  := '0' & unsigned(A);
137             temp2  := '0' & unsigned(A or B);
138             if (CI = '1') then
139                result := temp1 + temp2 + 1;
140             else
141                result := temp1 + temp2;
142             end if;
143             if (('0' = (temp1((N - 1)) xor temp2((N - 1)))) and
144                (temp1((N - 1)) /= result((N - 1)))) then
145                OVF <= '1';
146             end if;
147          when "11001" => -- ABn + (A or B)  + CI
148             temp1  := '0' & unsigned(A and (not B));
149             temp2  := '0' & unsigned(A or B);
150             if (CI = '1') then
151                result := temp1 + temp2 + 1;
152             else
153                result := temp1 + temp2;
154             end if;
155             if (('0' = (temp1((N - 1)) xor temp2((N - 1)))) and
156                (temp1((N - 1)) /= result((N - 1)))) then
157                OVF <= '1';
158             end if;
159          when "11010" => -- 0   + (A or B)  + CI
160             temp2  := '0' & unsigned(A or B);
161             if (CI = '1') then
162                result := temp2 + 1;
163             else
164                result := temp2;
165             end if;
166             if ((temp2((N - 1)) = '0') and
167                 (result((N - 1)) = '1')) then
168                OVF <= '1';
169             end if;
170          when "11011" => -- A   + A         + CI
171             temp1  := unsigned(A) & '0';
```

```
172              if (CI = '1') then
173                 result := temp1 + 1;
174              else
175                 result := temp1;
176              end if;
177              if (A((N - 1)) /= result((N - 1))) then
178                 OVF <= '1';
179              end if;
180           when "11100" => -- A   + B          + CI
181              temp1 := '0' & unsigned(A);
182              temp2 := '0' & unsigned(B);
183              if (CI = '1') then
184                 result := temp1 + temp2 + 1;
185              else
186                 result := temp1 + temp2;
187              end if;
188              if (('0' = (temp1((N - 1)) xor temp2((N - 1)))) and
189                 (temp1((N - 1)) /= result((N - 1)))) then
190                 OVF <= '1';
191              end if;
192           when "11101" => -- A   + AB         + CI
193              temp1 := '0' & unsigned(A);
194              temp2 := '0' & unsigned(A and B);
195              if (CI = '1') then
196                 result := temp1 + temp2 + 1;
197              else
198                 result := temp1 + temp2;
199              end if;
200              if (('0' = (temp1((N - 1)) xor temp2((N - 1)))) and
201                 (temp1((N - 1)) /= result((N - 1)))) then
202                 OVF <= '1';
203              end if;
204           when "11110" => -- A   + ABn        + CI
205              temp1 := '0' & unsigned(A);
206              temp2 := '0' & unsigned(A and (not B));
207              if (CI = '1') then
208                 result := temp1 + temp2 + 1;
209              else
210                 result := temp1 + temp2;
211              end if;
212              if (('0' = (temp1((N - 1)) xor temp2((N - 1)))) and
213                 (temp1((N - 1)) /= result((N - 1)))) then
214                 OVF <= '1';
215              end if;
216           when others  => -- 0, -1
217              result := (result'range => CI);
218        end case;
219
220        if (MODE = '1') then
221           COUT <= result(N);
222           Z    <= CONV_STD_LOGIC_VECTOR(result((N - 1) downto 0), N);
```

```
223          end if;
224       end process;
225    end RTL;
```

The simulation of the preceding VHDL code will be presented later. The VHDL code can be synthesized. Figure 12.2 shows part of the synthesized schematic. Note the many adders and subtracters. The area is 1,698 units.

12.3 IMPROVING THE DESIGN

The following VHDL code shows another architecture RTL1 for the ALU entity to compare with the architecture RTL VHDL code as shown earlier. The subtraction is done only in line 271. The addition is only performed in line 317. This allows the synthesis tool to better optimize the design. Figure 12.3 depicts the whole synthesized schematic. Note that the same component performs both addition and subtraction. The area is 695 units (compared to 1,698 units in architecture RTL). The timing delay of both architectures is about the same with 12 ns. Both are synthesized with the LSI lca300k library, the B9X9 wire load model, and the WCIND (worst-case industrial) operating condition.

It is easy to see that architecture RTL1 takes fewer lines. The overflow function is shared. The adding or subtracting of 1 based on carry input CI is directly used as in lines 271 and 317. In architecture RTL, additional if statement, as shown in lines 207 to 211, requires more coding. It also increases the difficulty of the synthesis tool to optimize the circuit. The area is much larger than necessary.

```
226    architecture RTL1 of ALU is
227    begin
228       proc : process (A, B, CI, MODE, OP)
229          variable temp1, temp2, result : unsigned(N downto 0);
230       begin
231          temp1  := (temp1'range => '0');
232          temp2  := (temp2'range => '0');
233          Z      <= (Z'range => '0');
234          COUT   <= '0';
235          OVF    <= '0';
236          result := (result'range => '0');
237          if (MODE = '0') then ------------ logical functions
238             case OP is
239                when "0000" => Z <= (Z'range => '0');
240                when "0001" => Z <= A;
241                when "0010" => Z <= B;
242                when "0011" => Z <= not A;
243                when "0100" => Z <= not B;
244                when "0101" => Z <= A and B;
245                when "0110" => Z <= A nand B;
246                when "0111" => Z <= A or B;
247                when "1000" => Z <= A nor B;
248                when "1001" => Z <= A xor B;
```

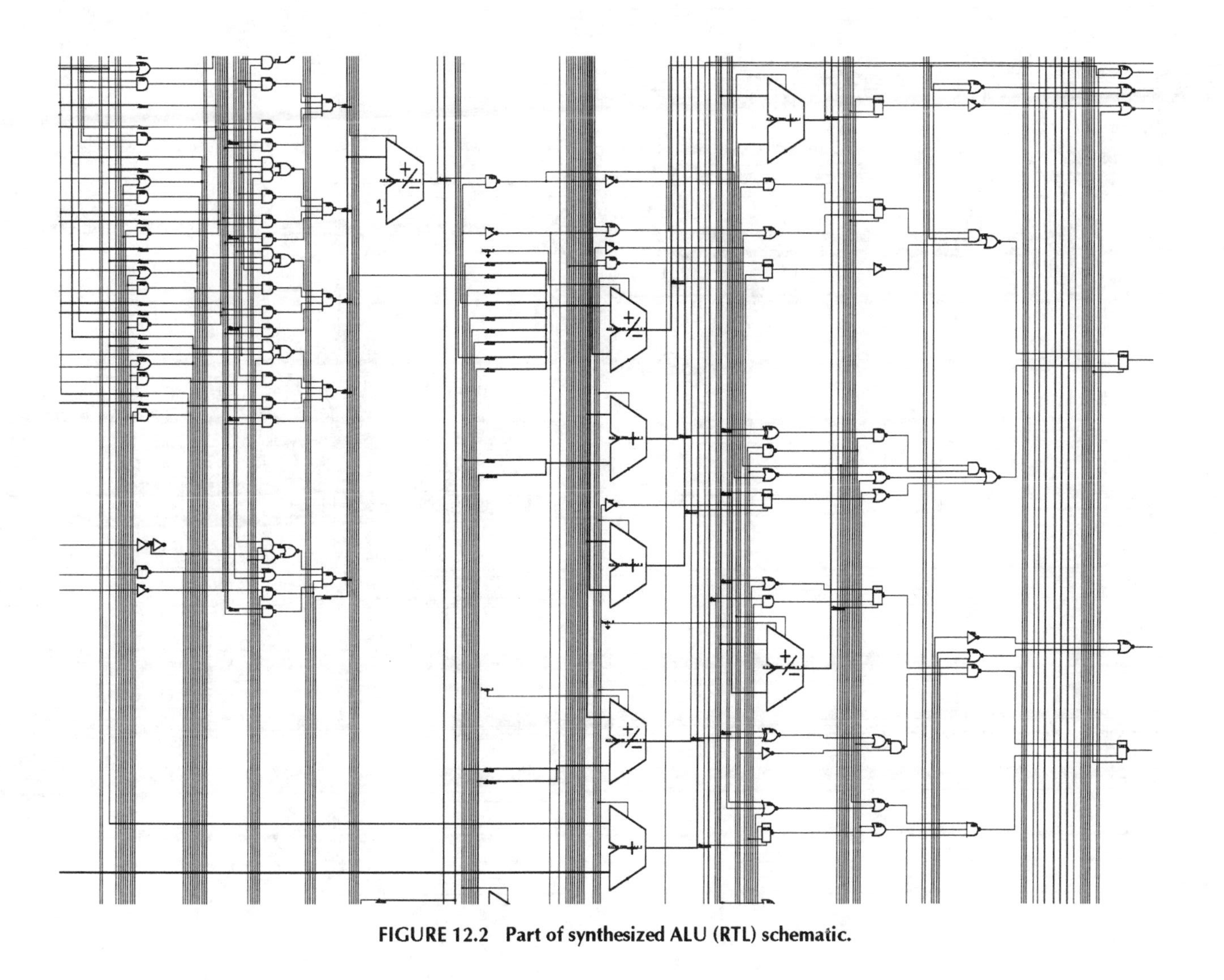

FIGURE 12.2 Part of synthesized ALU (RTL) schematic.

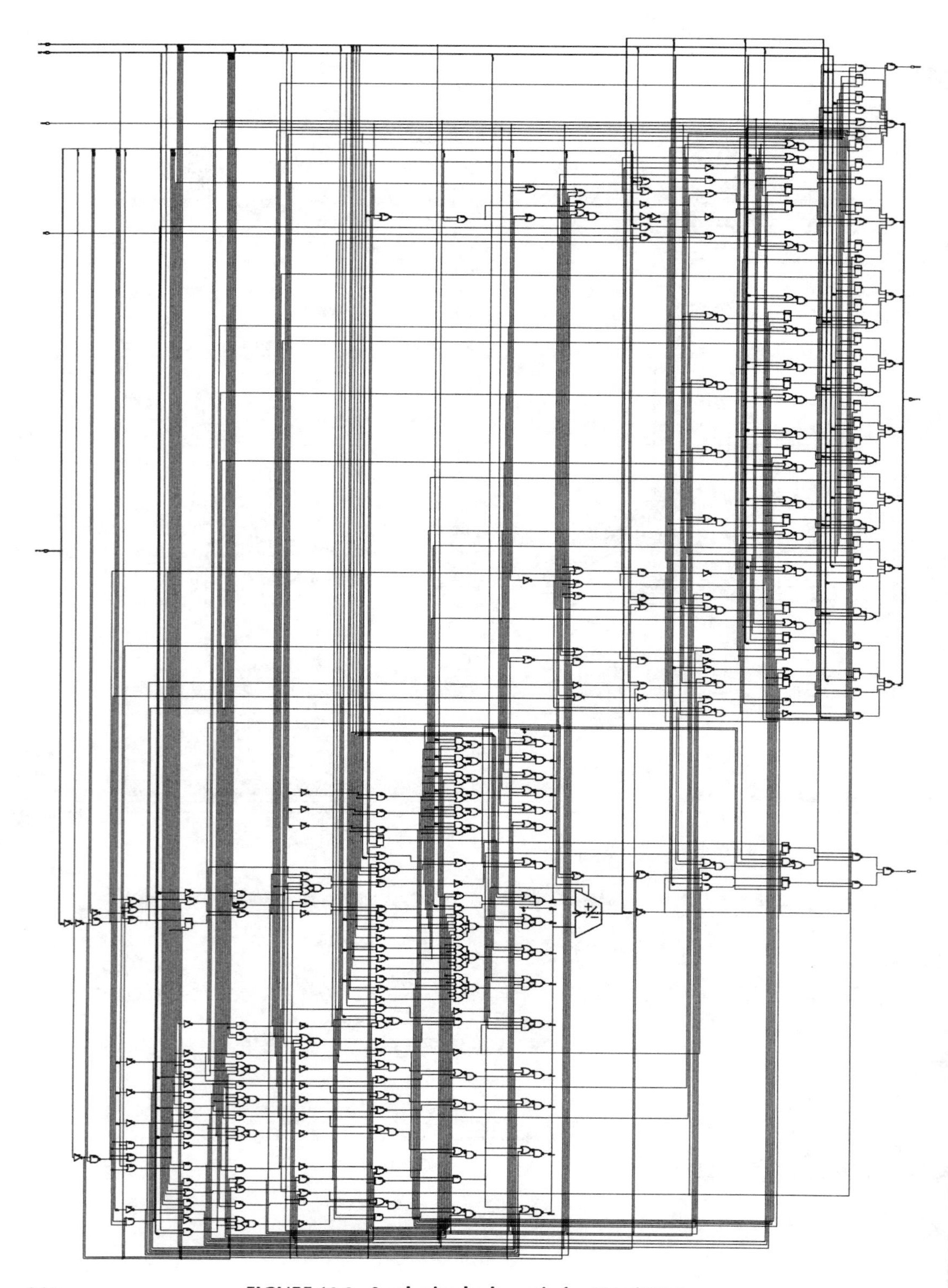

FIGURE 12.3 Synthesized schematic for ALU (RTL1).

```
249               when "1010" => Z <= (not A) or B;
250               when "1011" => Z <= not(A xor B);
251               when "1100" => Z <= A or (not B);
252               when "1101" => Z <= (not A) and B;
253               when "1110" => Z <= A and (not B);
254               when others => Z(0) <= '1';
255            end case;
256         elsif OP(3 downto 2) = "00" then -- minus operations
257            case OP(1 downto 0) is
258               when "00"   => -- A   - 0          - CI
259                  temp1    := '0' & unsigned(A);
260                  temp2    := (temp2'range => '0');
261               when "01"   => -- A   - B          - CI
262                  temp1    := '0' & unsigned(A);
263                  temp2    := '0' & unsigned(B);
264               when "10"   => -- AB  - 0          - CI
265                  temp1    := '0' & unsigned(A and B);
266                  temp2    := (temp2'range => '0');
267               when others => -- ABn - 0          - CI
268                  temp1    := '0' & unsigned(A and (not B));
269                  temp2    := (temp2'range => '0');
270            end case;
271            result := temp1 - temp2 - CI;
272            OVF   <= (temp1(N - 1) xor temp2(N - 1)) and
273                        (temp1(N - 1) xor result(N - 1));
274            COUT  <= result(N);
275            Z <= CONV_STD_LOGIC_VECTOR(result((N - 1) downto 0), N);
276         elsif (OP = "1111") then    ---------------- 0 and -1
277            Z     <= (Z'range => CI);
278            COUT  <= CI;
279            OVF   <= '0';
280         else  --------------------- plus operations
281            case OP is
282               when "0100" => -- A   + 0          + CI
283                  temp1    := '0' & unsigned(A);
284                  temp2    := (temp2'range => '0');
285               when "0101" => -- A   + (A or Bn) + CI
286                  temp1    := '0' & unsigned(A);
287                  temp2    := '0' & unsigned(A or ( not B));
288               when "0110" => -- AB  + (A or Bn) + CI
289                  temp1    := '0' & unsigned(A and B);
290                  temp2    := '0' & unsigned(A or (not B));
291               when "0111" => -- 0   + (A or Bn) + CI
292                  temp1    := (temp1'range => '0');
293                  temp2    := '0' & unsigned(A or (not B));
294               when "1000" => -- A   + (A or B)  + CI
295                  temp1    := '0' & unsigned(A);
296                  temp2    := '0' & unsigned(A or B);
297               when "1001" => -- ABn + (A or B)  + CI
298                  temp1    := '0' & unsigned(A and (not B));
```

```
299                     temp2     := '0' & unsigned(A or B);
300                  when "1010" => -- 0    + (A or B)   + CI
301                     temp1     := (temp1'range => '0');
302                     temp2     := '0' & unsigned(A or B);
303                  when "1011" => -- A    + A          + CI
304                     temp1     := '0' & unsigned(A);
305                     temp2     := '0' & unsigned(A);
306                  when "1100" => -- A    + B          + CI
307                     temp1     := '0' & unsigned(A);
308                     temp2     := '0' & unsigned(B);
309                  when "1101" => -- A    + AB         + CI
310                     temp1     := '0' & unsigned(A);
311                     temp2     := '0' & unsigned(A and B);
312                  when "1110" => -- A    + ABn        + CI
313                     temp1     := '0' & unsigned(A);
314                     temp2     := '0' & unsigned(A and (not B));
315                  when others => null; ------------ should not happen
316               end case;
317               result := temp1 + temp2 + CI;
318               OVF   <= (not (temp1(N - 1) xor temp2(N - 1))) and
319                         (temp1((N - 1)) xor result((N - 1)));
320               COUT   <= result(N);
321               Z <= CONV_STD_LOGIC_VECTOR(result((N - 1) downto 0), N);
322            end if;
323         end process;
324      end RTL1;
```

12.4 SIMULATE THE DESIGN WITH A TEST BENCH

Both RTL and RTL1 architectures of entity ALU, can be simulated. The following VHDL code shows a test bench to test both. Lines 9 to 17 declare the component ALU. Lines 19 to 31 declare constants and signals. Lines 34 to 43 instantiate two ALU components. Note that the outputs from both ALUs are mapped with signals Z, COUT, OVF, and (Z1, COUT1, OVF1), respectively. These signals are compared in lines 44 to 46.

The process statement in lines 47 to 79 generates values for A, B, CI, MODE, and OP. The values are changed with the control of the loop statements. Lines 52 to 55 assign values to A and B by converting integers Aint and Bint to a std_logic_vector. Both integers Aint and Bint can be regarded as stepping through the values between 0 and 255 randomly. It is easy to change them to have all the combinations. Lines 57 to 62 set up values for logical operation with selection signal S ranging from "0000" to "1111" (lines 59 to 62 loop statement). Signal S is mapped to ALU OP signal in lines 36 and 41. Lines 64 to 75 set up values for arithmetic operations. Another level of the loop statement is used to run two times with CI = '0' and CI = '1' respectively. Both ALUs input signals A, B, CI, MODE, and OP are mapped with the same signals. Lines 82 to 91 configure both ALUs, one with architecture RTL (line 85) and one with architecture RTL1 (line 88).

```
1     library IEEE;
2     use IEEE.std_logic_1164.all;
3     use IEEE.std_logic_arith.all;
4     entity TBALU is
5     end TBALU;
6
7     architecture BEH of TBALU is
8
9        component ALU
10       generic ( N  : in  integer := 8);
11       port (
12          A, B      : in  std_logic_vector(N-1 downto 0);
13          OP        : in  std_logic_vector(3 downto 0);
14          CI, MODE  : in  std_logic;
15          COUT, OVF : out std_logic;
16          Z         : out std_logic_vector(N-1 downto 0));
17       end component;
18
19       constant PERIOD : time := 20 ns;
20       constant N      : integer := 8;
21       signal   A      : std_logic_vector((N - 1) downto 0);
22       signal   B      : std_logic_vector((N - 1) downto 0);
23       signal   CI     : std_logic;
24       signal   MODE   : std_logic;
25       signal   S      : std_logic_vector(3 downto 0);
26       signal   COUT   : std_logic;
27       signal   OVF    : std_logic;
28       signal   Z      : std_logic_vector((N - 1) downto 0);
29       signal   COUT1  : std_logic;
30       signal   OVF1   : std_logic;
31       signal   Z1     : std_logic_vector((N - 1) downto 0);
32    begin
33
34       ALU0 : ALU
35          generic map (N => N)
36          port map ( A   => A,   B => B, OP     => S,
37                     CI  => CI,  MODE => MODE, COUT   => COUT,
38                     OVF => OVF, Z => Z);
39       ALU1 : ALU
40          generic map (N => N)
41          port map ( A   => A,   B => B, OP     => S,
42                     CI  => CI,  MODE => MODE, COUT   => COUT1,
43                     OVF => OVF1, Z => Z1);
44       assert COUT = COUT1 report "COUT not matched" severity NOTE;
45       assert OVF  = OVF1  report "OVF  not matched" severity NOTE;
46       assert Z    = Z1    report "Z    not matched" severity NOTE;
47       testproc : process
48          variable Aint : integer range 0 to 255 := 3;
49          variable Bint : integer range 0 to 255 := 230;
50       begin
51          for i in 0 to 1 loop -- M mode loop
```

```
52            Aint := (Aint + 5) mod 255;
53            Bint := (Bint + 53) mod 255;
54            A <= conv_std_logic_vector(Aint, 8);
55            B <= conv_std_logic_vector(Bint, 8);
56            if (i = 0) then
57               CI <= '0';
58               MODE <= '0';
59               for k in 0 to 15 loop -- S loop
60                  S <= CONV_STD_LOGIC_VECTOR(k, 4);
61                  wait for PERIOD;
62               end loop;
63            else
64               MODE <= '1';
65               for j in 0 to 1 loop -- CI loop
66                  if (j = 0) then
67                     CI    <= '0';
68                  else
69                     CI    <= '1';
70                  end if;
71                  for k in 0 to 15 loop -- S loop
72                     S <= CONV_STD_LOGIC_VECTOR(k, 4);
73                     wait for PERIOD;
74                  end loop;
75               end loop;
76            end if;
77         end loop;
78         wait for 2 * PERIOD;
79      end process;
80   end BEH;
81
82   configuration CFG_BEH_TBALU of TBALU is
83      for BEH
84         for alu0 : ALU
85            use entity work.ALU(RTL);
86         end for;
87         for alu1 : ALU
88            use entity work.ALU(RTL1);
89         end for;
90      end for;
91   end CFG_BEH_TBALU;
```

Figure 12.4 shows part of the simulation results for the logical operation (MODE = '0'). Figures 12.5 and 12.6 show some of the arithmetic operation results with CI = '0' and CI = '1', respectively. Note that the assertion statements do not generate any messages. The test bench can run for a long time to verify results automatically (assuming one is correct) without manually viewing the waveforms.

12.5 EXERCISES

12.1 If your synthesis tool does not take the arithmetic operators "+" and "–" to synthesize an adder-subtracter, write a VHDL code for such adder-subtracter and change

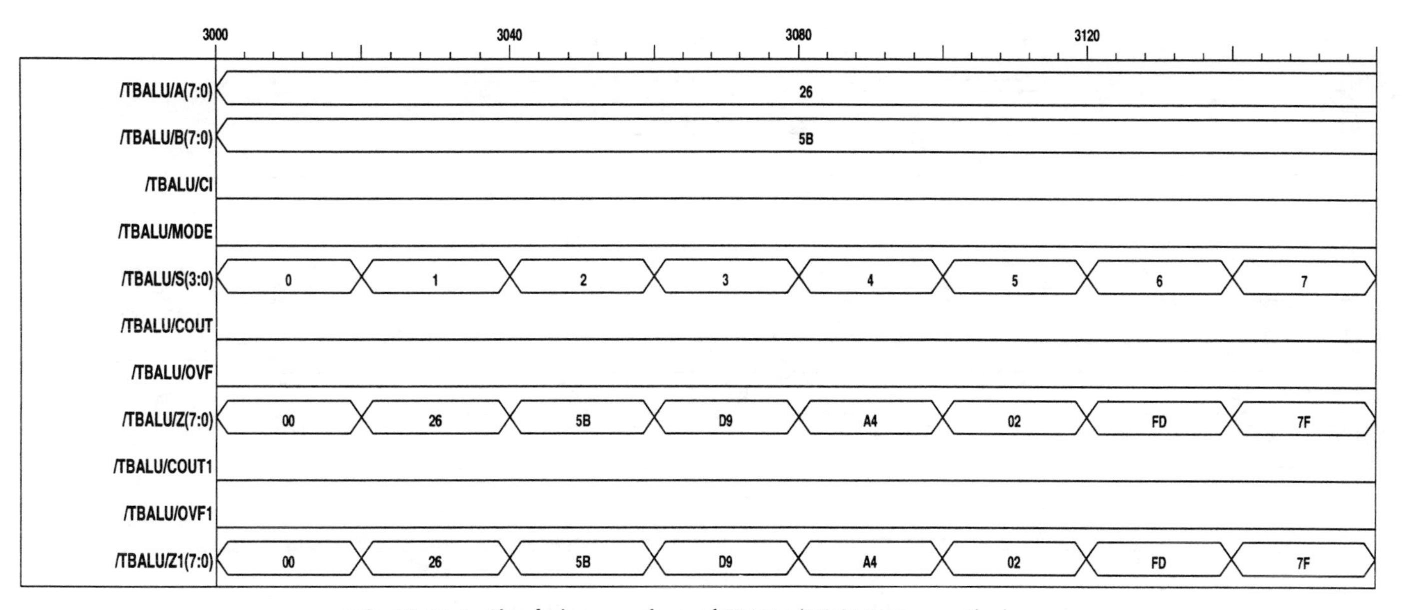

FIGURE 12.4 Simulation waveform of TBALU (BOOLEAN operation).

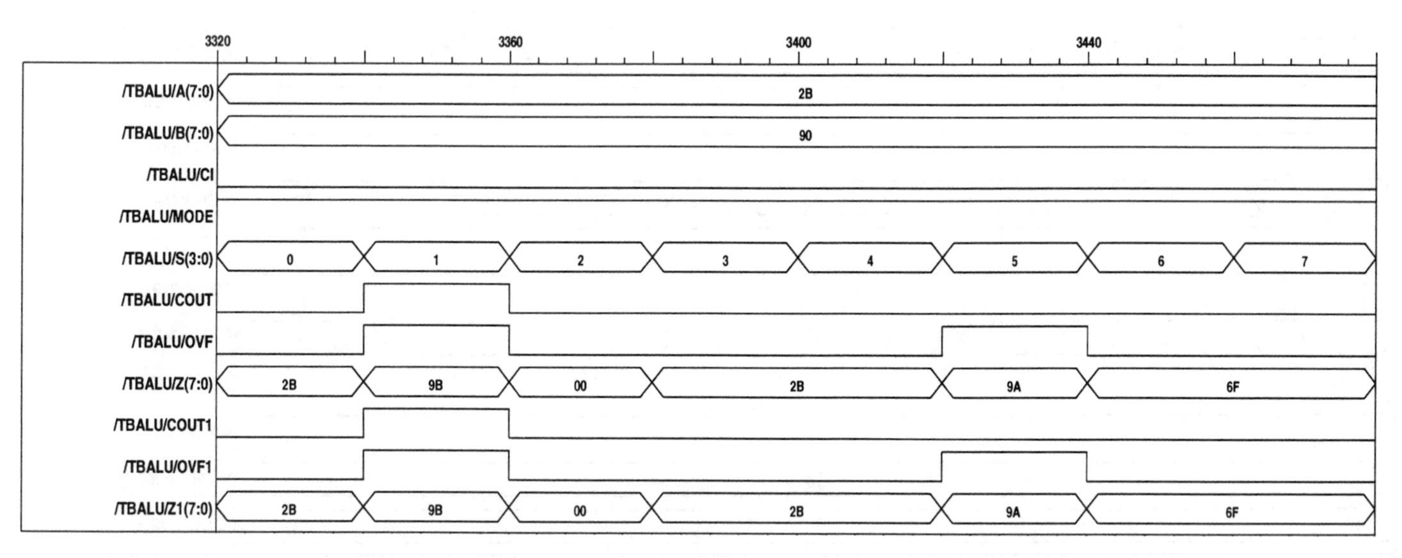

FIGURE 12.5 Simulation waveform of TBALU (arithmetic operation, CI = '0').

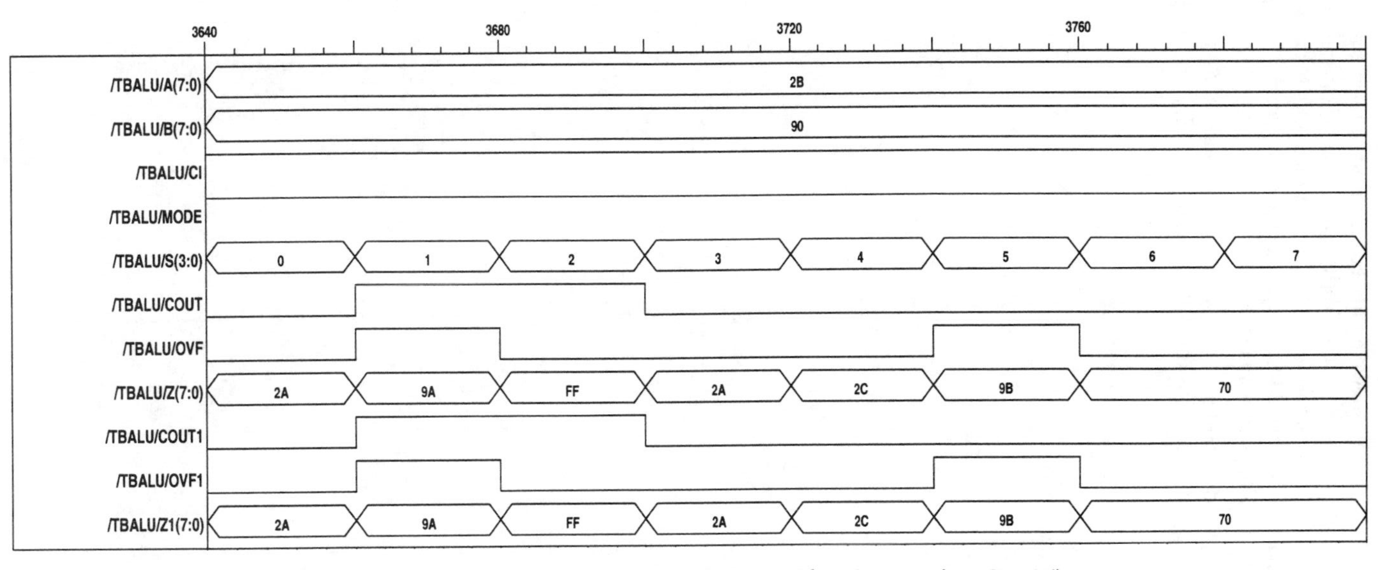

FIGURE 12.6 Simulation waveform of TBALU (arithmetic operation, CI = '1').

architecture RTL1 to include your adder-subtracter. Verify your design through simulation and synthesis.

12.2 Can the ALU design shown earlier be cascaded to form a bigger ALU (cascade two 8-bit ALU into a 16-bit ALU)? Verify your conclusion through simulation. From the timing and area point of view, what are the trade-offs between mapping the generic N with 16, and using two ALUs cascaded with each generic N mapped with 8? Verify your conclusions with your synthesis tool.

12.3 What improvements can you suggest to the timing of the ALU architecture RTL1? Does rearranging the operations (in truth table) help? Verify your results.

Chapter 13

A Design Project

The goal of this chapter is to take a small design project through the design phases of the design requirements, VHDL implementation, VHDL functional verification, test bench setup, VHDL synthesis, layout, and final verification with annotated layout timing.

13.1 DESIGN REQUIREMENTS

The design (we will call it CHIP) is to act as a target agent in a PCI (peripheral components interconnect) local bus. A PCI master agent is to initiate a bus command. A PCI target agent is to respond to a PCI master when the address in the bus matches the target agent's address space range. All the pin names and bus protocol are from PCI convention. Note that the purpose of this design project is to go through the design phases, not to be fully PCI compliant. We will make assumptions to limit the scope of the design.

CHIP has a memory space of 4096 words (32 bits per word). The first 1024 words are for a ROM. The second 1024 words are for a RAM. The third 1024 words occupy the first 256 bytes (16 words) of *PCI configuration registers*, with the rest not used. The fourth 1024 words have three registers with the rest not used. Figure 13.1 shows the CHIP memory space. The PCI address is 32-bit byte addressing. Each PCI agent will respond at the following two situations :

1. In the configuration mode, CHIP will respond when IDSEL is asserted high, and CBEN is set to "1010" or "1011" for configuration read or write cycle. This is the configuration phase when the system will find out each agent's memory space size, device specific information, and write base address to each agent so that each agent's memory space is relocatable. The system will write all '1' to the base register (x010) and read it back. The number of '1's in the most significant bits determines the memory size of the agent. For example, CHIP responds xFFFFC000 when the configuration register x010 is read after it is written all '1's. The register's least significant 14 bits are hardwired to '0'. There are only 18 flipflops for the most significant bits of the register.

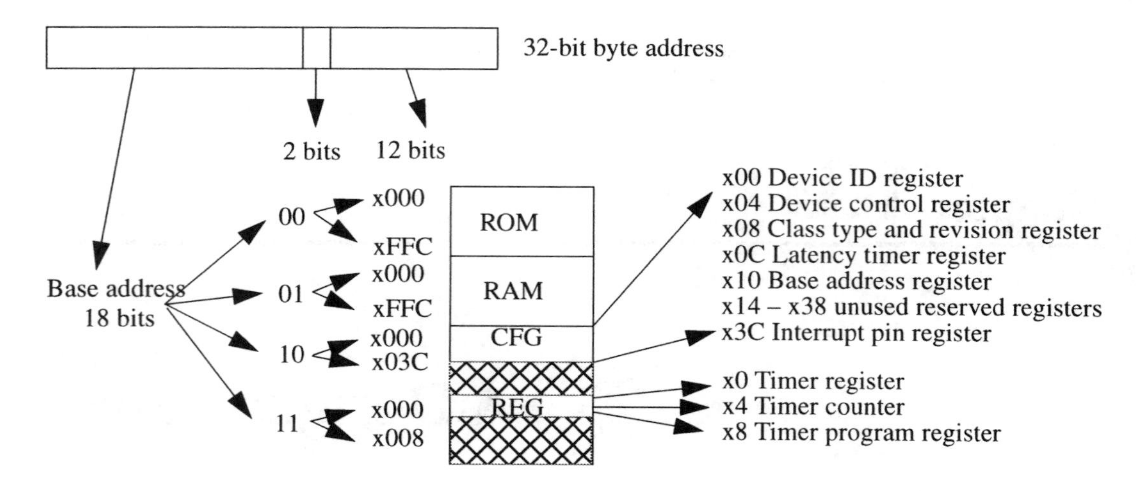

FIGURE 13.1 CHIP memory space.

2. In the normal operation mode, CHIP will respond to a memory write command (CBEn = "0111") and memory read command (CBEn = "0110") when the most significant 18-bit address matches the value in the base address register.

Figure 13.2 shows a PCI write cycle waveform. All PCI control signals are low active. The cycle starts with a PCI master asserting FRAMEn. The address is put on the AD (32-bit address and data mux bus) and the *memory write* bus command is put on CBEn (4-bit "0111") bus. CBEn bus has the bus command during the address phase when FRAMEn is first asserted. After that, CBEn is the byte enable to control the four byte lanes of 32-bit data. The master asserts IRDYn to indicate that the data is available in AD bus. The target asserts DEVSELn (device selected) after decoding the address matched with the base address register and the right bus command. The target also asserts TRDYn (target ready) to tell the master that the data is read or written successfully. The master deasserts FRAMEn to indicate the last data transfer. The target asserts the last TRDYn and deasserts DEVSELn. Then, the master deasserts IRDYn and the transaction cycle is complete.

In Figure 13.2, three data (TRDYn are asserted three times with IRDYn also asserted) are written to the RAM with starting address X"FEDC1000".

The PCI read cycle is about the same as a write cycle. Figure 13.3 shows a PCI read cycle timing diagram. The bus command is memory read (CBEn = "0110"). Whoever supplies the data to the AD bus would generate the even parity across AD and CBEn bus at signal PAR one clock cycle after the data is transferred (when both TRDYn and IRDYn are asserted). Note that the same three data values are read back after they are written.

To limit the scope of CHIP, the following conditions are assumed:

1. Only four PCI bus commands are implemented (configuration read "1010", configuration write "1011", memory read "0110", and memory write "0111").
2. IRDYn is always asserted during the data phase.

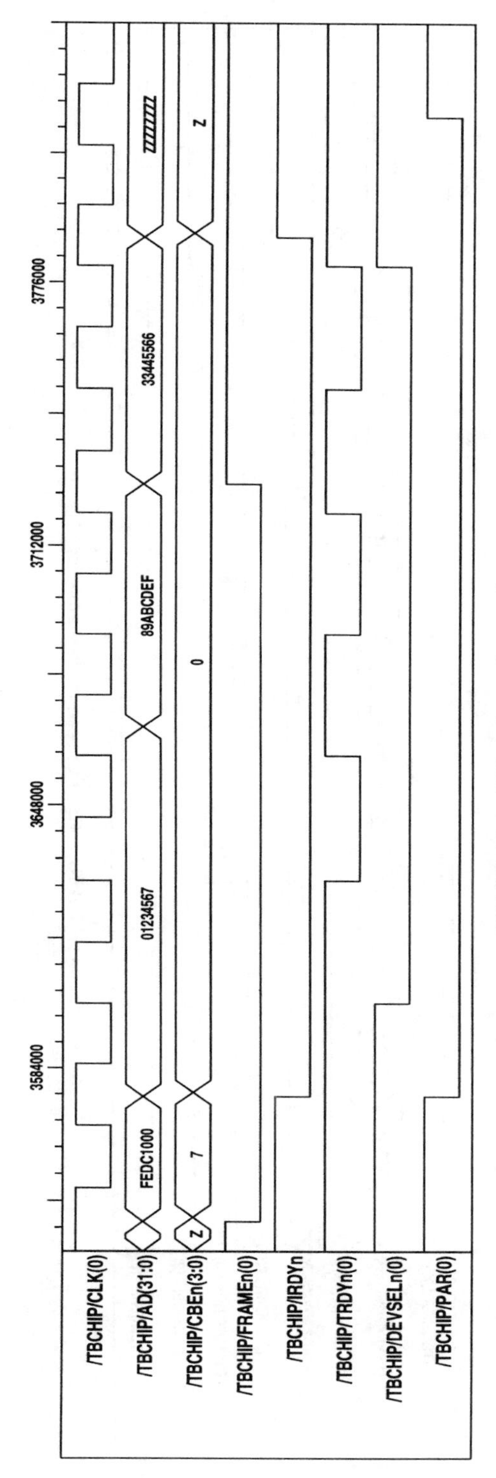

FIGURE 13.2 PCI write cycle waveform.

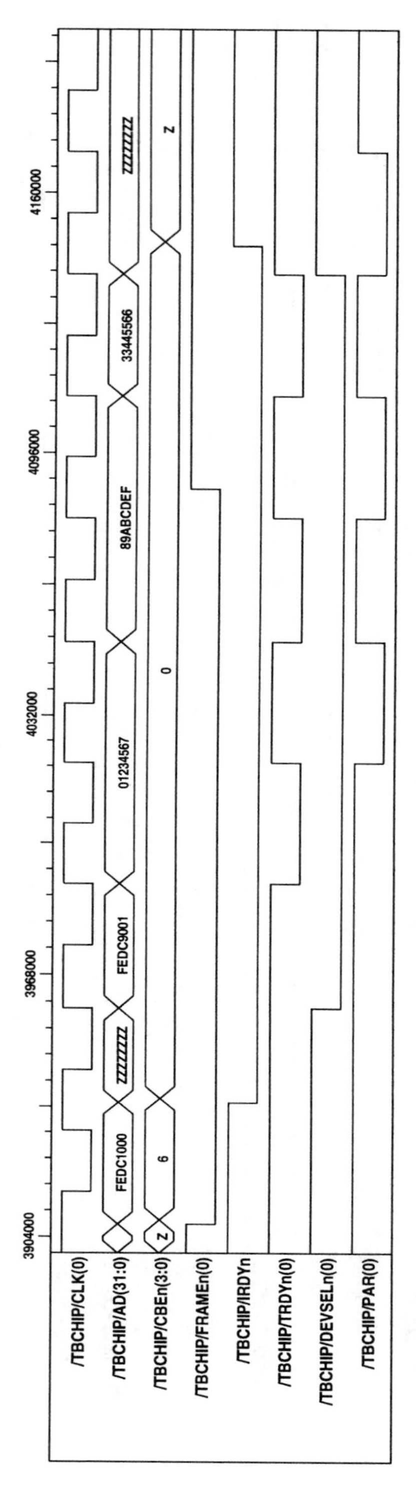

FIGURE 13.3 PCI read cycle waveform.

3. PERRn (data parity error), SERRn (address parity error), STOPn, LOCKn PCI signals are not implemented.
4. Medium speed response of DEVSELn is used such that DEVSELn is asserted right after the second rising edge of the clock after FRAMEn is asserted. This response speed is set up in the device control register.
5. CHIP always responds every two clock cycles.
6. When an address falls into the not used region, the write is not performed (memory or register not updated), and a random data is supplied for the read. The PCI bus acts normally.

For the 16 PCI configuration registers, most of the bits can be hardwired. The write to the unused reserved register would not update any register with normal return. The read to any of the unused reserved registers would return all '0's.

The first CHIP register is a 32-bit timer register TIMER_REG. The second CHIP register is a 32-bit timer counter TIMER_CNTR. The third CHIP register is an 8-bit timer program TIMER_PROG register. The TIMER_PROG is used to program how fast the TIMER_CNTR would decrease the value by 1. When PCI writes to the TIMER_CNTR, the value of the TIMER_REG would be parallel loaded to TIMER_CNTR. The TIMER_CNTR would decrease 1 in every TIMER_PROG + 2 clock cycles. If TIMER_PROG is 0, the TIMER_CNTR would count every clock cycle. When TIMER_CNTR reaches its terminal count, an interrupt is generated in INTAn pin. TIMER_CNTR is loaded with the value of TIMER_REG.

13.2 FUNCTIONAL VHDL IMPLEMENTATION

Before we present the VHDL code, we use the lowercase "n" at the end of the signal name to indicate a low active signal. Architecture name RTL indicates that the VHDL code would be synthesized to schematic. Architecture name BEH indicates the behavior functional model which is not to be synthesized.

The following VHDL code (file chip_e.vhd) shows the CHIP entity. Lines 1 to 4 reference library IEEE package std_logic_1164 and library EPOCH_LIB package COMPONENTS. CLK is the clock signal, RSTn is the asynchronous reset. The rest are for PCI pins. Note that the signals of single bit are declared as type std_logic_vector(0 downto 0). This is to be compatible with the pads in the EPOCH_LIB. Otherwise, std_logic can be used. Note that IRDYn is not a pin of the CHIP since we assumed IRDYn would be always asserted by the PCI master during the transfer of data. AD is a bidirectional bus. TRDYn, DEVSELn, and TAR are tristated output signals. INTAn is a straight output signal.

```
1     library IEEE;
2     library EPOCH_LIB;
3     use IEEE.std_logic_1164.all;
4     use EPOCH_LIB.COMPONENTS.all;
5     entity CHIP is
6        port (
7           CLK     : in    std_logic_vector(0 downto 0);
8           RSTn    : in    std_logic_vector(0 downto 0);
```

```
9          AD      : inout std_logic_vector(31 downto 0);
10         CBEn    : in    std_logic_vector(3 downto 0);
11         IDSEL   : in    std_logic_vector(0 downto 0);
12         FRAMEn  : in    std_logic_vector(0 downto 0);
13         TRDYn   : out   std_logic_vector(0 downto 0);
14         DEVSELn : out   std_logic_vector(0 downto 0);
15         PAR     : out   std_logic_vector(0 downto 0);
16         INTAn   : out   std_logic_vector(0 downto 0));
17    end CHIP;
```

At this stage, the design would be described from top to bottom. We assume the design CHIP has pads around a core from the physical silicon point of view. Figure 13.4 shows the relationship of CHIP, CORE, and four general types of pads (input, output, tristate output, and bidirectional). CORE has more signals than the CHIP due to the tristate buffer enable signals and the breaking of the bidirectional signals. The following VHDL code (a separate text file chip_beh.vhd) shows the architecture BEH of entity CHIP. Lines 4 to 21 declare the component CORE. Local signals for the connections between CORE and pads are declared in lines 33 to 40. Note that std_logic is used rather than std_logic_vector(0 downto 0) for a single bit signal.

```
1     library IEEE;
2     use IEEE.std_logic_1164.all;
3     architecture BEH of CHIP is
4        component CORE
5        port (
6           CLK          : in    std_logic;
7           RSTn         : in    std_logic;
8           ADi          : in    std_logic_vector(31 downto 0);
9           ADo          : out   std_logic_vector(31 downto 0);
10          AD_OEn       : out   std_logic_vector(1 downto 0);
11          CBEni        : in    std_logic_vector(3 downto 0);
12          IDSEL        : in    std_logic;
13          FRAMEni      : in    std_logic;
```

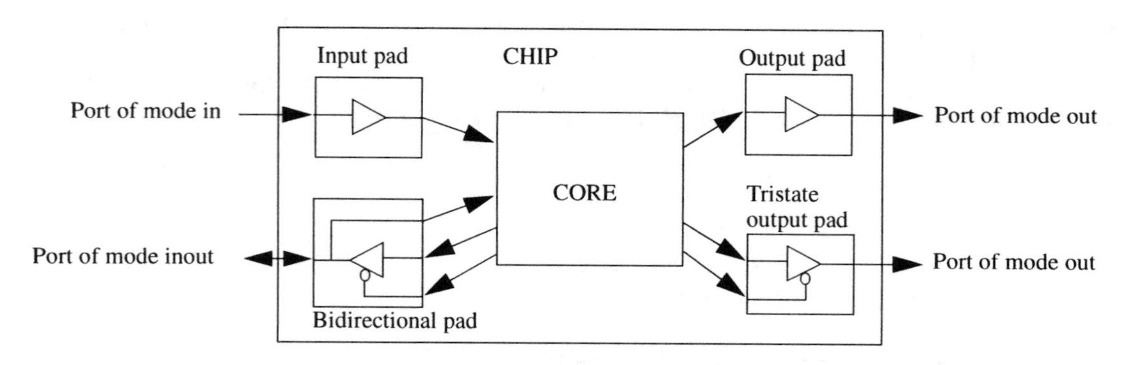

FIGURE 13.4 Physical chip, core, and pads.

```
14          TRDYno        : out   std_logic;
15          TRDYn_OEn     : out   std_logic;
16          DEVSELno      : out   std_logic;
17          DEVSELn_OEn   : out   std_logic;
18          PARo          : out   std_logic;
19          PAR_OEn       : out   std_logic;
20          INTAn         : out   std_logic);
21       end component;
22
23       function TO_STRONG (D : std_logic) return std_logic is
24          variable result : std_logic;
25       begin
26          if (D = 'L') then result := '0';
27          elsif (D = 'H') then result := '1';
28          else result := D;
29          end if;
30          return result;
31       end TO_STRONG;
32
33       signal ADi, ADo        : std_logic_vector(31 downto 0);
34       signal AD_OEn          : std_logic_vector(1 downto 0);
35       signal CBEni           : std_logic_vector(3 downto 0);
36       signal L_CLK, L_RSTn       : std_logic_vector(0 downto 0);
37       signal L_INTAn,  L_IDSEL   : std_logic_vector(0 downto 0);
38       signal FRAMEni, TRDYno     : std_logic_vector(0 downto 0);
39       signal DEVSELno, PARo      : std_logic_vector(0 downto 0);
40       signal TRDYn_OEn, DEVSELn_OEn, STOPn_OEn, PAR_OEn : std_logic;
41    begin
42       core0 : CORE
43       port map (
44          CLK        => L_CLK(0),       RSTn          => L_RSTn(0),
45          ADi        => ADi,            ADo           => ADo,
46          AD_OEn     => AD_OEn,         CBEni         => CBEni,
47          IDSEL      => L_IDSEL(0),     FRAMEni       => FRAMEni(0),
48          TRDYno     => TRDYno(0),      TRDYn_OEn     => TRDYn_OEn,
49          DEVSELno   => DEVSELno(0),    DEVSELn_OEn   => DEVSELn_OEn,
50          PARo       => PARo(0),        PAR_OEn       => PAR_OEn,
51          INTAn      => L_INTAn(0));
52          -------------------------------------------------------
53          L_CLK       <= CLK;    L_RSTn  <= RSTn;
54          L_IDSEL     <= IDSEL;  INTAn   <= L_INTAn;
55          ADi         <= AD;
56          AD          <= ADo when AD_OEn(0) = '0' else (others => 'Z');
57          CBEni       <= CBEn;
58          FRAMEni(0)  <= TO_STRONG(FRAMEn(0));
59          TRDYn(0)    <= TRDYno(0)    when TRDYn_OEn    = '0' else 'Z';
60          DEVSELn(0)  <= DEVSELno(0)  when DEVSELn_OEn  = '0' else 'Z';
61          PAR(0)      <= PARo(0)      when PAR_OEn      = '0' else 'Z';
62    end BEH;
```

Lines 42 to 51 instantiate the CORE component. Note that the mapping of input and output signals can use CHIP port signals directly. However, local signals are used because we want to use the actual instantiation of pad components later so that the code can be uniform and easily changed. Lines 53, 54, and 57 just tie the input and output signal directly, which will be replaced by input pads and output pads later. Lines 59 to 61 model the tristate output pads. Lines 55 and 56 model the bidirectional pad for the AD bus. Line 57 ties the FRAMEn signal with a function TO_STRONG declared in lines 23 to 31. The reason for using the function TO_STRONG and the modeling for a bidirectional pad were discussed in Chapter 10.

Now, we can go down one level to decompose CORE into three blocks (RAM, ROM, and other circuits). All the circuits not inside the RAM and ROM are grouped into one block SYN and synthesized. The following VHDL code shows the CORE VHDL code (core.vhd):

```
1     library IEEE;
2     library EPOCH_LIB;
3     use EPOCH_LIB.COMPONENTS.all;
4     use IEEE.std_logic_1164.all;
5     entity CORE is
6        port (
7           CLK          : in    std_logic;
8           RSTn         : in    std_logic;
9           ADi          : in    std_logic_vector(31 downto 0);
10          ADo          : out   std_logic_vector(31 downto 0);
11          AD_OEn       : out   std_logic_vector(1 downto 0);
12          CBEni        : in    std_logic_vector(3 downto 0);
13          IDSEL        : in    std_logic;
14          FRAMEni      : in    std_logic;
15          TRDYno       : out   std_logic;
16          TRDYn_OEn    : out   std_logic;
17          DEVSELno     : out   std_logic;
18          DEVSELn_OEn  : out   std_logic;
19          PARo         : out   std_logic;
20          PAR_OEn      : out   std_logic;
21          INTAn        : out   std_logic);
22    end CORE;
23    architecture RTL of CORE is
24       component ROM
25       generic (
26          N, WORDS, M : integer;
27          CODEFILE    : string := "rom.in";
28          CPB         : integer;
29          BUFFER_SIZE : string);
30       port (
31          A    : in  std_logic_vector(M-1 downto 0);
32          DOUT : out std_logic_vector(N-1 downto 0));
33       end component;
34       component HSRAM
35       generic ( N, WORDS, M, BPC, FLOORPLAN : integer);
36       port (
37          A   : in  std_logic_vector(M-1 downto 0);
38          DIN : in  std_logic_vector(N-1 downto 0);
39          WR  : in  std_logic;
```

```
40          DOUT: out std_logic_vector(N-1 downto 0));
41       end component;
42       component SYN
43       port (
44          CLK           : in    std_logic;
45          RSTn          : in    std_logic;
46          ADi           : in    std_logic_vector(31 downto 0);
47          ADo           : out   std_logic_vector(31 downto 0);
48          AD_OEn        : out   std_logic_vector(1 downto 0);
49          CBEni         : in    std_logic_vector(3 downto 0);
50          IDSEL         : in    std_logic;
51          FRAMEni       : in    std_logic;
52          TRDYno        : out   std_logic;
53          TRDYn_OEn     : out   std_logic;
54          DEVSELno      : out   std_logic;
55          DEVSELn_OEn   : out   std_logic;
56          PARo          : out   std_logic;
57          PAR_OEn       : out   std_logic;
58          INTAn         : out   std_logic;
59             -- signals to from RAM, ROM
60          RAMOUT        : in    std_logic_vector(31 downto 0);
61          ROMOUT        : in    std_logic_vector(31 downto 0);
62          ADDR          : out   std_logic_vector(9 downto 0);
63          RAM_WRn       : out   std_logic);
64       end component;
65       -------------------------------------------------------------
66       signal RAMOUT, ROMOUT : std_logic_vector(31 downto 0);
67       signal ADDR           : std_logic_vector(9 downto 0);
68       signal RAM_WRn        : std_logic;
69    begin
70       hsram0 : HSRAM
71          generic map (
72             N => 32, WORDS => 1024, M => 10, BPC => 8, FLOORPLAN => 0)
73          port map (
74             A => ADDR, DIN  => ADi, WR => RAM_WRn, DOUT => RAMOUT);
75       rom0 : ROM
76          generic map (
77             N => 32, WORDS=>1024, M=>10, CODEFILE=>"rom.in", CPB =>8,
78             BUFFER_SIZE => "2")
79          port map (
80             A => ADDR, DOUT => ROMOUT);
81       syn0 : SYN
82       port map (
83          CLK           => CLK,       RSTn          => RSTn,
84          ADi           => ADi,       ADo           => ADo,
85          AD_OEn        => AD_OEn,    CBEni         => CBEni,
86          IDSEL         => IDSEL,     FRAMEni       => FRAMEni,
87          TRDYno        => TRDYno,    TRDYn_OEn     => TRDYn_OEn,
88          DEVSELno      => DEVSELno,  DEVSELn_OEn   => DEVSELn_OEn,
89          PARo          => PARo,      PAR_OEn       => PAR_OEn,
90          INTAn         => INTAn,     RAMOUT        => RAMOUT,
```

```
91          ROMOUT        => ROMOUT,    ADDR          => ADDR,
92          RAM_WRn       => RAM_WRn);
93    end RTL;
```

The preceding CORE VHDL code is straightforward. It declares three components and instantiates each of them. Note that signals RAMOUT, ROMOUT, ADDR, and RAM_WRn are declared in lines 66 to 68 to connect the ROM, HSRAM, and SYN block. The SYN block contains all circuits of the chip other than the ROM and RAM. The SYN block will be described in VHDL and synthesized. The physical layout of ROM and RAM blocks will be generated from the layout tool directly. Note that these ROM and RAM models are very simple. They have no timing checks since we are now interested only in the functional behavior. The following VHDL code models the HSRAM (high speed RAM, file name hsram.vhd). Note that BPC (bit per column) and FLOORPLAN generic are used for the layout tool. They do not affect the RAM function.

```
1     library IEEE;
2     use IEEE.std_logic_1164.all;
3     use IEEE.std_logic_arith.all;
4     entity HSRAM is
5        generic (
6           N         : integer := 32;
7           WORDS     : integer := 1024;
8           M         : integer := 10;
9           BPC       : integer := 2;
10          FLOORPLAN : integer := 0);
11       port (
12          A   : in  std_logic_vector(M-1 downto 0);
13          DIN : in  std_logic_vector(N-1 downto 0);
14          WR  : in  std_logic;
15          DOUT: out std_logic_vector(N-1 downto 0));
16    end HSRAM;
17    architecture BEH of HSRAM is
18       subtype ARRAYWORD is std_logic_vector((N - 1) downto 0);
19       type ARRAYTYPE is array(Natural range <>) of ARRAYWORD;
20       signal MEM : ARRAYTYPE (0 to (WORDS - 1));
21    begin
22       DOUT <= MEM(conv_integer(unsigned(A))) after 3 ns;
23       check : process
24       begin
25          wait until WR'event and WR = '1';
26          MEM(conv_integer(unsigned(A))) <= DIN;
27       end process;
28    end BEH;
```

The ROM VHDL code follows. Line 9 CODEFILE generic is used to initialize the ROM values in VHDL code (lines 29 to 36) with the text file "rom.in". The same file is also used in the layout tool to generate the physical ROM layout. Line 10 CPB (column per bit) generic and line 11 BUFFER_SIZE generic are also used for the layout tool and do not affect the ROM function.

```
1    library IEEE;
2    use IEEE.std_logic_1164.all;
3    use IEEE.std_logic_arith.all;
4    use IEEE.std_logic_textio.all;
5    use STD.TEXTIO.all;
6    entity ROM is
7       generic (
8          N, WORDS, M : integer;
9          CODEFILE    : string := "rom.in";
10         CPB         : integer;
11         BUFFER_SIZE : string);
12      port (
13         A    : in  std_logic_vector(M-1 downto 0);
14         DOUT : out std_logic_vector(N-1 downto 0));
15   end ROM;
16   architecture BEH of ROM is
17   begin
18      ---------------------------------------------------
19      check : process(A)
20         file MEMIN   : text is in CODEFILE;
21         subtype ARRAYWORD is bit_vector((N - 1) downto 0);
22         type    ARRAYTYPE is array(Natural range <>) of ARRAYWORD;
23         variable address   : integer;
24         variable INLINE    : line;
25         variable MEM       : ARRAYTYPE (0 to (WORDS - 1));
26         variable datatemp : bit_vector(N-1 downto 0);
27         variable READDONE : boolean := FALSE;
28      begin
29         if (not READDONE) then
30            for i in MEM'range loop
31               readline(MEMIN, INLINE);
32               read(INLINE, datatemp);
33               MEM(i) := datatemp;
34            end loop;
35            READDONE := TRUE;
36         end if;
37         if (IS_X(A)) then
38            DOUT <= (others => 'X');
39            assert FALSE report "address not valid" severity NOTE;
40         else
41            address := CONV_INTEGER(unsigned(A));
42            DOUT    <= To_stdlogicvector(MEM(address));
43         end if;
44      end process;
45   end BEH;
```

Note that the memory array MEM in line 25 is a one-dimensional array of length WORDS generic. This allows the generic WORDS to be mapped with a smaller number for initial testing. If we just want to check the reading and writing interface to the ROM or RAM, we do not need the whole array. The large memory requirements for the whole array slow the run time. To ensure the address in line 41 falls inside the range of 0 to WORDS – 1, it can be changed to

```
address := CONV_INTEGER(unsigned(A)) mod WORDS;
```

Now we are ready to design the SYN block. Block SYN is decomposed into blocks PCILOGIC, TARFSM, MEMLOGIC, TREEC. and REGMUX. The function of each block is summarized as follows. The design of each block will be discussed in more detail after we have an overview of each block's function.

Block TREEC: It is used to generate two-stage inverters for the clock tree and the reset signal tree. It uses two generics, WIDE1 and BUF, of type integer. Generic WIDE1 indicates the number of inverters in the first stage, and generic BUF indicates the number of output signals required. For example, the following inverter tree, as shown in Figure 13.5, is generated with WIDE1 => 2, BUF => 8. The fanout of the first stage inverters is more or less balanced. Each inverter output is connected to the clock signal of other blocks to balance the loading of the clock signals. This clock tree is used because the layout tool does not generate the clock tree automatically. However, it will automatically buffer size the inverters and minimize the clock skew in the layout process. This block is used (instantiated) two times for the clock signal tree and the reset signal tree. The VHDL code for TREEC is left as an exercise.

Block PCILOGIC: This block monitors the FRAMEn signal, latches in the address, decodes the address and bus command (CBEn) to determine whether to respond or not, sends a signal to wake up the finite-state machine if it is required to respond, controls and increases the address counter for multiple words transaction, and generates parity for the read cycle.

Block TARFSM: This is a finite-state machine to respond to the PCI bus. It generates PCI related signals such as for DEVSELn and TRDYn.

Block MEMLOGIC: This block implements the TIMER_REG and TIMER_PROG registers and the TIMER_CNTR.

Block REGMUX: This block implements PCI configuration registers, and, based on the address, selects data from RAM, ROM, PCI configuration registers or other registers to respond to PCI bus during the read cycle.

With an overview of each block, the following shows the SYN VHDL code. This code is basically a collection of interconnections of the blocks through component instantiation statements. From the design point of view, this code may iterate up and down with the implementation of its lower-level blocks. Note that each line is more utilized to reduce the length of this code. In practice, you may have a set of coding guidelines.

```
1     library IEEE;
2     use IEEE.std_logic_1164.all;
3     entity SYN is
4        port (
5           CLK         : in   std_logic;
6           RSTn        : in   std_logic;
7           ADi         : in   std_logic_vector(31 downto 0);
8           CBEni       : in   std_logic_vector(3 downto 0);
9           IDSEL       : in   std_logic;
10          FRAMEni     : in   std_logic;
```

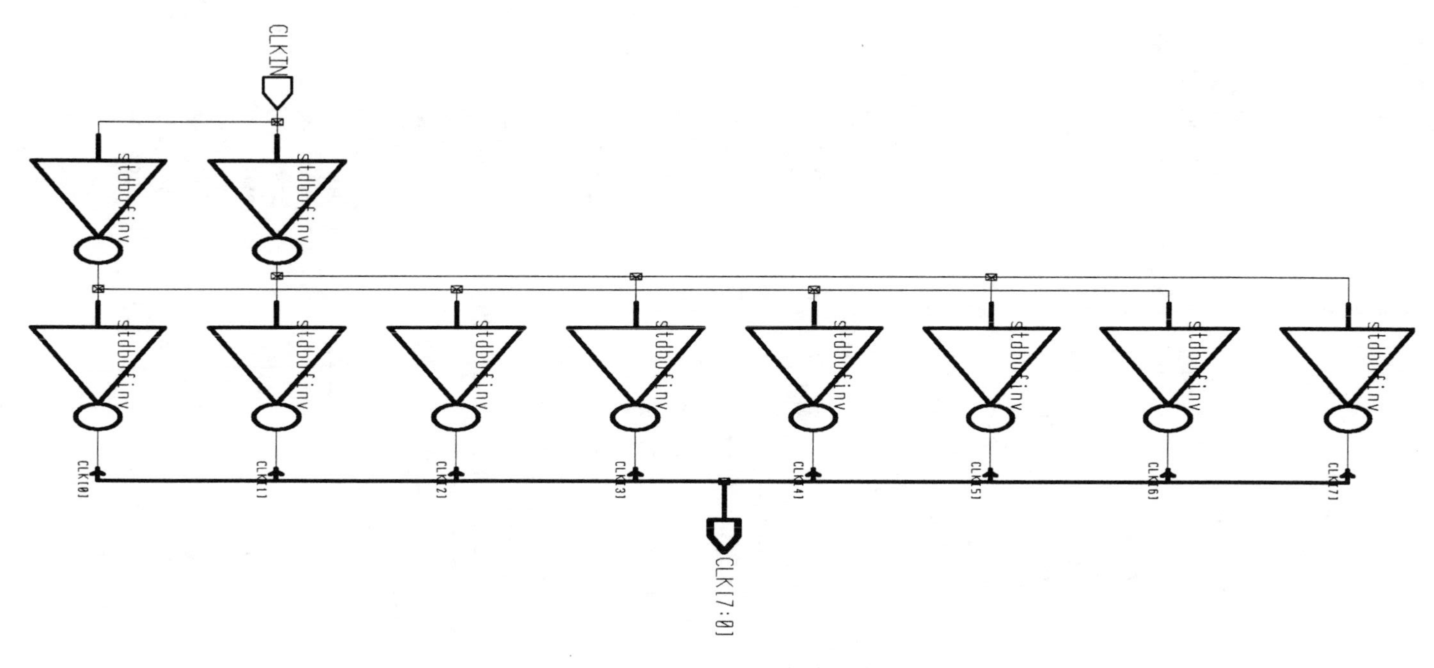

FIGURE 13.5 Inverter tree for clock and reset signals.

```
11         ADo           : out   std_logic_vector(31 downto 0);
12         AD_OEn        : out   std_logic_vector(1 downto 0);
13         TRDYno        : out   std_logic;
14         TRDYn_OEn     : out   std_logic;
15         DEVSELno      : out   std_logic;
16         DEVSELn_OEn   : out   std_logic;
17         PARo          : out   std_logic;
18         PAR_OEn       : out   std_logic;
19         INTAn         : out   std_logic;
20         RAMOUT        : in    std_logic_vector(31 downto 0);
21         ROMOUT        : in    std_logic_vector(31 downto 0);
22         ADDR          : out   std_logic_vector(11 downto 2);
23         RAM_WRn       : out   std_logic);
24   end SYN;
25   architecture RTL of SYN is
26      component PCILOGIC
27      port (
28         CLK, CLK1, RSTn, RSTn1  : in  std_logic;
29         IDSEL, FRAMEni          : in  std_logic;
30         RESPOND_EN, TAR_ACTIVE  : in  std_logic;
31         ADDR_CNTR_EN, TRDYno    : in  std_logic;
32         ADi, ADof               : in  std_logic_vector(31 downto 0);
33         CBEni                   : in  std_logic_vector(3 downto 0);
34         BASEADDR                : in std_logic_vector(31 downto 14);
35         ADDR                    : out std_logic_vector(11 downto 2);
36         ADDR13_12               : out std_logic_vector( 1 downto 0);
37         PARo, PAR_OEn, TAR_WRITE: out std_logic;
38         HIT, CFG_CYCLE          : out std_logic);
39      end component;
40      component TARFSM
41      port (
42         CLK, RSTn, FRAMEni, HIT, TAR_WRITE : in  std_logic;
43         DEVSELno, TRDYn_OEn, TRDYno        : out std_logic;
44         ADDR_CNTR_EN, WRn                  : out std_logic;
45         AD_OEn              : out std_logic_vector(1 downto 0);
46         DEVSELn_OEn, TLATCH_EN, TAR_ACTIVE : out std_logic);
47      end component;
48      component MEMLOGIC
49      port (
50         CLK, CLK1, CLK2         : in  std_logic;
51         RSTn, RSTn1, RSTn2, WRn : in  std_logic;
52         CFG_CYCLE               : in  std_logic;
53         ADDR13_12               : in std_logic_vector(13 downto 12);
54         ADDR                    : in std_logic_vector(11 downto  2);
55         ADi                     : in std_logic_vector(31 downto  0);
56         RAM_WRn, CFG_WRn, INTAn : out std_logic;
57         TIMER_REG, TIMER_CNTR   : out std_logic_vector(31 downto 0);
58         TIMER_PROG              : out std_logic_vector( 7 downto 0));
59      end component;
60      component REGMUX
61      port (
62         CLK, CLK1, RSTn, RSTn1: in  std_logic;
63         CFG_WRn, TLATCH_EN    : in  std_logic;
```

```
64          ADDR                   : in  std_logic_vector(11 downto 2);
65          ADDR13_12              : in  std_logic_vector(13 downto 12);
66          ADi, RAMOUT, ROMOUT    : in  std_logic_vector(31 downto 0);
67          TIMER_REG, TIMER_CNTR  : in  std_logic_vector(31 downto 0);
68          TIMER_PROG             : in  std_logic_vector( 7 downto 0);
69          RESPOND_EN             : out std_logic;
70          BASEADDR               : out std_logic_vector(31 downto 14);
71          ADo                    : out std_logic_vector(31 downto 0));
72       end component;
73       component TREEC
74       generic(
75          WIDE1  : integer range 1 to 10 := 4;
76          BUF    : integer range 1 to 100 := 20);
77       port(
78          CLKIN  : in  std_logic; -- input clock
79          CLK    : out std_logic_vector(BUF-1 downto 0));
80       end component;
81       ---------------------------------------------------------
82       constant NUMBER : integer := 8;
83       signal CLOCK  : std_logic_vector(NUMBER - 1 downto 0);
84       signal RESETn : std_logic_vector(NUMBER - 1 downto 0);
85       signal RESPOND_EN, CFG_CYCLE, CFG_WRn   : std_logic;
86       signal HIT, TRDYno_TEMP, ADDR_CNTR_EN, TLATCH_EN : std_logic;
87       signal WRn, TAR_ACTIVE, TAR_WRITE     : std_logic;
88       signal BASEADDR                     : std_logic_vector(31 downto 14);
89       signal ADDR13_12                    : std_logic_vector(13 downto 12);
90       signal ADDR_TEMP                    : std_logic_vector(11 downto 2);
91       signal ADo_TEMP                     : std_logic_vector(31 downto 0);
92       signal TIMER_REG, TIMER_CNTR        : std_logic_vector(31 downto 0);
93       signal TIMER_PROG                   : std_logic_vector( 7 downto 0);
94    begin
95       TRDYno <= TRDYno_TEMP;
96       ADo    <= ADo_TEMP;
97       ADDR   <= ADDR_TEMP;
98       ---------------------------------------------------------
99       clktree : TREEC
100      generic map (
101         WIDE1        => 2,              BUF          => NUMBER)
102      port map (
103         CLKIN        => CLK,            CLK          => CLOCK);
104      rsttree : TREEC
105      generic map (
106         WIDE1        => 2,              BUF          => NUMBER)
107      port map (
108         CLKIN        => RSTn,           CLK          => RESETn);
109      pcilogic0 : PCILOGIC
110      port map (
111         CLK          => CLOCK(0),       CLK1         => CLOCK(1),
112         RSTn         => RESETn(0),      RSTn1        => RESETn(1),
113         IDSEL        => IDSEL,          FRAMEni      => FRAMEni,
114         RESPOND_EN   => RESPOND_EN,     TAR_ACTIVE   => TAR_ACTIVE,
115         ADDR_CNTR_EN => ADDR_CNTR_EN,   TRDYno       => TRDYno_TEMP,
```

```
116          ADi          => ADi,          ADof         => ADo_TEMP,
117          CBEni        => CBEni,        BASEADDR     => BASEADDR,
118          ADDR         => ADDR_TEMP,    ADDR13_12    => ADDR13_12,
119          PARo         => PARo,         PAR_OEn      => PAR_OEn,
120          TAR_WRITE    => TAR_WRITE,    HIT          => HIT,
121          CFG_CYCLE    => CFG_CYCLE );
122       tarfsm0 : TARFSM
123       port map (
124          CLK          => CLOCK(2),     RSTn         => RESETn(2),
125          FRAMEni      => FRAMEni,      HIT          => HIT,
126          TAR_WRITE    => TAR_WRITE,    DEVSELno     => DEVSELno,
127          TRDYn_OEn    => TRDYn_OEn,    TRDYno       => TRDYno_TEMP,
128          ADDR_CNTR_EN => ADDR_CNTR_EN, WRn          => WRn,
129          AD_OEn       => AD_OEn,       DEVSELn_OEn  => DEVSELn_OEn,
130          TLATCH_EN    => TLATCH_EN,    TAR_ACTIVE   => TAR_ACTIVE);
131       memlogic0 : MEMLOGIC
132       port map (
133          CLK          => CLOCK(3),     CLK1         => CLOCK(4),
134          CLK2         => CLOCK(5),     RSTn         => RESETn(3),
135          RSTn1        => RESETn(4),    RSTn2        => RESETn(5),
136          WRn          => WRn,          CFG_CYCLE    => CFG_CYCLE,
137          ADDR13_12    => ADDR13_12,    ADDR         => ADDR_TEMP,
138          ADi          => ADi,          RAM_WRn      => RAM_WRn,
139          CFG_WRn      => CFG_WRn,      INTAn        => INTAn,
140          TIMER_REG    => TIMER_REG,    TIMER_CNTR   => TIMER_CNTR,
141          TIMER_PROG   => TIMER_PROG );
142       regmux0 : REGMUX
143       port map (
144          CLK          => CLOCK(6),     CLK1         => CLOCK(7),
145          RSTn         => RESETn(6),    RSTn1        => RESETn(7),
146          CFG_WRn      => CFG_WRn,      TLATCH_EN    => TLATCH_EN,
147          ADDR         => ADDR_TEMP,    ADDR13_12    => ADDR13_12,
148          ADi          => ADi,          RAMOUT       => RAMOUT,
149          ROMOUT       => ROMOUT,       TIMER_REG    => TIMER_REG,
150          TIMER_CNTR   => TIMER_CNTR,   TIMER_PROG   => TIMER_PROG,
151          RESPOND_EN   => RESPOND_EN,   BASEADDR     => BASEADDR,
152          ADo          => ADo_TEMP );
153    end RTL;
```

The preceding SYN VHDL code describes the same information as the schematic in Figure 13.6. More signals are declared in lines 83 to 93 to connect among blocks that are not declared as port signals. Component TREEC is instantiated two times in lines 99 to 108. Other than PCI related signals declared as the port signals, input port signals RAMOUT and ROMOUT are the data coming from the output of the HSRAM and ROM, respectively, and the address ADDR (10 bits to address 1024 words) goes out to address the HSRAM and ROM. Output port signal RAM_WRn is the write-enable low active pulse for writing the HSRAM. In line 90, the ADDR_TEMP signal is declared. It is used in the port mapping in lines 137 and 147. It is then assigned to ADDR output signal in line 97. Similarly, TRDYno_TEMP is declared in line 86 and used in lines 115, 127, and 95. ADo_TEMP is declared in line 91 and used in lines 116, 152, and 96.

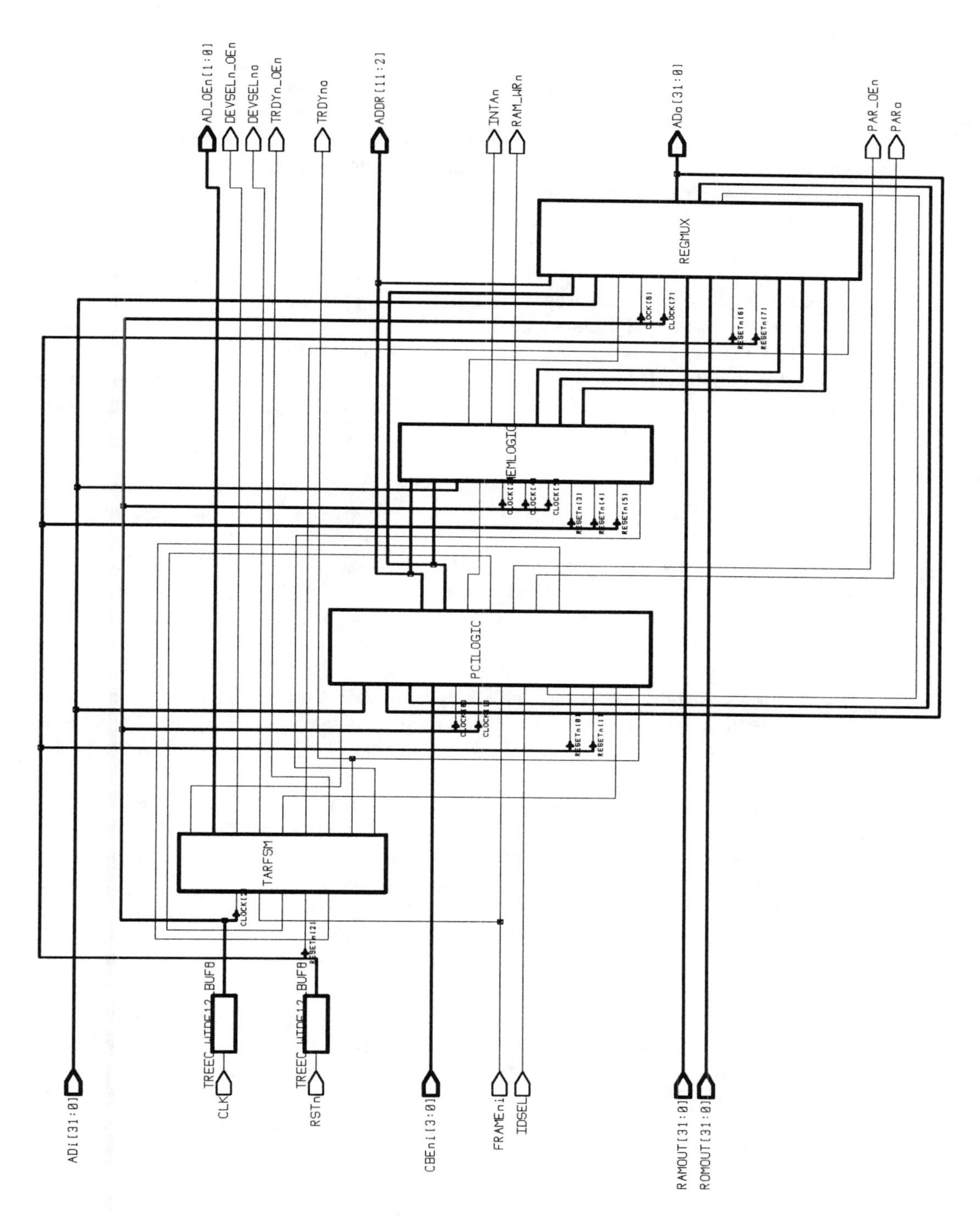

FIGURE 13.6 SYN block diagram.

The block TARFSM is an important block which generates many control signals for other blocks. The waveform in Figure 13.7 summarizes the operation of TARFSM for the write cycle. TARFSM is normally in the IDLE state. PCILOGIC block gets the address when FRAMEn is asserted low. It decodes the address and CBEn and drives signal HIT high. TARFSM wakes up when HIT = '1' and moves to state PREPARE. It asserts TAR_ACTIVE signal to indicate the PCI target finite state machine is active. The PCI signal DEVSELn should be asserted '0' also. During the TRANSFER state, TRDYn is asserted, signal WRn is asserted, and the address counter is enabled (ADDR_CNTR_EN) to be increased by 1. The address is ready in the PREPARE state. WRn signal is used to enable writing to registers. WRn is also used to generate the RAM_WRn write-enable pulse of half clock cycle wide. When FRAMEn is deasserted to '1', the last data is transferred. It moves to the turnaround state TURN_AR, and then goes back to IDLE state again.

The read cycle is similar to the write cycle. CHIP should drive AD bus to respond with data. AD bus is released by the PCI master after the address is put in the AD bus. AD bus is tristated after that until AD is driven by CHIP when CHIP starts to assert DEVSELn. WRn is deasserted high during the whole read cycle. Signal TLATCH_EN enables the data to be clocked into flipflops to go out to AD bus during the read cycle (it is low for the whole write cycle). The other signals are the same as in the write cycle. Figure 13.8 shows the read cycle waveform. Note the same three data values are read back.

The following VHDL code shows the implementation of the TARFSM block. Lines 5 to 9 define the input and output port signals. FRAMEni is the FRAMEn input signal. HIT comes from the PCILOGIC block to wake up TARFSM state machine. TAR_WRITE signal from the PCILOGIC block indicates a write cycle when it is '1' (a read cycle when it is '0'). DEVSELno and DEVSELn_OEn are the output signals that go to the tristate output pad for DEVSELn. TRDYno and TRDYn_OEn are the signal pairs that go to tristate pad for TRDYn. In practice, related signals are better to be declared together. AD_OEn (2-bit) is used to control the bidirectional AD bus pads. Each 1 bit drives 16 pads. ADDR_CNTR_EN goes to PCILOGIC block to enable the address counter. WRn goes to MEMLOGIC block to enable writing to registers and to generate a half clock wide RAM_WRn signal for writing RAM. TAR_ACTIVE signal indicates that the TARFSM state machine is active. TLATCH_EN goes to the PCILOGIC block to enable the clocking of data to go out to the AD bus during the read cycle. Line 12 declares the type STATE_TYPE for the states of the state machine. Line 13 declares the signals STATE and NEXT_STATE based on the STATE_TYPE. Lines 14 and 15 declare couple signals to communicate between concurrent statements.

To make the response time faster for the output signals after the rising edge of the clock, all output signals are flipflop outputs. Lines 17 to 42 generate all flipflops in this block. Lines 45 to 74 are another process statement to generate combinational circuits to go to the data input of flipflops and to determine the state transitions. Lines 47 to 49 set the default values. Note that signals DEVSELno, DEVSELn_OEn, and TRDYn_OEn are the same in lines 22 to 24 and lines 33 to 35. They can be combined so that one flipflop is used rather than three flipflops. The state transition in the case statement (lines 50 to 73) is straightforward. Note that the state machine is a MEALY FSM such that the state machine output (for example, TLATCH_EN_comb in line 69) depends both on input signals (for example, FRAMEni) and the state value.

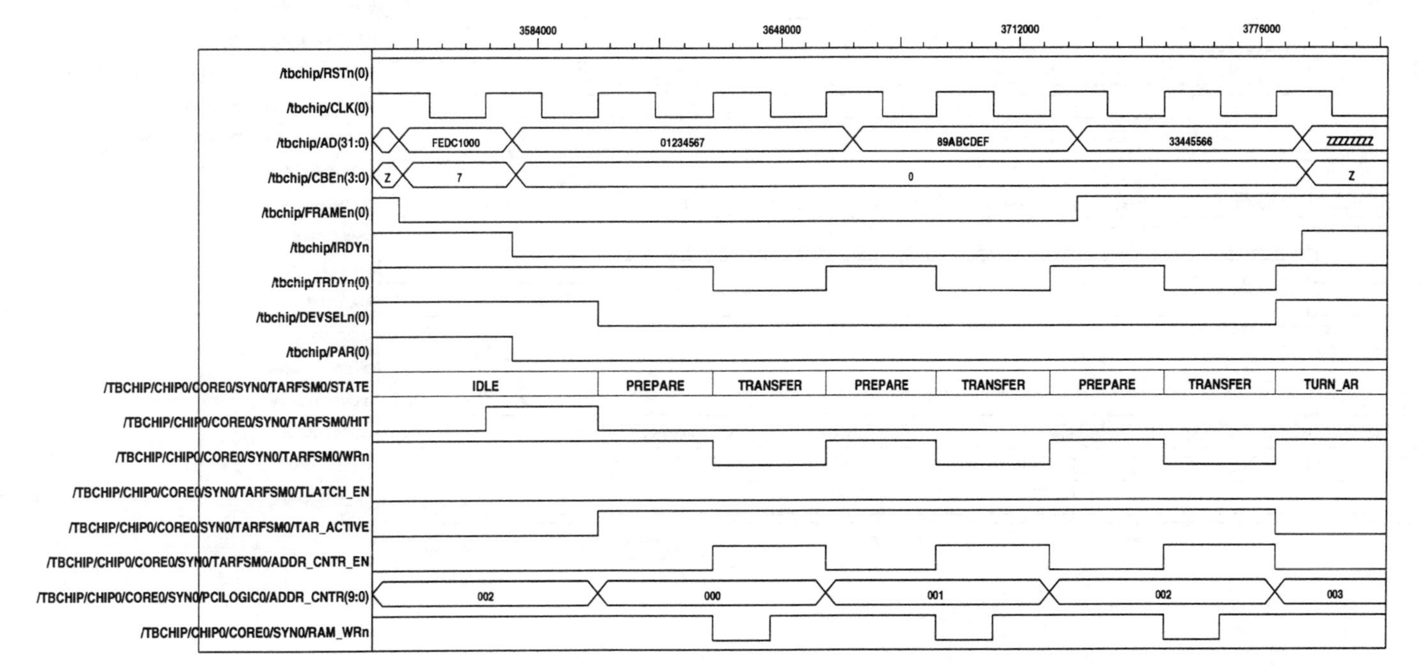

FIGURE 13.7 Write cycle detailed waveform.

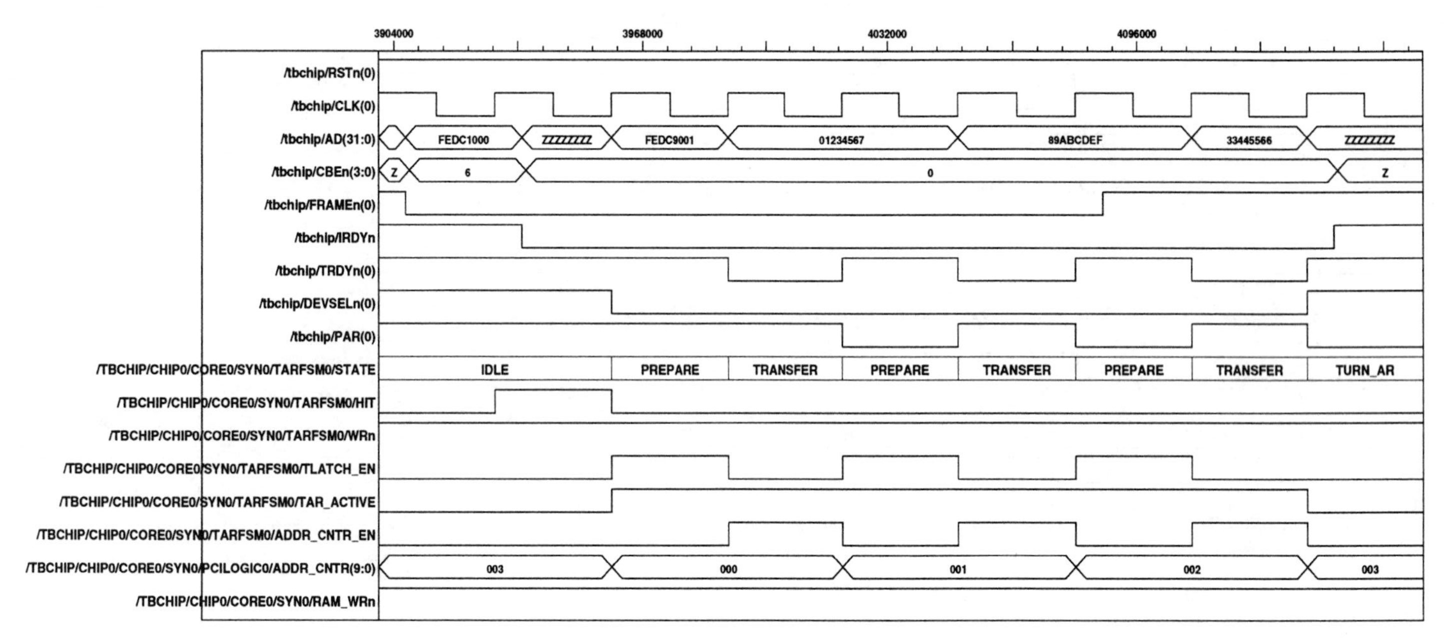

FIGURE 13.8 Read cycle detailed waveform.

```
1     library IEEE;
2     use IEEE.std_logic_1164.all;
3     entity TARFSM is
4        port (
5           CLK, RSTn, FRAMEni, HIT, TAR_WRITE : in  std_logic;
6           DEVSELno, TRDYn_OEn, TRDYno        : out std_logic;
7           ADDR_CNTR_EN, WRn                  : out std_logic;
8           AD_OEn               : out std_logic_vector(1 downto 0);
9           DEVSELn_OEn, TLATCH_EN, TAR_ACTIVE : out std_logic);
10    end TARFSM;
11    architecture RTL of TARFSM is
12       type state_type is (IDLE, PREPARE, TRANSFER, TURN_AR);
13       signal STATE, NEXT_STATE : state_type;
14       signal TAR_ACTIVE_comb, READ_comb, TLATCH_En_comb : std_logic;
15       signal TRDYno_comb                                : std_logic;
16    begin
17       ffs_gen : process(CLK, RSTn)
18       begin
19          if (RSTn = '0') then
20             STATE        <= IDLE;
21             TAR_ACTIVE   <= '0';
22             DEVSELno     <= '1';
23             DEVSELn_OEn  <= '1';
24             TRDYn_OEn    <= '1';
25             TLATCH_EN    <= '0';
26             TRDYno       <= '1';
27             WRn          <= '1';
28             ADDR_CNTR_EN <= '0';
29             AD_OEn       <= "11";
30          elsif (CLK'event and CLK = '1') then
31             STATE        <= NEXT_STATE;
32             TAR_ACTIVE   <= TAR_ACTIVE_comb;
33             DEVSELno     <= not TAR_ACTIVE_comb;
34             DEVSELn_OEn  <= not TAR_ACTIVE_comb;
35             TRDYn_OEn    <= not TAR_ACTIVE_comb;
36             TLATCH_EN    <= TLATCH_EN_comb;
37             TRDYno       <= TRDYno_comb;
38             WRn          <= TRDYno_comb or (not TAR_WRITE);
39             ADDR_CNTR_EN <= not TRDYno_comb;
40             AD_OEn       <= READ_comb & READ_comb;
41          end if;
42       end process;
43       READ_comb <= TAR_ACTIVE_comb nand (not TAR_WRITE);
44
45       comb : process(STATE , FRAMEni, HIT, TAR_WRITE)
46       begin
47          TRDYno_comb      <= '1';
48          TLATCH_EN_comb   <= '0';
49          TAR_ACTIVE_comb  <= '0';
50          case STATE  is
51             when IDLE  =>
52                if (HIT = '1') then
```

```
53                    TAR_ACTIVE_comb <= '1';
54                    NEXT_STATE      <= PREPARE;
55                    TLATCH_EN_comb  <= not TAR_WRITE;
56                 else
57                    NEXT_STATE      <= IDLE;
58                 end if;
59              when PREPARE   =>
60                 TRDYno_comb     <= '0';
61                 TAR_ACTIVE_comb <= '1';
62                 NEXT_STATE      <= TRANSFER;
63              when TRANSFER  =>
64                 if (FRAMEni = '1') then
65                    NEXT_STATE <= TURN_AR;
66                 else
67                    TAR_ACTIVE_comb <= '1';
68                    NEXT_STATE      <= PREPARE;
69                    TLATCH_EN_comb  <= not TAR_WRITE;
70                 end if;
71              when TURN_AR   =>
72                 NEXT_STATE      <= IDLE;
73           end case;
74        end process;
75     end RTL;
```

Now, we come to the PCILOGIC block. The following shows the PCILOGIC VHDL entity. Lines 8 to 18 declare input and output port signals. CLK and CLK1 (RSTn and RSTn1) comes from the clock signal tree (reset signal tree) so that about 32 flipflops are connected to one clock (one reset) signal (from the output of the two-stage inverters tree). IDSEL, FRAMEni, CBEni, ADi are from input pads directly. IDSEL = '1', CBEn = "1010" (or CBEn = "1011"), and AD(1 downto 0) = "00" are used to indicate a PCI configuration read (or write) cycle. RESPOND_EN and BASEADDR are from the PCI configuration registers. TAR_ACTIVE, ADDR_CNTR_EN, TRDYno are from block TARFSM. ADof is the data for driving AD bus during the read cycle. It is also used to generate an even parity bit PARo one clock cycle after the TRDYn is asserted. PAR_OEn controls the tristate PAR output pad. TAR_WRITE indicates a write cycle when it is '1' ('0' indicates a read cycle). HIT signal goes to TARFSM block to wake up the TARFSM state machine. CFG_CYCLE indicates a PCI configuration cycle (either read or write). The PCI configuration registers can only be updated when CFG_CYCLE is asserted high.

```
1     library IEEE, SYNOPSYS;
2     use IEEE.std_logic_1164.all;
3     use IEEE.std_logic_arith.all;
4     use SYNOPSYS.attributes.all;
5     use work.PACK.all;
6     entity PCILOGIC is
7        port (
8           CLK, CLK1, RSTn, RSTn1    : in  std_logic;
9           IDSEL, FRAMEni            : in  std_logic;
10          RESPOND_EN, TAR_ACTIVE    : in  std_logic;
```

```
11          ADDR_CNTR_EN, TRDYno      : in  std_logic;
12          ADi, ADof                 : in  std_logic_vector(31 downto 0);
13          CBEni                     : in  std_logic_vector(3 downto 0);
14          BASEADDR                  : in  std_logic_vector(31 downto 14);
15          ADDR                      : out std_logic_vector(11 downto 2);
16          ADDR13_12                 : out std_logic_vector( 1 downto 0);
17          PARo, PAR_OEn, TAR_WRITE: out std_logic;
18          HIT, CFG_CYCLE            : out std_logic);
19     end PCILOGIC;
```

The waveform in Figure 13.9 illustrates the implementation of PCILOGIC block. The input ADi, CBEni, IDSEL, FRAMEni are all clocked every clock (rising edge) into ADif, CBEnif, IDSELf, and FRAMEnif flipflops. As shown in the waveform, they are exactly one clock behind. DECODE signal is generated exactly one clock cycle wide to indicate the address decoding phase. FRAMEnif is clocked into FRAMEnif_half at the falling edge of the clock. A half cycle wide ALE pulse is generated to enable the address latch (TAR_ADDR) to latch in the address in ADif and to latch the bus command in CBEnif to TAR_CBEn. Based on the TAR_ADDR, TAR_CBEn, RESPOND_EN, IDSELf, signal HIT is generated which wakes up TAR-FSM state machine.

To generate the even parity, ADof (for PCI, TAR_CBEn should be included) is used with flipflop output PARo. PAR_OEn is the delay of TRDYno during the read cycle.

The PCILOGIC block circuits are described in Figure 13.10. The rectangles denote flipflops signals. FRAMEnif_half is clocked at the falling edge of the clock. Shaded rectangles represent latches. The rounded corner block is for the decoding and other combinational circuits.

Internal signals are declared in lines 23 to 28. Lines 36 to 43 generate 32-bit flipflops ADif to clock in ADi every clock so that the data can be obtained correctly with the very minimum hold time (0 hold time) for the AD bus. CBEnif (lines 48 and 57) clocks in the CBEni input for the same reason. ADif has 32 bits already. The rest of the flipflops are put into another process (lines 44 to 78) using another clock signal CLK1 and reset signal RSTn1. Lines 49 to 51 and lines 57 to 59 generate FRAMEnif, FRAMEnif_delay, IDSELf flipflops. FRAMEnif_half is generated in a separate process due to its falling edge of the clock in lines 79 to 86.

```
20     architecture RTL of PCILOGIC is
21        attribute async_set_reset of RSTn1   : signal is "true";
22        attribute async_set_reset of RSTn    : signal is "true";
23        signal ADDR_CNTR              : std_logic_vector(9 downto 0);
24        signal ADif, TAR_ADDR         : std_logic_vector(31 downto 0);
25        signal CBEnif, TAR_CBEn       : std_logic_vector(3 downto 0);
26        signal DECODE, CFG, CBEn67, ALE, IDSELf         : std_logic;
27        signal FRAMEnif, FRAMEnif_delay, FRAMEnif_half : std_logic;
28        signal ADDR_MATCH, CFG_CYCLE_FF                 : std_logic;
29     begin
30        ADDR       <= ADDR_CNTR when CFG_CYCLE_FF = '0' else
31                         "000000" & ADDR_CNTR(3 downto 0);
32       ADDR13_12<= TAR_ADDR(13 downto 12) when CFG_CYCLE_FF='0' else "10";
```

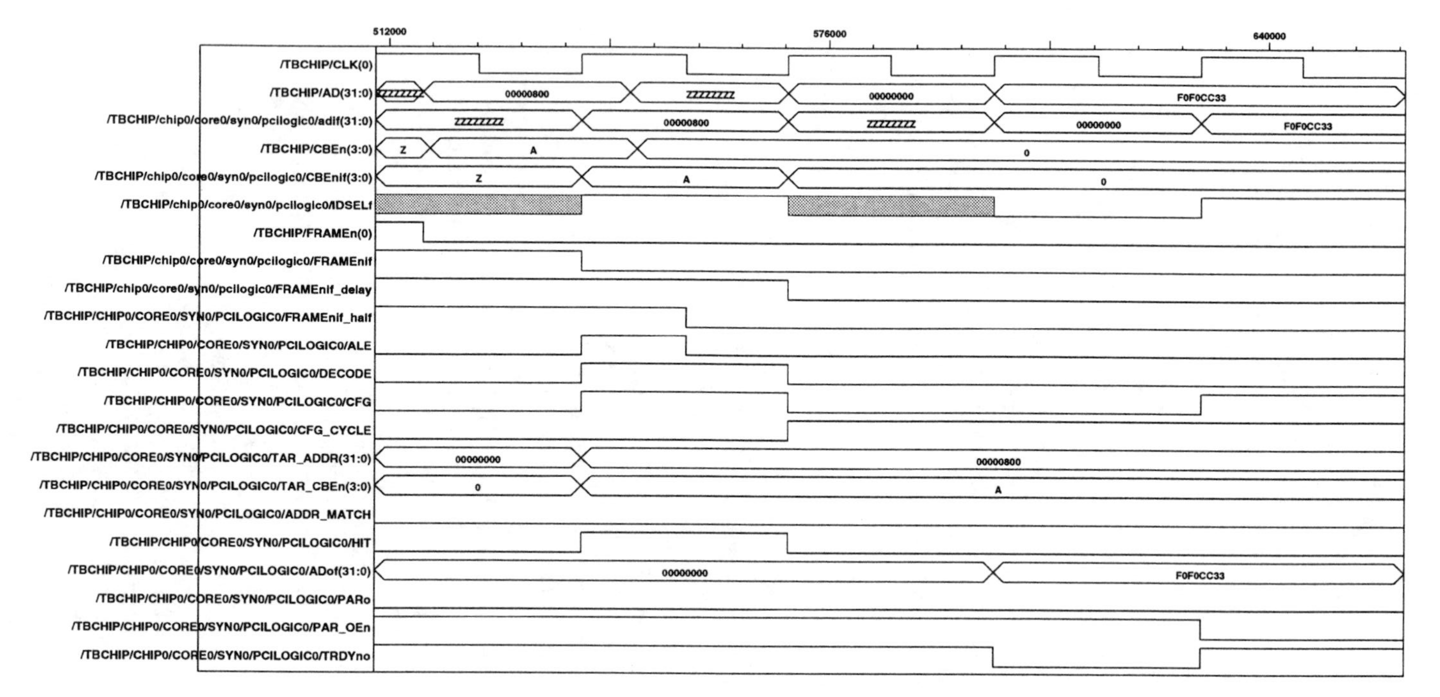

FIGURE 13.9 PCILOGIC block signals waveform.

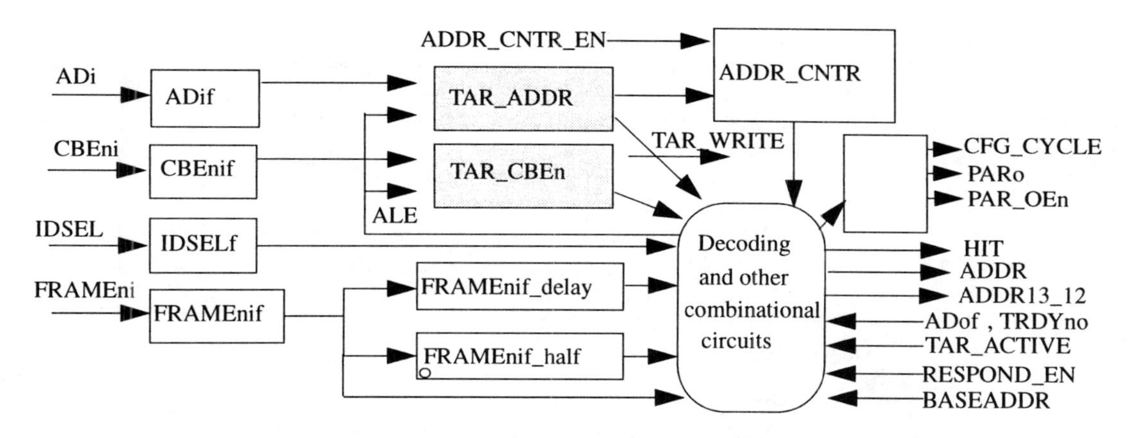

FIGURE 13.10 PCILOGIC block circuit diagram.

```
33      TAR_WRITE <= TAR_CBEn(0);
34      CFG_CYCLE <= CFG_CYCLE_FF;
35      ----------------------------------------------------------
36      ffs0 : process(RSTn, CLK)
37      begin
38         if (RSTn = '0') then
39            ADif     <= (ADif'range => '0');
40         elsif (CLK'event and CLK = '1') then
41            ADif     <= ADi;
42         end if;
43      end process;
44      ffs1 : process(RSTn1, CLK1)
45         variable TEMP : std_logic;
46      begin
47         if (RSTn1 = '0') then
48            CBEnif          <= "0000";
49            FRAMEnif        <= '1';
50            FRAMEnif_delay  <= '1';
51            IDSELf          <= '0';
52            PARo            <= '0';
53            PAR_OEn         <= '1';
54            ADDR_CNTR       <= "0000000000";
55            CFG_CYCLE_FF    <= '0';
56         elsif (CLK1'event and CLK1 = '1') then
57            CBEnif          <= CBEni;
58            FRAMEnif        <= FRAMEni;
59            FRAMEnif_delay  <= FRAMEnif;
60            IDSELf          <= IDSEL;
61            TEMP := ADof(0);
62            for i in 31 downto 1 loop
63               TEMP := TEMP xor ADof(i);
64            end loop;
65            PARo            <= TEMP;
66            PAR_OEn         <= TRDYno or TAR_CBEn(0);
67            if (DECODE = '1') then
68               ADDR_CNTR    <= TAR_ADDR(11 downto 2);
```

```
69              elsif (ADDR_CNTR_EN = '1') then
70                 ADDR_CNTR     <= unsigned(ADDR_CNTR) + 1;
71              end if;
72              if (CFG and DECODE) = '1' then
73                 CFG_CYCLE_FF <= '1';
74              elsif (TAR_ACTIVE = '0') then
75                 CFG_CYCLE_FF    <= '0';
76              end if;
77           end if;
78        end process;
79        ffs2 : process(RSTn1, CLK1)
80        begin
81           if (RSTn1 = '0') then
82              FRAMEnif_half <= '1';
83           elsif (CLK1'event and CLK1 = '0') then
84              FRAMEnif_half <= FRAMEnif;
85           end if;
86        end process;
87        DECODE  <= FRAMEnif nor (not FRAMEnif_delay);
88        ALE     <= FRAMEnif nor (not FRAMEnif_half);
89        LATCHR (RSTn, ADif,    ALE, TAR_ADDR);
90        LATCHR (RSTn, CBEnif, ALE, TAR_CBEn);
91        ---------------------------------------------------------
92        CBEn67<='1' when TAR_CBEn = "0111" or TAR_CBEn = "0110" else '0';
93        CFG     <= '1' when (TAR_CBEn = "1010" or TAR_CBEn = "1011") and
94                (IDSELf = '1') and (TAR_ADDR(1 downto 0) = "00") else '0';
95        ADDR_MATCH<='1' when TAR_ADDR(31 downto 14) = BASEADDR else '0';
96        HIT<=DECODE and ((ADDR_MATCH and CBEn67 and RESPOND_EN) or CFG);
97     end RTL;
```

Line 87 generates one clock cycle wide DECODE pulse. Line 88 generates half clock cycle wide ALE pulse to enable the latches. The HIT is determined in lines 92 to 96. Lines 89 and 90 are concurrent procedure call statements which infer latches. The procedure LATCHR is declared in package PACK which is referenced in line 5. Lines 61 to 64 generate the even parity. Note that we assume the CBEn is "0000" during the data phase. Otherwise, TAR_CBEn should be included in the parity generation. Lines 72 to 76 set up CFG_CYCLE when CFG is asserted in the DECODE phase. It is cleared when the cycle is ended. Lines 67 to 71 parallel load the address counter in the DECODE phase and increase by 1 when the signal ADDR_CNTR_EN is set by the TARFSM block. Lines 30 to 32 generate address to go out to HSRAM, ROM, and registers. The address is split into a 10-bit word (32-bit) address ADDR due to 1024 words for each section of the memory space. ADDR13_12 is used to select one of four sections of the memory. The address for the configuration cycle would be changed since we make the PCI configuration registers to a fixed address inside CHIP which should be independent of the PCI bus, otherwise, the address would be the output of the address counter. Note that lines 21 and 22 specify attributes to tell the synthesis tool that signal RSTn and RSTn1 are asynchronous reset signals which should be connected to asynchronous reset input of flipflops and latches. The attribute is defined in package attributes of library SYNOPSYS, which are referenced in lines 1 and 4.

The major function for the MEMLOGIC block is to implement registers TIMER_REG, TIMER_CNTR, and TIMER_PROG. Input signal WRn comes from TARFSM block to indicate a write operation. Input signal CFG_CYCLE comes from PCILOGIC block to indicate a PCI configuration cycle. ADDR13_12 and ADDR are the addresses for decoding to reference the right memory space. ADi is directly from the AD bus to be written to registers. Output signal RAM_WRn is a half clock cycle wide pulse to write the HSRAM. Output signal CFG_WRn goes to the REGMUX block to enable writing to PCI configuration registers. Registers TIMER_REG, TIMER_CNTR, and TIMER_PROG output signals go to block REGMUX to be selected to go to AD bus for the read cycle. INTAn is the output interrupt signal that is asserted when TIMER_CNTR reaches its terminal count (time out). These port signals are declared in lines 6 to 14.

```
1    library IEEE;
2    use IEEE.std_logic_1164.all;
3    use IEEE.std_logic_arith.all;
4    entity MEMLOGIC is
5       port (
6          CLK, CLK1, CLK2           : in  std_logic;
7          RSTn, RSTn1, RSTn2, WRn   : in  std_logic;
8          CFG_CYCLE                 : in  std_logic;
9          ADDR13_12                 : in  std_logic_vector(13 downto 12);
10         ADDR                      : in  std_logic_vector(11 downto  2);
11         ADi                       : in  std_logic_vector(31 downto  0);
12         RAM_WRn, CFG_WRn, INTAn   : out std_logic;
13         TIMER_REG, TIMER_CNTR     : out std_logic_vector(31 downto 0);
14         TIMER_PROG                : out std_logic_vector( 7 downto 0));
15   end MEMLOGIC;
```

Lines 17 to 25 declare a synchronous 32-bit down counter with parallel load (PL) and counter enable (CIN). It is asynchronously preset to all '1's, initially. If PL = '1', the counter is loaded with DIN input. DOUT indicates the counter value. TC is asserted when the counter reaches its terminal count. Local signals are declared in lines 27 to 31.

Based on the memory map and CFG_CYCLE, REG_WRn, and CFG_WRn are generated in lines 33 to 35. Line 36 generates RAM_WRn_full signal to indicate a full cycle wide write pulse. It is then delayed by half clock in lines 38 to 45 to get the signal RAM_WRn_delay. RAM_WRn is then generated in line 37 to get the half clock wide RAM_WRn signal.

Line 47 generates a signal REG_ADDR to ensure ADDR(11 downto 4) are all '0' for writing to the three registers. Lines 48 to 58 infer the TIMER_REG register. It is written with the value ADi when REG_ADDR is '1', and the lower 2 bits of ADDR are "00", and REG_WRn is asserted '0'. Line 59 connects the TIMER_REG_FF signal to the output port TIMER_PROG.

```
16   architecture RTL of MEMLOGIC is
17      component SCNTEDP32
```

```
18      port (
19         CLK   : in  std_logic;
20         RSTn  : in  std_logic;
21         CIN   : in  std_logic; -- tie to count enable
22         PL    : in  std_logic;
23         DIN   : in  std_logic_vector(31 downto 0);
24         TC    : out std_logic;
25         DOUT  : out std_logic_vector(31 downto 0));
26      end component;
27      signal REG_WRn,RAM_WRn_full,RAM_WRn_delay, REG_ADDR : std_logic;
28      signal TIMER_CNTR_EN, TIMER_CNTR_PL, TIMER_CNTR_TC : std_logic;
29      signal TIME_OUT : std_logic;
30      signal TIMER_REG_FF,TIMER_CNTR_FF:std_logic_vector(31 downto 0);
31   signal TIMER_PROG_FF,TIMER_PROG_CNTR:std_logic_vector(7 downto 0);
32   begin
33      REG_WRn      <= WRn or (ADDR13_12(13) nand ADDR13_12(12)) ;
34      CFG_WRn      <= WRn or (not CFG_CYCLE) or
35                         (ADDR13_12(13) nand (not ADDR13_12(12)));
36      RAM_WRn_full <= WRn or ((not ADDR13_12(13)) nand ADDR13_12(12));
37      RAM_WRn      <= RAM_WRn_full or (not RAM_WRn_delay);
38      delay_half : process (CLK1, RSTn1)
39      begin
40         if (RSTn1 = '0') then
41            RAM_WRn_delay <= '1';
42         elsif (CLK1'event and CLK1 = '0') then
43            RAM_WRn_delay <= RAM_WRn_full;
44         end if;
45      end process;
46      -------------------------------------------------------------
47      REG_ADDR <= '1' when ADDR(11 downto 4) = "00000000" else '0';
48      cntr_init : process (CLK, RSTn)
49      begin
50         if (RSTn = '0') then
51            TIMER_REG_FF <= (TIMER_REG_FF'range => '1');
52         elsif (CLK'event and CLK = '1') then
53            if (REG_ADDR = '1') and (ADDR(3 downto 2) = "00") and
54               (REG_WRn = '0') then
55               TIMER_REG_FF <= ADi;
56            end if;
57         end if;
58      end process;
59      TIMER_REG <= TIMER_REG_FF;
60      cntr_prog : process (CLK1, RSTn1)
61      begin
62         if (RSTn1 = '0') then
63            TIMER_PROG_FF   <= (TIMER_PROG_FF'range => '0');
64            TIMER_PROG_CNTR <= (TIMER_PROG_CNTR'range => '0');
65            TIME_OUT        <= '1';
66            TIMER_CNTR_EN   <= '0';
67         elsif (CLK1'event and CLK1 = '1') then
```

```
68              if (REG_ADDR = '1') and (ADDR(3 downto 2) = "10") and
69                 (REG_WRn = '0') then
70                 TIMER_PROG_FF <= ADi(7 downto 0);
71              end if;
72              if (TIMER_CNTR_PL = '1') or (TIMER_CNTR_EN = '1') then
73                 TIMER_PROG_CNTR <= TIMER_PROG_FF;
74              else
75                 TIMER_PROG_CNTR <= unsigned(TIMER_PROG_CNTR) - 1;
76              end if;
77              if (TIMER_PROG_CNTR = "00000000") then
78                 TIMER_CNTR_EN <= '1';
79              else
80                 TIMER_CNTR_EN <= '0';
81              end if;
82              TIME_OUT         <= not TIMER_CNTR_TC;
83           end if;
84        end process;
85        INTAn          <= TIME_OUT;
86        TIMER_PROG     <= TIMER_PROG_FF;
87        TIMER_CNTR_PL <= '1' when (TIME_OUT = '0') or
88           (REG_ADDR='1' and ADDR13_12="11" and ADDR(3 downto 2)="01"
89            and REG_WRn = '0') else '0';
90        ----------------------------------------------------------
91        timer0 : SCNTEDP32
92        port map (
93           CLK   => CLK2,           RSTn  => RSTn2,
94           CIN   => TIMER_CNTR_EN,  PL    => TIMER_CNTR_PL,
95           DIN   => TIMER_REG_FF,   TC    => TIMER_CNTR_TC,
96           DOUT  => TIMER_CNTR);
97     end RTL;
```

Lines 60 to 84 implement TIMER_PROG register and a TIMER_PROG_CNTR. TIMER_PROG is implemented the same way as TIMER_REG in lines 68 to 70. It has only 8 bits and the lower 2 bits of the address are "10". The register is connected to output port in line 86. TIMER_PROG_CNTR is an 8-bit down counter. It is first loaded with the value of TIMER_PROG_FF (same value as TIMER_PROG, but TIMER_PROG output port cannot be read here). It is then counted down every clock cycle unless it is parallel loaded when the TIMER_CNTR is parallel loaded or the TIMER_PROG_CNTR reaches its terminal count. Lines 77 to 81 generate a signal TIMER_CNTR_EN to enable the counting of TIMER_CNTR. It is a flipflop output when TIMER_PROG_CNTR reaches the 0 count. It enables the TIMER_CNTR to count down and also to parallel load the TIMER_PROG_CNTR again. The TIMER_PROG_CNTR is described in lines 72 to 76.

The TIMER_CNTR is implemented with a component instantiation statement in lines 91 to 96. Note that the DIN input of the component SCNTEDP32 is mapped to TIMER_REG. The down counter is parallel loaded when it reaches its terminal count or it is written from a PCI write cycle. TIME_OUT signal is asserted '0' in line 82 when TIMER_CNTR_TC is asserted '1'. TIME_OUT signal is then connected to INTAn output port and also used in line 87 to 89 to generate the TIMER_CNTR_PL parallel load signal which is mapped to the CIN input of component SCNTEDP32 in line 94. Note that when the TIMER_CNTR register is written from a PCI write cycle, TIMER_CNTR is parallel loaded with the value in TIMER_REG, not the value of ADi.

Three clock signals CLK, CLK1, and CLK2, and three reset signals RSTn, RSTn1, RSTn2, are used in this block so that a clock (reset) signal will not connect to more than 32 flipflops.

The waveform in Figure 13.11 shows the relationship of counter and register values. Note that TIMER_PROG has value of 1. The TIMER_CNTR is counted down every three clock cycles. When TIMER_CNTR has value 0 and it is enabled to count down, TIMER_CNTR_TC is asserted '1', which asserted the TIME_OUT signal '0' at the next clock.

The REGMUX block implements the PCI configuration registers and the selection of HSRAM, ROM, configuration registers, TIMER_PROG, TIMER_CNTR, or TIMER_PROG registers to go to AD bus for the read cycle. The following VHDL code implements the REGMUX block.

Lines 6 to 15 declare input and output port signals. CFG_WRn signal is used to enable writing to the PCI configuration registers. TLATCH_EN comes from TARFSM block to enable clocking in the data to AD bus during the read cycle. Address ADDR and ADDR13_12 are used for decoding and selection of memory space. ADi comes from AD bus directly. RAMOUT, ROMOUT, TIMER_REG, TIMER_CNTR, TIMER_PROG come from HSRAM, ROM, and MEMLOGIC blocks. Output signal RESPOND_EN is a register bit in the configuration register to determine whether CHIP should respond to a PCI bus transaction or not. Output signal BASEADDR is the CHIP base address which goes to PCILOGIC for PCI address decoding. ADo goes to AD bus for read data.

Most of the PCI configuration registers are hardwired and are declared as constants from lines 18 to 29. Refer to the PCI specification document for more details. The least significant 14 bits of the base address register are hardwired to '0' because CHIP memory space has 4096 32-bit words from the PCI bus point of view. The only PCI configuration registers here are RESPOND_EN and the 18 most significant bits of the base address register. These are generated in lines 46 to 59. To write to these registers, CFG_WRn should be asserted and the address should match. Lines 38 to 44 are used to form full 32-bit registers (even though many of them are hardwired).

```
1     library IEEE;
2     use IEEE.std_logic_1164.all;
3     use work.PACK.all;
4     entity REGMUX is
5        port (
6           CLK, CLK1, RSTn, RSTn1: in  std_logic;
7           CFG_WRn, TLATCH_EN    : in  std_logic;
8           ADDR                  : in  std_logic_vector(11 downto 2);
9           ADDR13_12             : in  std_logic_vector(13 downto 12);
10          ADi, RAMOUT, ROMOUT   : in  std_logic_vector(31 downto 0);
11          TIMER_REG, TIMER_CNTR : in  std_logic_vector(31 downto 0);
12          TIMER_PROG            : in  std_logic_vector( 7 downto 0);
13          RESPOND_EN            : out std_logic;
14          BASEADDR              : out std_logic_vector(31 downto 14);
```

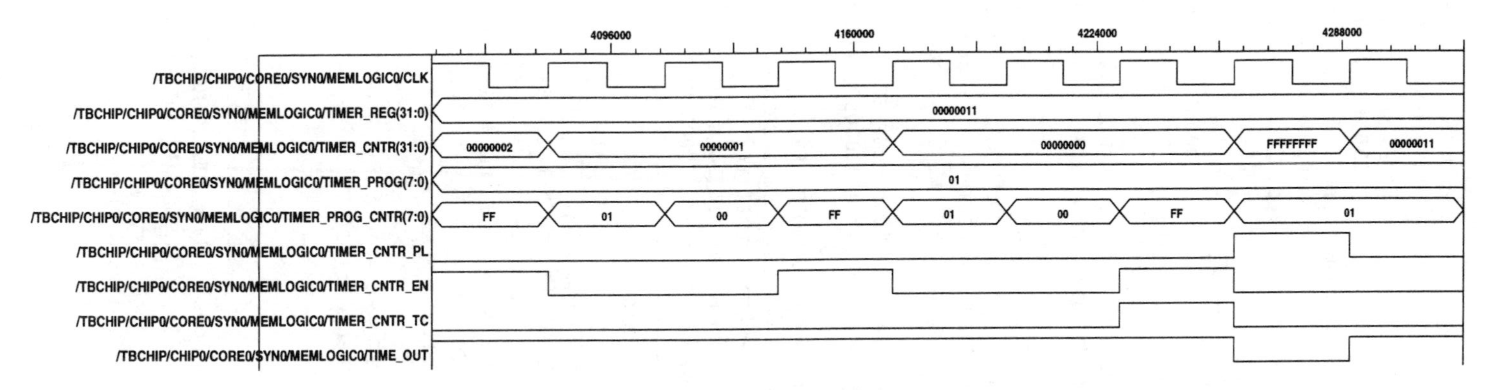

FIGURE 13.11 Waveform for the TIMER_CNTR.

```
15          ADo                        : out std_logic_vector(31 downto 0));
16     end REGMUX;
17     architecture RTL of REGMUX is
18        constant CHIP_ID     : std_logic_vector(15 downto 0)
19                             := "1111000011110000";
20        constant COMP_ID     : std_logic_vector(15 downto 0)
21                             := "1100110000110011";
22        constant CLASS_CODE  : std_logic_vector(23 downto 0)
23                             := "010101011010101010001000";
24        constant REVISION    : std_logic_vector( 7 downto 0)
25                             := "00000001";
26        constant BASEREG13_0 : std_logic_vector(13 downto 0)
27                             := "00000000000000";
28        constant INT_PIN     : std_logic_vector( 7 downto 0)
29                             := "00000001";
30        signal   DEVID_REG, DEVCTL_REG : std_logic_vector(31 downto 0);
31        signal   REVID_REG, BASE_REG   : std_logic_vector(31 downto 0);
32        signal   INT_PIN_REG           : std_logic_vector(31 downto 0);
33        signal   BASEREG31_14          : std_logic_vector(31 downto 14);
34        signal   PROG_REG, REGOUT      : std_logic_vector(31 downto 0);
35        signal   CFGOUT, DATAOUT       : std_logic_vector(31 downto 0);
36        signal   RESPOND_EN_FF         : std_logic;
37     begin
38        DEVID_REG    <= CHIP_ID & COMP_ID;
39        DEVCTL_REG   <= "000000101000000000000000000000" & RESPOND_EN_FF
                          & '0';
40        RESPOND_EN   <= RESPOND_EN_FF;
41        REVID_REG    <= CLASS_CODE & REVISION;
42        BASEADDR     <= BASEREG31_14;
43        BASE_REG     <= BASEREG31_14 & BASEREG13_0;
44        INT_PIN_REG <= "0000000000000000" & INT_PIN & "00000000";
45        ---------------------------------------------------------------
46        ff_gen : process (RSTn, CLK)
47        begin
48           if (RSTn = '0') then
49              BASEREG31_14  <= "111111111111111111";
50              RESPOND_EN_FF <= '0'; -- default no respond
51           elsif (CLK'event and CLK = '1') then
52              if (CFG_WRn = '0') and (ADDR(5 downto 2) = "0001") then
53                 RESPOND_EN_FF  <= ADi(1);
54              end if;
55              if (CFG_WRn = '0') and ADDR(5 downto 2) = "0100" then
56                 BASEREG31_14(31 downto 14) <= ADi(31 downto 14);
57              end if;
58           end if;
59        end process;
60        ---------------------------------------------------------------
61        cfg_gen : process (ADDR, DEVID_REG, DEVCTL_REG, REVID_REG,
62                          BASE_REG, INT_PIN_REG)
63        begin
64           case ADDR(5 downto 2) is
```

```
65              when "0000" => CFGOUT <= DEVID_REG;
66              when "0001" => CFGOUT <= DEVCTL_REG;
67              when "0010" => CFGOUT <= REVID_REG;
68              when "0100" => CFGOUT <= BASE_REG;
69              when "1111" => CFGOUT <= INT_PIN_REG;
70              when others => CFGOUT <= (CFGOUT'range => '0');
71           end case;
72        end process;
73        PROG_REG <= "000000000000000000000000" & TIMER_PROG;
74        KMUX31(ADDR(3 downto 2), TIMER_REG, TIMER_CNTR, PROG_REG,
               REGOUT);
75        KMUX41(ADDR13_12, ROMOUT, RAMOUT, CFGOUT, REGOUT, DATAOUT);
76        ado_gen : process (RSTn1, CLK1)
77        begin
78           if (RSTn1 = '0') then
79              ADo <= (ADo'range => '0');
80           elsif (CLK1'event and CLK1 = '1') then
81              if (TLATCH_En = '1') then
82                 ADo <= DATAOUT;
83              end if;
84           end if;
85        end process;
86     end RTL;
```

The PCI configuration registers are selected (1 of 16) with the address bits ADDR(5 downto 2) as shown in lines 61 to 72. Note that reading the unused and reserved registers should return all '0' as implemented in line 70.

Line 73 makes the full 32-bit for the TIMER_PROG register. Line 74 selects one of the three registers with a 3-to-1 multiplexer concurrent procedure call. Line 75 selects one of four 1024 memory blocks (ROM, HSRAM, PCI configuration registers, and CHIP registers) with ADDR13_12 address. The procedures KMUX31 and KMUX41 are declared in package PACK, which is referenced in line 3. The output of the KMUX41 is going to 32-bit flipflops in lines 76 to 85. The signal ADo is going to the AD bus pads and also to the PCILOGIC block for parity generation.

13.3 VHDL TEST BENCH

At this point, the VHDL code for CHIP is completed. A test bench can be set up to easily test the CHIP functions. Figure 13.12 shows a possible test bench for testing CHIP. PCI ARBITOR arbitrates among PCI master agents to determine who should get the PCI bus by asserting an individual GNTn signal to each PCI master agent. In Figure 13.12, two PCI master agents (PCIKC) are shown. PCIMTR gets the input signals from the PCI bus and generates a text file to record the address and data in the PCI bus. The output text file can be read much easier than looking at the simulation waveforms.

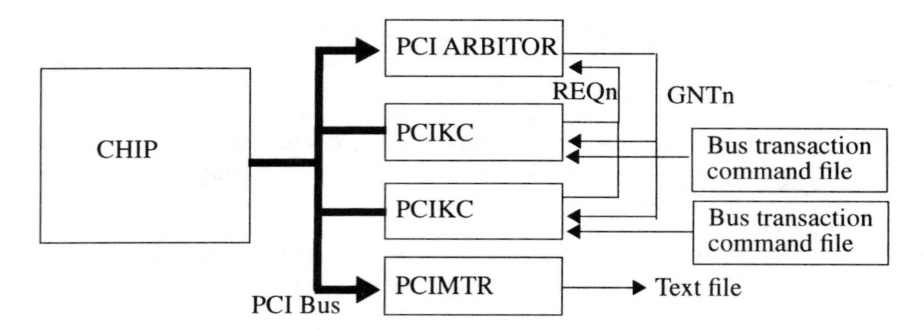

FIGURE 13.12 Test bench for the CHIP.

Each PCI bus master can read in a bus transaction command file. For example, the following shows a bus command file. Character "#" is used for a comment line. Line 31 is a read transaction with 16 words to the address 00000800 with bus command 'A' as configuration read. Byte enable CBEn is "0000" in the data phase and IRDYn has 0 wait states. PCI0 reads in the bus command file, generates signals to PCI bus, and performs the bus protocol. This makes writing the test vectors much easier than writing simulator commands to assign values to signals at particular simulation times.

```
1     #-----------------------------------------------------
2     #  PCI Bus Command File
3     #
4     #  1. I n idle n clock cycles
5     #  2. write transaction
6     #     2.a w n address pci_cmd pci_bye_enable X
7     #         data data data data  (max 4 words per line)
8     #     2.b W n address pci_cmd
9     #         data byte_enable X (1 word per line)
10    #  3. read transaction
11    #     3.a r n addresss pci_cmd pci_byte_enable X
12    #     3.b R n address pci_cmd
13    #         byte_enable X
14    #  4. S - wait forever in the driver
15    #  5. # comment line
16    #
17    #  command format :
18    #  1. First letter indicate the command type
19    #  2. address, data, byte_enable are in HEX format
20    #  3. n is in INTEGER format
21    #  4. X indicates number of wait state to control IRDYn
22    #-----------------------------------------------------
23    #  base register 18 bits
24    #       ROM    FC00_0000
25    #       RAM    FC00_1000
26    #       CFG    FC00_2000
27    #       REG    FC00_3000
28    #-----------------------------------------------------
29    I 10
```

```
30    # configuration read, should respond
31    r 16 00000800 A 0 0
32    # configuration write to base register
33    w 1 00000810 B 0 0
34    FEDC3FFF
35    # memory        read, should NOT respond, master abort
36    r 1 FEDC0000 6 0 0
37    # configuration write to enable respond_en
38    w 1 00000C04 B 0 0
39    00000002
40    # Register write and read
41    w 3 FEDC3000 7 0 0
42    00000011 00112233 00000001
43    r 3 FEDC3000 6 0 0
44    # ROM read
45    r 2 FEDC0000 6 0 0
46    # RAM write and read
47    w 3 FEDC1000 7 0 0
48    01234567 89ABCDEF 33445566
49    r 3 FEDC1000 6 0 0
50    S
```

The test bench for the CHIP follows. Lines 6 to 78 declare the components. Component PCIMTR has a generic output text file name and input PCI bus signals. Component PCIKC has many generics to control the timing of PCI bus signals generated. It has a generic input file name for the bus command file and an output file (not used in this test bench). The ARBITOR block generates GNTn signal based on the REQn signal. Lines 80 to 91 declare local signals to connect these blocks. A half clock period constant HALF_CLOCK is defined in line 92.

```
1     library IEEE;
2     use IEEE.std_logic_1164.all;
3     entity TBCHIP is
4     end TBCHIP;
5     architecture BEH of TBCHIP is
6        component PCIMTR
7        generic (
8           outfile  : string := "monitor.out");
9        port (
10          CLK      : in    std_logic;
11          AD       : in    std_logic_vector(31 downto 0);
12          CBEn     : in    std_logic_vector(3 downto 0);
13          FRAMEn   : in    std_logic;
14          IRDYn    : in    std_logic;
15          TRDYn    : in    std_logic);
16       end component;
17       component CHIP
18       port (
19          CLK     : in    std_logic_vector(0 downto 0);
20          RSTn    : in    std_logic_vector(0 downto 0);
21          AD      : inout std_logic_vector(31 downto 0);
```

```
22         CBEn     : in    std_logic_vector(3 downto 0);
23         IDSEL    : in    std_logic_vector(0 downto 0);
24         FRAMEn   : in    std_logic_vector(0 downto 0);
25         TRDYn    : out   std_logic_vector(0 downto 0);
26         DEVSELn  : out   std_logic_vector(0 downto 0);
27         PAR      : out   std_logic_vector(0 downto 0);
28         INTAn    : out   std_logic_vector(0 downto 0));
29      end component;
30      component PCIKC
31      generic (
32         T_AD      : time    := 7.0 ns;
33         T_CBEn    : time    := 8.0 ns;
34         T_FRAMEn  : time    := 7.0 ns;
35         T_IRDYn   : time    := 7.0 ns;
36         T_TRDYn   : time    := 7.0 ns;
37         T_STOPn   : time    := 7.0 ns;
38         T_LOCKn   : time    := 7.0 ns;
39         T_PAR     : time    := 7.0 ns;
40         T_PERRn   : time    := 7.0 ns;
41         T_SERRn   : time    := 7.0 ns;
42         T_DEVSELn: time     := 7.0 ns;
43         PERIOD    : time    := 30.0 ns;
44         T_WAIT    : integer   := 2; -- default # of target wait states
45         CID       : std_logic_vector(3 downto 0);
46         REV_PAR   : std_logic := '1'; -- reverse parity
47         DATASIZE  : integer:= 32;  -- size of the data array
48         ADDRSIZE  : integer:= 5;  -- size of the addr
49         infile    : string := "pci.in";
50         outfile   : string := "pci.out");
51      port (
52         CLK       : in    std_logic;
53         RSTn      : in    std_logic;
54         IDSEL     : in    std_logic;
55         REQn      : out   std_logic;
56         GNTn      : in    std_logic;
57         AD        : inout std_logic_vector(31 downto 0);
58         CBEn      : inout std_logic_vector(3 downto 0);
59         FRAMEn    : inout std_logic;
60         IRDYn     : inout std_logic;
61         TRDYn     : inout std_logic;
62         STOPn     : inout std_logic;
63         LOCKn     : inout std_logic;
64         PAR       : inout std_logic;
65         PERRn     : inout std_logic;
66         SERRn     : inout std_logic;
67         DEVSELn   : inout std_logic);
68      end component;
69      component ARBITOR
70      generic (
71         N      : integer := 2 ;     -- number of ports
72         T_DLY : time     := 1.0 ns  );
73      port (
74         CLK  : in  std_logic;
```

```
75          RSTn : in  std_logic;
76          REQn : in  std_logic_vector((N-1) downto 0);
77          GNTn : out std_logic_vector((N-1) downto 0));
78       end component;
79       ---------------------------------------------------
80       signal CLK       : std_logic_vector(0 downto 0);
81       signal RSTn      : std_logic_vector(0 downto 0);
82       signal AD        : std_logic_vector(31 downto 0);
83       signal CBEn      : std_logic_vector(3 downto 0);
84       signal FRAMEn    : std_logic_vector(0 downto 0);
85       signal TRDYn     : std_logic_vector(0 downto 0);
86       signal DEVSELn   : std_logic_vector(0 downto 0);
87       signal PAR       : std_logic_vector(0 downto 0);
88       signal INTAn     : std_logic_vector(0 downto 0);
89       signal IRDYn, STOPn, PERRn, SERRn, LOCKn : std_logic;
90       signal REQn      : std_logic_vector(3 downto 0);
91       signal GNTn      : std_logic_vector(3 downto 0);
92       constant HALF_PERIOD : time := 15 ns;
93    begin
94       RSTn(0) <= '0', '1' after 5 * HALF_PERIOD;
95       REQn(3 downto 1) <= "111";
96       clk_gen : process
97       begin
98          CLK(0) <= '1'; wait for HALF_PERIOD;
99          CLK(0) <= '0'; wait for HALF_PERIOD;
100      end process;
101      pcikc0 :  PCIKC
102      generic map (
103         PERIOD   => HALF_PERIOD * 2,  T_WAIT   => 0,
104         CID      => "0000",            REV_PAR  => '0',
105         DATASIZE => 32,                infile   => "pci0.in",
106         outfile  => "pci0.out")
107      port map (
108         CLK      => CLK(0),            RSTn     => RSTn(0),
109         IDSEL    => AD(30),            REQn     => REQn(0),
110         GNTn     => GNTn(0),           AD       => AD,
111         CBEn     => CBEn,              FRAMEn   => FRAMEn(0),
112         IRDYn    => IRDYn,             TRDYn    => TRDYn(0),
113         STOPn    => STOPn,             LOCKn    => LOCKn,
114         PAR      => PAR(0),            PERRn    => PERRn,
115         SERRn    => SERRn,             DEVSELn  => DEVSELn(0));
116      pcimtr0 : PCIMTR
117      port map (
118         CLK      => CLK(0),            AD       => AD,
119         CBEn     => CBEn,              FRAMEn   => FRAMEn(0),
120         IRDYn    => IRDYn,             TRDYn    => TRDYn(0));
121      arbitor0 : ARBITOR
122      generic map (
123         N        => 4,                 T_DLY    => 2.0 ns)
124      port map (
125         CLK      => CLK(0),            RSTn     => RSTn(0),
126         REQn     => REQn,              GNTn     => GNTn);
127      chip0 : CHIP
```

```
128      port map (
129         CLK      => CLK,                 RSTn      => RSTn,
130         AD       => AD,                  CBEn      => CBEn,
131         IDSEL    => AD(11 downto 11),    FRAMEn    => FRAMEn,
132         TRDYn    => TRDYn,               DEVSELn   => DEVSELn,
133         PAR      => PAR,                 INTAn     => INTAn);
134   end BEH;
```

Line 94 generates the reset signal. Lines 96 to 100 generate the clock signal. Note that CLK and RSTn are declared as a bus signal of 1 bit wide to be consistent with CHIP block. The rest of the code is for the component instantiation of each block. In this test bench, only one PCIKC block is instantiated to make the test bench shorter. Line 95 ties 3 of the 4-bit REQn signals to '1' because only one PCIKC block is used. The test bench is configured with the following VHDL code. Note that CHIP is hierarchically configured from lines 12 to 52.

```
1     configuration CFG_TBCHIP of TBCHIP is
2        for BEH
3           for all : PCIMTR
4              use entity WORK.PCIMTR(BEH);
5           end for;
6           for all : PCIKC
7              use entity WORK.PCIKC(BEH);
8           end for;
9           for all : ARBITOR
10             use entity WORK.ARBITOR(BEH);
11          end for;
12          for all : CHIP
13             use entity WORK.CHIP(BEH);
14             for BEH
15                for all : CORE
16                   use entity WORK.CORE(RTL);
17                   for RTL
18                      for all : ROM
19                         use entity WORK.ROM(BEH);
20                      end for;
21                      for all : HSRAM
22                         use entity WORK.HSRAM(BEH);
23                      end for;
24                      for all : SYN
25                         use entity WORK.SYN(RTL);
26                         for RTL
27                            for all : TREEC
28                               use configuration WORK.CFG_RTL_TREEC;
29                            end for;
30                            for all : TARFSM
31                               use entity WORK.TARFSM(RTL);
32                            end for;
33                            for all : MEMLOGIC
34                               use entity WORK.MEMLOGIC(RTL);
```

```
35                                   for RTL
36                                      for all : SCNTEDP32
37                                      use configuration WORK.CFG_SCNTEDP32;
38                                      end for;
39                                   end for;
40                                end for;
41                                for all : PCILOGIC
42                                   use entity WORK.PCILOGIC(RTL);
43                                end for;
44                                for all : REGMUX
45                                   use entity WORK.REGMUX(RTL);
46                                end for;
47                             end for;
48                          end for;
49                       end for;
50                    end for;
51                 end for;
52              end for;
53           end for;
54        end CFG_TBCHIP;
```

The test bench runs with a text file generated from PCIMTR block as follows. At line 1, address 00000800 with bus command A (configuration read) is started at time 540 ns. The 16 data from the configuration registers follows. Note that the data comes out every 60 ns (two clock periods). Line 18 indicates writing to the base address configuration register. Line 20 reads a value from the ROM, but no data comes out. This is as it should be since the RESPOND_EN bit in the PCI configuration register is not set. PCIKC block will perform master abort to terminate the PCI bus transaction. Line 21 enables the RESPOND_EN bit which enables CHIP to respond with PCI memory read (CBEn = "0110") and memory write (CBEn = "0111") commands. The writing and reading of HSRAM, ROM, and CHIP registers afterward shows that CHIP functions correctly. Note that in line 25, TIMER_CNTR is written with value 00112233, but reads back 0000000D. This is because TIMER_CNTR is loaded with TIMER_REG value 00000011 and starts to count down. Value 0000000D is read. Of course, more tests can be run by adding more bus commands to the bus command file or having different bus command files to test different functions.

Note that the IDSEL signal of CHIP is connected to AD(11 downto 11) in the test bench (line 131). This is why address 00000800 is used in the first configuration read transaction.

The implementation of the ARBITOR block can vary depending on the arbitration algorithm. PCIMTR block VHDL code is not shown here. They are left as exercises. PCIKC block is not presented here because it is much more involved (it is close to implementing the whole PCI bus protocol).

```
1    00000800  A  ---- PCI read  ----    Time: 540 NS
2      F0F0CC33  0  Time: 630 NS
3      02800000  0  Time: 690 NS
4      55AA8801  0  Time: 750 NS
```

```
5       00000000  0  Time: 810 NS
6       FFFFC000  0  Time: 870 NS
7       00000000  0  Time: 930 NS
8       00000000  0  Time: 990 NS
9       00000000  0  Time: 1050 NS
10      00000000  0  Time: 1110 NS
11      00000000  0  Time: 1170 NS
12      00000000  0  Time: 1230 NS
13      00000000  0  Time: 1290 NS
14      00000000  0  Time: 1350 NS
15      00000000  0  Time: 1410 NS
16      00000000  0  Time: 1470 NS
17      00000100  0  Time: 1530 NS
18    00000810  B  ---- PCI write ----     Time: 1680 NS
19      FEDC3FFF  0  Time: 1770 NS
20    FEDC0000  6  ---- PCI read  ----     Time: 1920 NS
21    00000C04  B  ---- PCI write ----     Time: 2310 NS
22      00000002  0  Time: 2400 NS
23    FEDC3000  7  ---- PCI write ----     Time: 2550 NS
24      00000011  0  Time: 2640 NS
25      00112233  0  Time: 2700 NS
26      00000001  0  Time: 2760 NS
27    FEDC3000  6  ---- PCI read  ----     Time: 2910 NS
28      00000011  0  Time: 3000 NS
29      0000000D  0  Time: 3060 NS
30      00000001  0  Time: 3120 NS
31    FEDC0000  6  ---- PCI read  ----     Time: 3270 NS
32      FEDC9000  0  Time: 3360 NS
33      FEDC9001  0  Time: 3420 NS
34    FEDC1000  7  ---- PCI write ----     Time: 3570 NS
35      01234567  0  Time: 3660 NS
36      89ABCDEF  0  Time: 3720 NS
37      33445566  0  Time: 3780 NS
38    FEDC1000  6  ---- PCI read  ----     Time: 3930 NS
39      01234567  0  Time: 4020 NS
40      89ABCDEF  0  Time: 4080 NS
41      33445566  0  Time: 4140 NS
```

13.4 SYNTHESIS AND LAYOUT

Blocks TREEC, SCNTEDP32, PCILOGIC, MEMLOGIC, REGMUX, and TARFSM are synthesized with SYNOPSYS Design Compiler. The schematics are not shown. These blocks are connected in the SYN block. A structural VHDL code is generated (2587 lines). This file is referenced as SYN.synout.vhd.

The pads are instantiated directly with VHDL code. The architecture BEH of entity CHIP was presented earlier, which describes the simple behavior of all pads. The following VHDL code (file name chip_str.vhd) shows the architecture STR of entity CHIP which instantiates all the pads (from the library EPOCH_LIB) for the chip CHIP and the CORE block. Note that the signals for the pads are declared as a bus (std_logic_vector). This is the reason that CLK (RSTn, and so on) single bit port signals are declared as std_logic_vector(0 downto 0) in CHIP entity VHDL code.

```
1     library IEEE;
2     library EPOCH_LIB;
3     use IEEE.std_logic_1164.all;
4     use EPOCH_LIB.COMPONENTS.all;
5     architecture STR of CHIP is
6        constant SLIM_FLAG         : integer := 1;
7        constant LEVEL_SHIFTING    : integer := 0;
8        constant SCHMITT_TRIGGER   : integer := 1;
9        constant PULL_TYPE         : STRING  := "None";
10       constant YPITCH            : STRING  := "4MA";
11       constant OUTDRIVE          : STRING  := "4MA";
12       constant PTYPE             : STRING  := "REGULAR";
13       component CORE
14       port (
15          CLK           : in    std_logic;
16          RSTn          : in    std_logic;
17          ADi           : in    std_logic_vector(31 downto 0);
18          ADo           : out   std_logic_vector(31 downto 0);
19          AD_OEn        : out   std_logic_vector(1 downto 0);
20          CBEni         : in    std_logic_vector(3 downto 0);
21          IDSEL         : in    std_logic;
22          FRAMEni       : in    std_logic;
23          TRDYno        : out   std_logic;
24          TRDYn_OEn     : out   std_logic;
25          DEVSELno      : out   std_logic;
26          DEVSELn_OEn   : out   std_logic;
27          PARo          : out   std_logic;
28          PAR_OEn       : out   std_logic;
29          INTAn         : out   std_logic);
30       end component;
31       signal ADi, ADo        : std_logic_vector(31 downto 0);
32       signal AD_OEn          : std_logic_vector(1 downto 0);
33       signal CBEni           : std_logic_vector(3 downto 0);
34       signal L_CLK, L_RSTn         : std_logic_vector(0 downto 0);
35       signal L_INTAn,  L_IDSEL     : std_logic_vector(0 downto 0);
36       signal FRAMEni, TRDYno       : std_logic_vector(0 downto 0);
37       signal DEVSELno, PARo        : std_logic_vector(0 downto 0);
38       signal TRDYn_OEn, DEVSELn_OEn, STOPn_OEn, PAR_OEn : std_logic;
39    begin
40       core0 : CORE
41       port map (
42          CLK          => L_CLK(0),      RSTn          => L_RSTn(0),
43          ADi          => ADi,           ADo           => ADo,
44          AD_OEn       => AD_OEn,        CBEni         => CBEni,
45          IDSEL        => L_IDSEL(0),    FRAMEni       => FRAMEni(0),
46          TRDYno       => TRDYno(0),     TRDYn_OEn     => TRDYn_OEn,
```

```
47        DEVSELno     => DEVSELno(0),  DEVSELn_OEn  => DEVSELn_OEn,
48        PARo         => PARo(0),      PAR_OEn      => PAR_OEn,
49        INTAn        => L_INTAn(0));
50     cgnd1_inst: padgnd generic map (9, SLIM_FLAG, "core_only","4ma")
51        port map (cgnds(1));
52     vdd1_inst: padvdd generic map (10, SLIM_FLAG, "rail_only","4ma")
53        port map (vdds(1));
54     rstn_inst : padin_unbuf generic map (11, 11, SLIM_FLAG, YPITCH)
55        port map (RSTn, L_RSTn);
56     clk_inst : padin_unbuf generic map (12, 12, SLIM_FLAG, YPITCH)
57        port map (CLK, L_CLK);
58     idsel_inst : padin_unbuf generic map (13, 13, SLIM_FLAG, YPITCH)
59        port map (IDSEL, L_IDSEL);
60     gnd1_inst: padgnd generic map (14, SLIM_FLAG, "rail_only","4ma")
61        port map (gnds(1));
62     ad31_28 : padbiuni generic map (18, 15, SLIM_FLAG, "8ma",
63           LEVEL_SHIFTING, SCHMITT_TRIGGER, PULL_TYPE)
64        port map (AD_OEn(1), ADo(31 downto 28), AD(31 downto 28),
65                  ADi(31 downto 28));
66     gnd2_inst: padgnd generic map (19, SLIM_FLAG, "rail_only","4ma")
67        port map (gnds(2));
68     ad27_24 : padbiuni generic map (23, 20, SLIM_FLAG, "8ma",
69           LEVEL_SHIFTING, SCHMITT_TRIGGER, PULL_TYPE)
70        port map (AD_OEn(1), ADo(27 downto 24), AD(27 downto 24),
71                  ADi(27 downto 24));
72     cvdd1_inst: padvdd generic map (24, SLIM_FLAG,"core_only","4ma")
73        port map (cvdds(1));
74     cgnd2_inst: padgnd generic map (25, SLIM_FLAG,"core_only","4ma")
75        port map (cgnds(2));
76     intan_inst : padout generic map (26, 26, SLIM_FLAG, OUTDRIVE)
77        port map (L_INTAn, INTAn);
78     gnd3_inst: padgnd generic map (27, SLIM_FLAG, "rail_only","4ma")
79        port map (gnds(3));
80     ad23_20 : padbiuni generic map (31, 28, SLIM_FLAG, "8ma",
81           LEVEL_SHIFTING, SCHMITT_TRIGGER, PULL_TYPE)
82        port map (AD_OEn(1), ADo(23 downto 20), AD(23 downto 20),
83                  ADi(23 downto 20));
84     vdd2_inst: padvdd generic map (32, SLIM_FLAG, "rail_only","4ma")
85        port map (vdds(2));
86     cben_inst : padin_buf
87       generic map (36, 33,SLIM_FLAG,LEVEL_SHIFTING,SCHMITT_TRIGGER,
88           PULL_TYPE, YPITCH)
89        port map (CBEn, CBEni);
90     framen_inst : padin_buf
91       generic map (37, 37,SLIM_FLAG,LEVEL_SHIFTING,SCHMITT_TRIGGER,
92           "Up", YPITCH)
93        port map (FRAMEn, FRAMEni);
94     gnd4_inst: padgnd generic map (38, SLIM_FLAG, "rail_only","4ma")
95        port map (gnds(4));
96     trdyn_inst : padout_tri
97        generic map (39, 39, SLIM_FLAG, OUTDRIVE)
```

```
98        port map (TRDYn_OEn, TRDYno, TRDYn);
99     cvdd2_inst: padvdd generic map (40, SLIM_FLAG,"core_only","4ma")
100       port map (cvdds(2));
101    cgnd3_inst: padgnd generic map (41, SLIM_FLAG,"core_only","4ma")
102       port map (cgnds(3));
103    ad19_16 : padbiuni generic map (45, 42, SLIM_FLAG, "8ma",
104           LEVEL_SHIFTING, SCHMITT_TRIGGER, PULL_TYPE)
105       port map (AD_OEn(1), ADo(19 downto 16), AD(19 downto 16),
106                 ADi(19 downto 16));
107    gnd5_inst: padgnd generic map (46, SLIM_FLAG, "rail_only","4ma")
108       port map (gnds(5));
109    ad15_12 : padbiuni generic map (50, 47, SLIM_FLAG, "8ma",
110           LEVEL_SHIFTING, SCHMITT_TRIGGER, PULL_TYPE)
111       port map (AD_OEn(0), ADo(15 downto 12), AD(15 downto 12),
112                 ADi(15 downto 12));
113    vdd3_inst: padvdd generic map (51, SLIM_FLAG, "rail_only","4ma")
114       port map (vdds(3));
115    devseln_inst : padout_tri
116       generic map (52, 52, SLIM_FLAG, OUTDRIVE)
117       port map (DEVSELn_OEn, DEVSELno, DEVSELn);
118    gnd6_inst: padgnd generic map (53, SLIM_FLAG, "rail_only","4ma")
119       port map (gnds(6));
120    par_inst : padout_tri
121       generic map (54, 54, SLIM_FLAG, OUTDRIVE)
122       port map (PAR_OEn, PARo, PAR);
123    gnd7_inst: padgnd generic map (55, SLIM_FLAG, "rail_only","4ma")
124       port map (gnds(7));
125    cvdd3_inst: padvdd generic map (56, SLIM_FLAG,"core_only","4ma")
126       port map (cvdds(3));
127    cgnd4_inst: padgnd generic map (57, SLIM_FLAG,"core_only","4ma")
128       port map (cgnds(4));
129    ad11_8 : padbiuni generic map (61, 58, SLIM_FLAG, "8ma",
130           LEVEL_SHIFTING, SCHMITT_TRIGGER, PULL_TYPE)
131       port map (AD_OEn(0), ADo(11 downto 8), AD(11 downto 8),
132                 ADi(11 downto 8));
133    vdd4_inst: padvdd generic map (62, SLIM_FLAG, "rail_only","4ma")
134       port map (vdds(4));
135    ad7_6 : padbiuni generic map (64, 63, SLIM_FLAG, "8ma",
136           LEVEL_SHIFTING, SCHMITT_TRIGGER, PULL_TYPE)
137       port map (AD_OEn(0), ADo(7 downto 6), AD(7 downto 6),
138                 ADi(7 downto 6));
139    ad5_4 : padbiuni generic map (2, 1, SLIM_FLAG, "8ma",
140           LEVEL_SHIFTING, SCHMITT_TRIGGER, PULL_TYPE)
141       port map (AD_OEn(0), ADo(5 downto 4), AD(5 downto 4),
142                 ADi(5 downto 4));
143    gnd8_inst: padgnd generic map (3, SLIM_FLAG, "rail_only", "4ma")
144       port map (gnds(8));
145    ad3_0 : padbiuni generic map (7, 4, SLIM_FLAG, "8ma",
146           LEVEL_SHIFTING, SCHMITT_TRIGGER, PULL_TYPE)
147       port map (AD_OEn(0), ADo(3 downto 0), AD(3 downto 0),
```

```
148                  ADi(3 downto 0));
149      cvdd4_inst: padvdd generic map (8, SLIM_FLAG, "core_only","4ma")
150         port map (cvdds(4));
151   end STR;
```

Files SYN.synout.vhd, core.vhd, chip_e.vhd, and chip_str.vhd are read into the layout tool (Cascade Design Automation EPOCH Silicon Compiler). The layout is generated as shown in the plot with ES2 CMOS 0.7 micron double metal technology. The chip size is 187 by 210 mils. The HSRAM appears at the top. The ROM is shown in the bottom-right corner. The rest of them are 1100 standard cells.

13.5 LAYOUT BACKANNOTATION AND VERIFICATION

The layout tool generates actual timing delay (1066 lines) with a structural VHDL code of simulation primitives (14722 lines) with architecture name CDA_STRUCTURAL. The following VHDL configures the test bench to use the actual post-layout delay VHDL code (line 13):

```
1     configuration CFG_TBCHIPSTR of TBCHIP is
2        for BEH
3           for all : PCIMTR
4              use entity WORK.PCIMTR(BEH);
5           end for;
6           for all : PCIKC
7              use entity WORK.PCIKC(BEH);
8           end for;
9           for all : ARBITOR
10             use entity WORK.ARBITOR(BEH);
11          end for;
12          for all : CHIP
13             use entity WORK.CHIP(CDA_STRUCTURAL);
14          end for;
15       end for;
16    end CFG_TBCHIPSTR;
```

The test bench is run with another text file generated from the PCIMTR block. By comparing the text file generated from the functional VHDL code, (using UNIX diff command), we see that they are identical!

Now, let's examine the CHIP response time for PCI signals. Figure 13.13 shows the simulation waveform. The AD bus, TRDYn, DEVSELn response time is within the PCI specification of 11 ns (after the rising edge of the clock). The simulation environment is done with the worst case 4.5V, and 125 degrees C. The 32-bit down counter is traced, which shows TIMER_D (data goes to D input of the counter flip-flops) with enough setup time in this region. The worst case for the TIMER_CNTR is changing from all bits with '0' to all bits with '1'. Its waveform is shown in Figure 13.14. When TIMER_CNTR is changing from all '0' to all '1', it has a setup time of about 5 ns relative to the clock signal CLK outside of the chip CHIP. It would be a lot more than 5 ns setup time inside the chip.

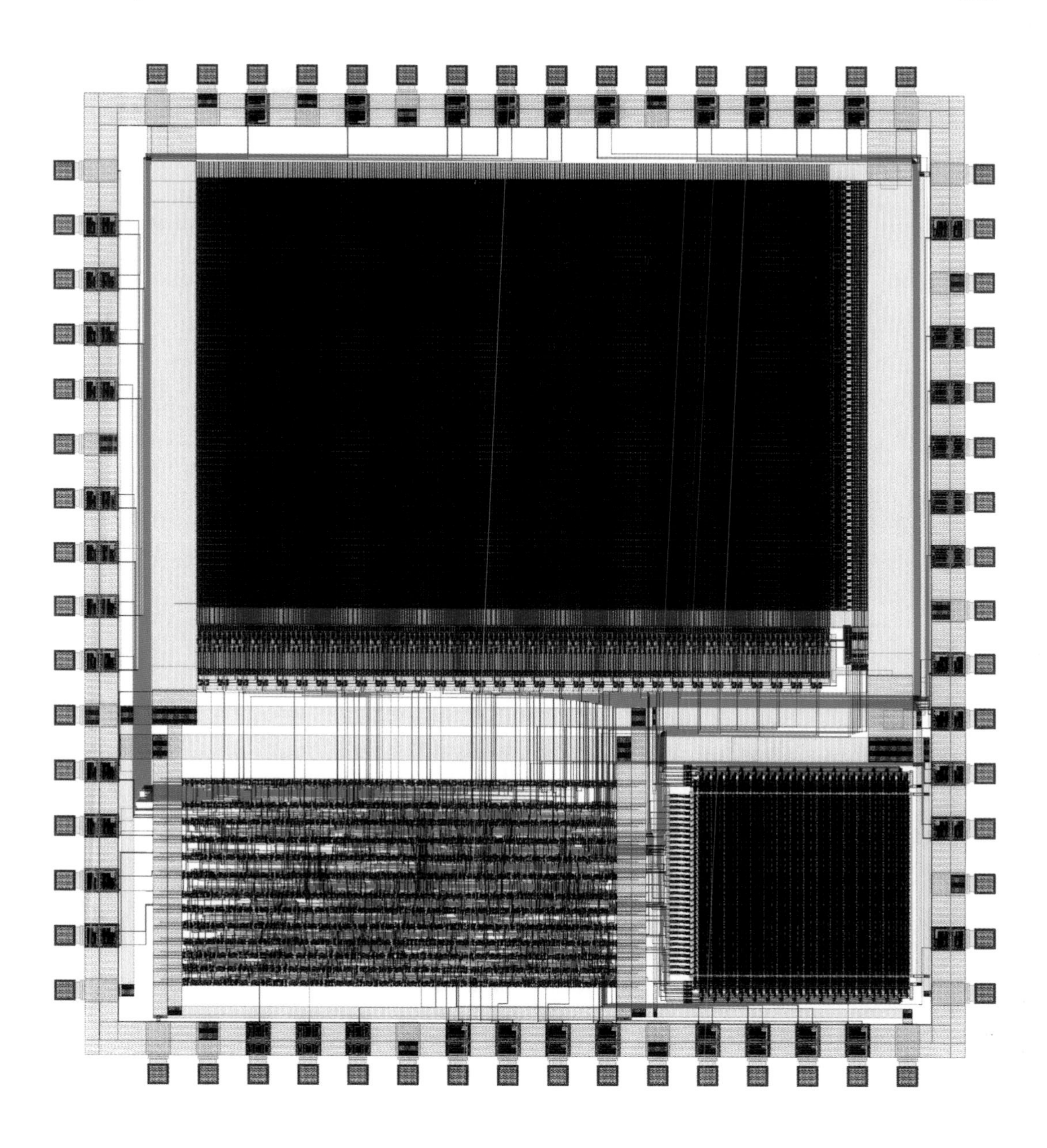

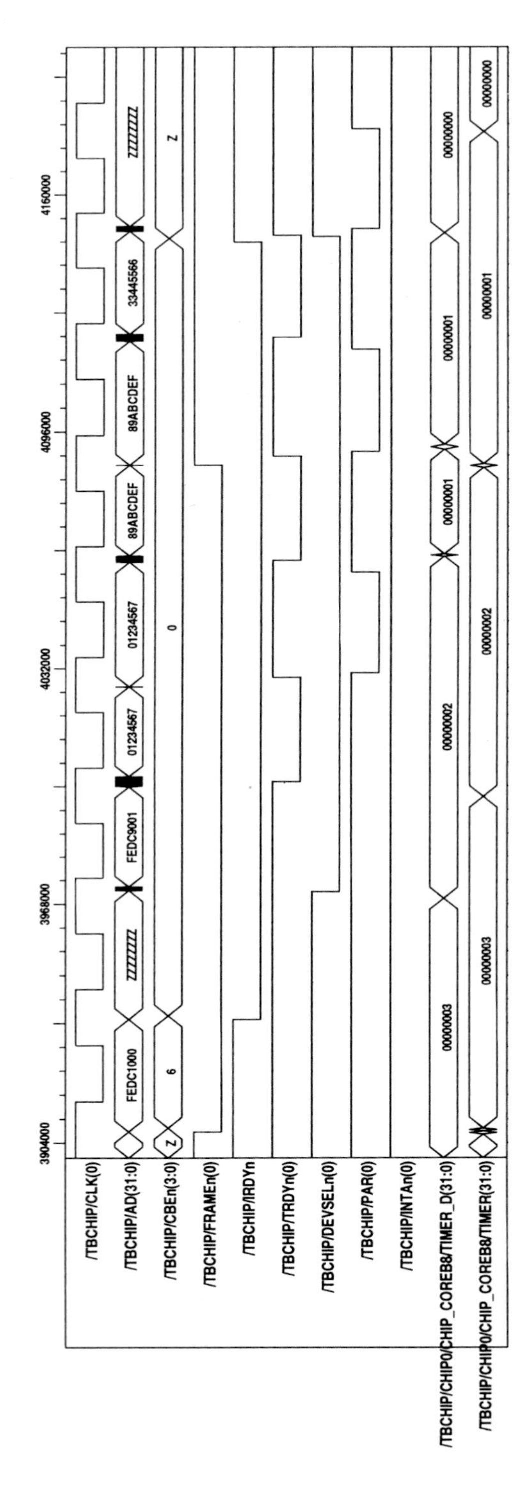

FIGURE 13.13 CHIP actual response time after layout.

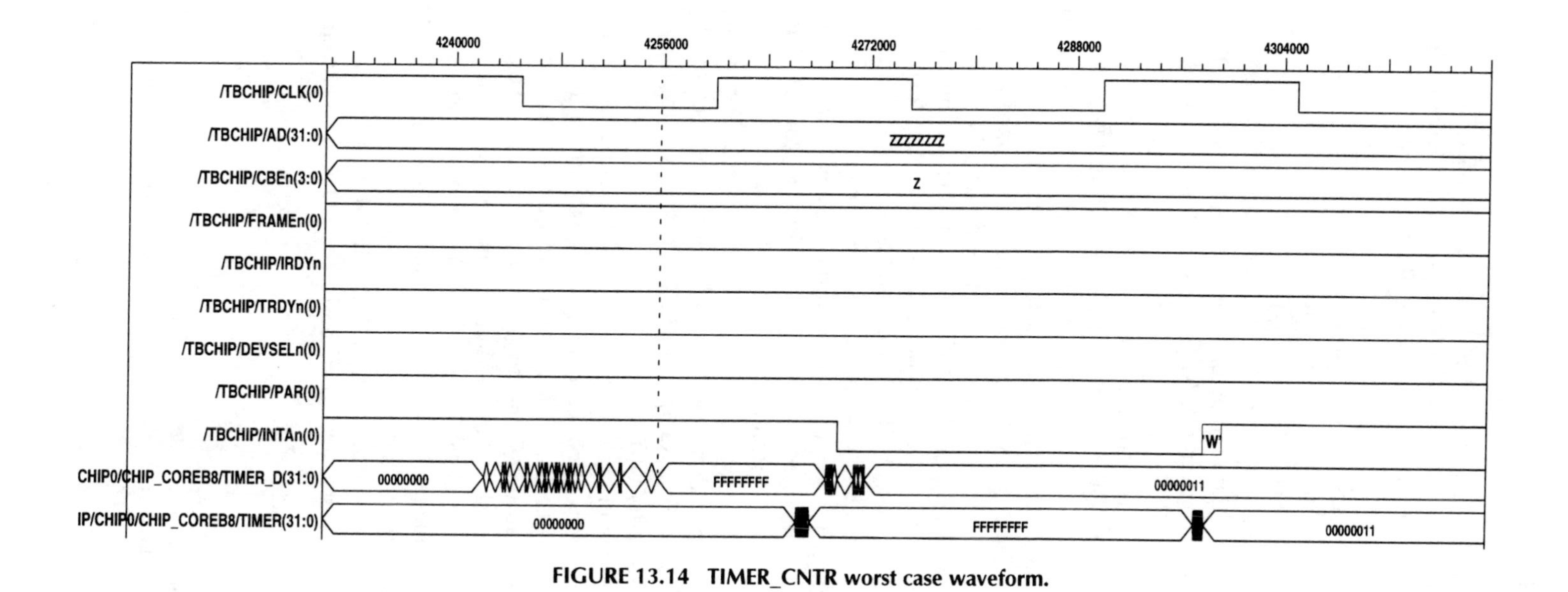

FIGURE 13.14 TIMER_CNTR worst case waveform.

13.6 VHDL PARTITIONING

Another VHDL design issue is the VHDL partitioning. For example, how do we determine which functions (circuits) should be inside a VHDL entity? There are no "absolutes." However, the following are some of guidelines:

1. Logics, to achieve the related function, were put together unless they were too big for the efficient synthesis. For example, PCILOGIC, TARFSM, and MEMLOGIC blocks.
2. When there was a need to break a function into more blocks, register boundaries were the cutting line for easy specification of synthesis timing constraints. For example, REGMUX block gets the RAMOUT, ROMOUT, and registers TIMER_REG, TIMER_CNTR, and TIMER_PROG into the same block with the configuration registers. TIMER_REG, TIMER_CNTR, and TIMER_PROG are not muxed in the MEMLOGIC block (combinational output) and then go to the REGMUX block. All the selection functions are done in the REGMUX block because there are many hardwired signals for the registers. They can be better optimized in the same block.
3. Fewer VHDL hierarchy levels were preferred to reduce the structural VHDL coding and synthesis efforts. For example, as shown in Figure 13.15, block TOP is partitioned into four lower-level blocks A, B, C, and D. The partition X requires less VHDL coding and less effort of synthesis compared to partition Y. Partition Y has two middle-level blocks that do not gain any functions. Partition X also reduces the effort of design changes. As an example, an input signal is required to be added into block B from TOP. Partition X is easier to modify than Partition Y.
4. The reusable portion is placed in one block. For example, the middle-level block in partition Y would make sense if C and D together could be reused as one block. For example, TREEC block is reusable to generate different two-stage inverter buffers.
5. The physical layout partition was also a good criteria for the partition. Some of the logics might have a physical place in the die for timing and floorplanning consideration. If they stayed in the same block, it would be easier to specify attributes and to generate structural VHDL separately for layout tools. This also simplified the layout process when only minor changes were needed in a given block, allowing the other blocks to remain undisturbed. Most of the layout was kept the same with only the changed block needing to be recompiled and layout.
6. Instantiated blocks such as RAM, ROM, register file, and datapaths stayed in the same block such as HSRAM and ROM in the design of CHIP.

13.7 EXERCISES

13.1 Write the VHDL code for the TREEC block. The TREEC component declaration is shown in the SYN VHDL code. Synthesize TREEC VHDL code verifying that a tree of inverters is generated correctly.

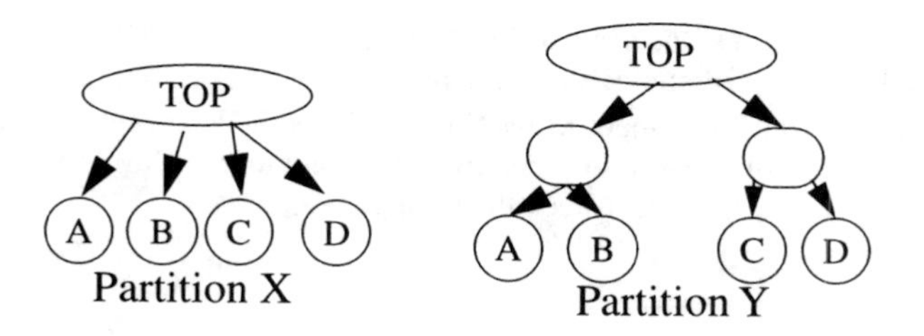

FIGURE 13.15 VHDL hierarchy partitioning.

13.2 Write a configuration VHDL code to configure TREEC.

13.3 The VHDL code for package PACK is not shown. Write the package PACK to have procedures LATCHR, KMUX31, and KMUX41. Try to overload the procedures so that the same procedure name can take both the single bit input and an unconstrained bus input.

13.4 In the MEMLOGIC block, a counter SCNTEDP32 is used, but the VHDL code is not shown. Write a VHDL code to implement SCNTEDP32. Simulate and synthesize the VHDl code. Determine the counter's maximum running speed. Modify the VHDL code so that a generic N of integer type can be passed in, to generate a N-bit down counter.

13.5 In the MEMLOGIC block, TIMER_CNTR has enabled every TIMER_PROG + 2 clock cycles. What should be changed in the MEMLOGIC VHDL code so that TIMER_CNTR can be enabled for every TIMER_PROG + 1 clock cycles?

13.6 In the CHIP test bench, a block PCIMTR is used to generate address and data traces text files. The VHDL code is not shown. Write a VHDL code for the PCIMTR block. (Hint: data is transferred when both IRDYn and TRDYn are asserted.)

13.7 In the CHIP test bench, the ARBITOR block VHDL code is not shown. Write VHDL code for the ARBITOR block.

13.8 In the design of CHIP, CBEn is assumed to be "0000" in the data phase. What needs to be changed to allow combinations of CBEn values in the data phase?

13.9 In the design of CHIP, PCI signal IRDYn is assumed asserted for the whole transaction. What needs to be changed to allow the PCI master agent to deassert IRDYn (master is not ready and the target should wait) for any number of clock cycles after seeing TRDYn is asserted?

13.10 In the design of CHIP, each data transfer is done with two clock cycles. What needs to be changed if the data transfer is done in one clock cycle?

13.11 Design project: Obtain a TI Databook. Write a VHDL code for SN54ALS616 16 bits error detection and correction (EDAC) circuits. Write a VHDL test bench to fully verify its functions. Synthesize the VHDL code to generate a netlist. Check the timing. Layout the design if you have access to layout tools.

13.12 Design project: Do the same as in Exercise 13.11 for SN54ALS632A 32 bits EDAC circuits.

13.13 Design project: Do the same as in Exercise 13.11 for INTEL 8259 programmable interrupt controller.

13.14 Design project: Do the same as in Exercise 13.11 for INTEL 8254 programmable interval timer.

13.15 Design project: Do the same as in Exercise 13.11 for a UART circuit such as AMI 8250 asynchronous communication element.

13.16 Design project: Implement M by N parallel Booth-Wallace Tree multiplier with VHDL code where M and N are positive even numbers. Develop a VHDL test bench. Synthesize VHDL code. Check the timing.

Chapter 14

VHDL'93

In 1993, IEEE approved the new VHDL standard, and a *Language Reference Manual* was published. The purpose of this chapter is to address most of the VHDL'93 important updates. The *Language Reference Manual* (LRM) is recommended for more detailed and complete study.

14.1 MORE REGULAR SYNTAX

In VHDL'87, the declaration syntax is not uniform. For example, the following syntax is for the entity and component declarations. In the first line of each declaration, the entity declaration has the required reserved word **is**, while the component declaration does not have it. In the last line of each closing declaration, an optional entity simple name can be added in the entity declaration, while a reserved word **component** is used to close the component declaration. The same problem exists in packages, architectures, subprograms, and other declarations. This places extra burden on the user to remember the exact syntax of each declaration.

entity_declaration ::= **entity** identifier **is**

---

end [entity_simple_name] ;

component_declaration ::= **component** identifier

---

end component ;

To maintain upward portability and to make the syntax more uniform, VHDL'93 updated the syntax which is shown here. Only the beginning and the closing of the declarations are shown below because the middle portion of the declarations remains the same. The added portions of VHDL'93 are underlined.

entity_declaration ::= **entity** identifier **is**

---

end [**entity**] [entity_simple_name] ;

component_declaration ::= **component** identifier [**is**]
---
end component [component_simple_name];

architecture_body ::= **architecture** identifier **of** entity_name **is**
---
end [**architecture**] [architecture_simple_name] ;

configurartion_declaration ::= **configuration** identifier **of** entity_name **is**
---
end [**configuration**] [configuration_simple_name] ;

function_body_declaration::= function_declaration **is**
---
end [**function**] [function designator] ;

procedure_body_declaration::= procedure_declaration **is**
---
end [**procedure**] [procedure designator] ;

package_declaration ::= **package** identifier **is**
---
end package [identifier] ;

package_body ::= **package body** identifier **is**
---
end [**package body**] [identifier] ;

It is not easy to remember what is required and what is optional. The only common rule of the preceding declarations is that the identifier at the end of the closing declaration is optional. The easiest way is to write all the required and optional portions of reserved words to match the beginning and the closing of the declarations. The following VHDL code shows a package example with VHDL'93 syntax in lines 7 and 22. Statements inside the package declaration (lines 2 to 6) and inside the package body (lines 9 to 21) will be discussed later. The added syntax (mostly optional) for VHDL'93 is underlined.

```
1     package V93PACK is
2        shared variable NUMBER : integer := 10;
3        pure function PLUS1(DIN : integer) return integer;
4        impure function MINUS1 (DIN : integer) return integer;
```

```
5         procedure TIME2 (signal DIN : in integer;
6            signal DOUT : out integer);
7      end package V93PACK;
8      package body V93PACK is
9         pure   function PLUS1 (DIN : integer) return integer is
10        begin
11           oplabelb : return DIN + 1;
12        end function PLUS1;
13        impure function MINUS1 (DIN : integer) return integer is
14        begin
15           oplabelc : return NUMBER - DIN;
16        end function MINUS1;
17        procedure TIME2 (signal DIN : in integer;
18           signal DOUT : out integer) is
19        begin
20           oplabeld : DOUT <= DIN * 2 after 2 ns;
21        end procedure TIME2;
22     end package body V93PACK;
```

14.2 SEQUENTIAL STATEMENTS

VHDL'93 allows an optional label for each sequential statement. The following shows *if statements* and *case statements* syntax with an optional label in the beginning of the statement and the optional repeated label after the reserved words **end if** (**case**) at the closing of the statements.

if_statment ::= [label :]   **if** condition **then**

---

end if [label] ;

case_statement ::= [label :]   **case** expression **is**

---

end case [label] ;

A new report statement is added in VHDL'93 which is similar to the assert statement without the beginning reserved word **assert** and BOOLEAN expression portion. This statement is useful to generate a message without the assert FALSE portion as in VHDL'87. The default severity level of the report statement is NOTE rather than ERROR in the assert statement.

report_statement ::= [label :] **report** expression [**severity** expression] ;

The VHDL'93 signal assignment statement syntax is changed as follows :

signal_assignment_statement ::= [label :] target [delay_mechanism] waveform;

delay_mechanism ::= **transport** | [**reject** time_expression] **inertial**

waveform ::= waveform_element { , waveform_element } | **unaffected**

The waveform element is the same as in VHDL'87. Reserved word **inertial** can be optional or explicitly specified to indicate inertial delay. The inertial delay has another optional reject clause (reserved word **reject** followed by a time expression). The time expression specifies a pulse rejection limit. If the reject clause is not explicitly specified, the pulse rejection limit is specified by the time expression associated with the first waveform element. An error occurs when the pulse rejection limit is either negative or greater than the timing expression associated with the first waveform element. Note that when the rejection clause is specified, the reserved word **inertial** is required (not optional). The reserved words **transport** and **inertial** for the delay mechanism can be either in a concurrent or a sequential signal assignment statement.

The following VHDL code shows examples of using reserved words **inertial**, **transport**, and **reject**. Figure 14.1 shows the simulation waveform. Note that signals X, Y, and Z have exactly the same waveform. Signal T has a pulse of 5 ns due to the transport delay of A nand B. Signal W has the same waveform as signal T since the pulse is wider than 3 ns rejection pulse width (the 5 ns pulse will not be rejected). Signal U has the same waveform as X, Y, and Z since the rejection pulse width is wider than 5 ns and thereby rejecting the 5 ns pulse.

```
1     entity DELAY93 is
2     end entity DELAY93;
3     architecture BEH of DELAY93 is
4        signal A, B, T, X, Y, Z, W, U : bit;
5     begin
6        p0 : process (A, B) is
7        begin
8           T <= transport A nand B after 10 ns;
9           X <= A nand B after 10 ns;
10          Z <= reject 10 ns inertial A nand B after 10 ns;
11          W <= reject  3 ns inertial A nand B after 10 ns;
12       end process p0;
13       Y <= inertial A nand B after 10 ns;
14       U <= reject  6 ns inertial A nand B after 10 ns;
15       p1 : process is
16       begin
17          A <= '0', '1' after 20 ns, '0' after 40 ns,
18                '1' after 60 ns;
19          B <= '0', '1' after 30 ns, '0' after 35 ns,
20                '1' after 50 ns;
21          wait for 80 ns;
22       end process p1;
23    end architecture BEH;
```

14.3 CONCURRENT STATEMENTS

The reserved word **unaffected** can only appear in a concurrent signal assignment statement. It does not introduce a transaction by putting the same value into the signal driver. This allows the signal to maintain its old value without evaluating the new value even though the new value may be the same. This improves the simulation performance.

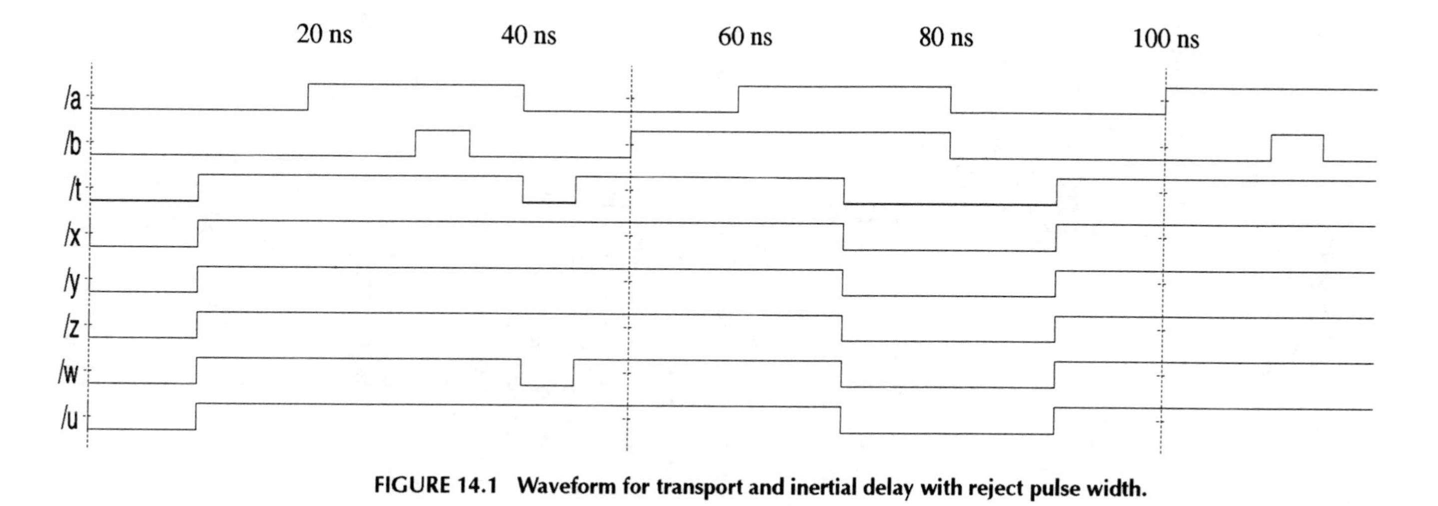

FIGURE 14.1 Waveform for transport and inertial delay with reject pulse width.

The syntax of the concurrent process, block, and generate statements is modified in VHDL'93 as follows:

process_statement ::=
[process_label :] [**postponed**] **process** [(sensitivity_list)] [**is**]
declaration
begin
sequential statements
end [postponed] process [process_label] ;

block_statement ::= block_label : **block** [(guard_expression)] [**is**]
declaration
begin
concurrent statements
end block [block_label] ;

generate_statement ::=
generate_label : **for** generate_parameter_specification | **if** condition **generate**
declaration
begin
concurrent statements
end generate [generate_label] ;

Reserved word **is** is optional in the process and block statements before the reserved word **begin**. An optional declaration and the reserved word **begin** are added in the *generate statement* making the syntax more regular.

Use of reserved word **postponed** can be optional in *concurrent statements* that have equivalent *process statements* (concurrent procedure call, concurrent signal assignment, concurrent assert statements). The optional reserved word **postponed** can be added before the reserved word **process,** as shown in the preceding process statement syntax. A postponed process is only activated in the last delta of the simulation cycle. For example, the following VHDL code example shows the difference between a postponed process and a regular process. Reserved word **postponed** is used in line 14. Both architectures BEH and BEH1 are about the same, except in lines 8 and 14.

```
1     entity POST is
2        signal A, A1 : integer;
3     end entity POST;
4     architecture BEH of POST is
5     begin
6        A  <= 0, 1 after 2 ns, 2 after 4 ns;
7        A1 <= A + 1;
8        assert A1 = A + 1 report "A1 not equal to A + 1" severity NOTE;
9     end architecture BEH;
10    architecture BEH1 of POST is
11    begin
12       A  <= 0, 1 after 2 ns, 2 after 4 ns;
```

```
13        A1 <= A + 1;
14        postponed assert A1 = A + 1 report "A1 not equal to A + 1";
15    end architecture BEH1;
```

In line 8 of architecture BEH, the *concurrent assert statement* will be activated and evaluated whenever there is an event on A or A1. When the *assert statement* is evaluated, there exists a delta time such that A1 is not equal to A + 1, which results in a printed message. In line 14 of architecture BEH1, the reserved word **postponed** ensures that the assert statement is evaluated only at the last delta delay before advancing the simulation time. A1 is always equal to A + 1 except at time 0 in the initialization cycle. The following messages are from the actual simulation of the preceding VHDL'93 code. Only one message is generated in architecture BEH1, while four messages are generated from architecture BEH.

```
# Loading ./work.post(beh)
# ** Note: A1 not equal to A + 1
#   Time: 0 ns  Iteration: 0  Instance:/
# ** Note: A1 not equal to A + 1
#   Time: 0 ns  Iteration: 1  Instance:/
# ** Note: A1 not equal to A + 1
#   Time: 2 ns  Iteration: 0  Instance:/
# ** Note: A1 not equal to A + 1
#   Time: 4 ns  Iteration: 0  Instance:/

# Loading ./work.post(beh1)
# ** Error: A1 not equal to A + 1
#   Time: 0 ns  Iteration: 0  Instance:/
```

It is important to know the restrictions of using the reserved word **postponed**. The corresponding process could not introduce more delta delay since it is assumed to be the last delta in that simulation cycle before advancing the simulation time. This implies that the corresponding process could not assign a value to a signal with 0 time delay, or that it must simply wait for 0 ns. The use of the postponed process may improve the simulation speed; however, its use requires caution. For example, a *concurrent signal assignment statement* "postponed A <= B;" results in error, while "postponed A <= B after 0.001 ns;" would be fine.

14.4 NEW RESERVED WORDS

In VHDL'93, the following reserved words are added:

group	impure	inertial	literal	postponed
pure	reject	rol	ror	shared
sla	sll	sra	srl	unaffected
xnor				

Reserved word **xnor** is a BOOLEAN operator, which is the inverted result of the operator **xor**. For example, in the VHDL'87, a signal assignment statement Z <= **not** (A **xor** B); can be written in VHDL'93 as Z <= A **xnor** B;.

Reserved words **pure** and **impure** are used in function declarations. A **pure** function returns the same value every time when the same parameters are passed in. An **impure** function may return different values for each function call, even though the same parameters are passed in. The predefined function NOW, which returns the current simulation time, is an example of an impure function. Depending on when it is called, the returned values are normally different. The function declaration syntax is then changed in VHDL'93 to have the optional reserved word **pure** or **impure** before the reserved word **function**. The V93PACK VHDL code in Section 14.1 shows examples of a pure and an impure function declaration and function body. Lines 3 and 4 declare the pure function PLUS1 and the impure function MINUS1. The function bodies are specified in lines 9 to 12 and lines 13 to 16 respectively. Note that the reserved words **function** and **procedure** are optional in the closing of the subprogram body such as in lines 12, 16, and 21.

Reserved word **shared** is used to declare a global variable. A shared variable is not the same as the local variable used inside a process and subprograms. A shared (global) variable can be read and updated by many processes and subprograms. It cannot be declared in a region (inside a process or a subprogram body) where a local variable can be declared. For example, line 2 of V93PACK package declares a **shared** variable NUMBER in a package.

A new reserved word **inertial** is added so that there is the option to explicitly specify it in the signal assignment statement in the same way as the reserved word **transport**. The default still refers to inertial delay if none of the reserved words **inertial** or **transport** is specified.

Reserved words **rol**, **ror**, **sla**, **sll**, **sra**, **srl** are *rotate left*, *rotate right*, *shift left arithmetic*, *shift left logical, shift right arithmetic, and shift right logical* operators, respectively. Each of them takes a left operand of one-dimensional array with element of type BIT or BOOLEAN and a right operand of type INTEGER to indicate how many positions of shifting and rotating. If the right integer operand is a negative number, the other direction is used for shifting or rotating. For the basic shift operation of the **sll** (**srl**) operator, the T'left value is used for the target right (left) most element where T is the base type of the element. This basic operation is repeated the same number of times as the right operand integer. For example, TARGET := "10110011" **sll** 1;, the value of TARGET is "01100110". The left-most bit of '1' is dropped, the right-most source 7 bits go to the left most 7 bits of TARGET, the right-most bit of TARGET is replaced with BIT'left which is '0'. For the **sla** (**sra**) operator, the right- (left) most bit of the left operand is used to replace the right (left) most bit of the target. For example, TARGET := "10110011" **sla** 2;, the value of TARGET is "11001111".

In VHDL'93, a new reserved word **group** is introduced to group many related items and to attach a single value for annotation. The syntax of the group template declaration and the group declaration is as follows:

group_template_declaration ::=

group group_template_name **is** (entity_class [<>] {, entity_class [<>] }) ;

group_declaration ::= **group** group_template_name (group_constituent_list);

group_constituent_list ::= name | character_literal {, name | character_literal }
entity_class ::= **entity** | **architecture** | **configuration** | **procedure** | **function** | **package** | **type** | **subtype** | **constant** | **signal** | **variable** | **component** | **label** | **literal** | **units** | **group** | **file**

The following VHDL code shows an example of defining a group. Line 4 declares signals. Line 5 defines a group template which has a pair of signal entity classes. Lines 6 to 9 are group declarations of signal pairs. Line 10 is another group template with an unconstrained group as the entity class. Line 11 is a group declaration based on the group template PATH_GROUP with four groups (PATH1, PATH2, PATH3, PATH4). Line 12 is an attribute declaration. Line 13 is the attribute specification which annotates a delay of 2 ns to group PATHS.

```
1     entity GROUPEX is
2     end entity GROUPEX;
3     architecture RTL of GROUPEX is
4        signal DA, DB, DZ : bit_vector(3 downto 0);
5        group SIG2SIG is (signal, signal);
6        group PATH1 : SIG2SIG (DA(3), DZ(3));
7        group PATH2 : SIG2SIG (DB(2), DZ(2));
8        group PATH3 : SIG2SIG (DA(1), DZ(1));
9        group PATH4 : SIG2SIG (DB(0), DZ(0));
10       group PATH_GROUP is (group <>);
11       group PATHS       : PATH_GROUP (PATH1, PATH2, PATH3, PATH4);
12       attribute PATH_DELAY : time;
13       attribute PATH_DELAY of PATHS : group is 2 ns;
14    begin
15       gate3 : DZ(3) <= DA(3) nand DB(3);
16       gate2 : DZ(2) <= DA(2) nor  DB(2);
17       gate1 : DZ(1) <= DA(1) xor  DB(1);
18       gate0 : DZ(0) <= DA(0) xnor DB(0);
19    end architecture RTL;
```

Note that the reserved word **literal** is used as one of the entity class.

14.5 PREDEFINED STANDARD PACKAGE

The predefined package STANDARD has been updated. The following VHDL code shows part of the package STANDARD that differs from VHDL'87. The type CHARACTER is extended to have 256 characters. The first 128 characters are the same.

```
1     package STANDARD is
2        type CHARACTER is (
3           nul, soh, stx, etx, eot, enq, ack, bel,
4           bs,  ht,  lf,  vt,  ff,  cr,  so,  si,
5           dle, dc1, dc2, dc3, dc4, nak, syn, etb,
6           can, em,  sub, esc, fsp, gsp, rsp, usp,
7
```

```
8        ' ', '!', '"', '#', '$', '%', '&', ''',
9        '(', ')', '*', '+', ',', '-', '.', '/',
10       '0', '1', '2', '3', '4', '5', '6', '7',
11       '8', '9', ':', ';', '<', '=', '>', '?',
12
13       '@', 'A', 'B', 'C', 'D', 'E', 'F', 'G',
14       'H', 'I', 'J', 'K', 'L', 'M', 'N', 'O',
15       'P', 'Q', 'R', 'S', 'T', 'U', 'V', 'W',
16       'X', 'Y', 'Z', '[', '\', ']', '^', '_',
17
18       '`', 'a', 'b', 'c', 'd', 'e', 'f', 'g',
19       'h', 'i', 'j', 'k', 'l', 'm', 'n', 'o',
20       'p', 'q', 'r', 's', 't', 'u', 'v', 'w',
21       'x', 'y', 'z', '{', '|', '}', '~', del,
22
23       c128, c129, c130, c131, c132, c133, c134, c135,
24       c136, c137, c138, c139, c140, c141, c142, c143,
25       c144, c145, c146, c147, c148, c149, c150, c151,
26       c152, c153, c154, c155, c156, c157, c158, c159,
27
28       ' ', '¡', '¢', '£', '¤', '¥', '|', '§',
29       '¨', '©', 'ª', '«', '¬', '-', '®', '¯',
30       '°', '¿', '2', '3', '´', 'u', '¶', '·',
31       '¸', '1', 'º', '»', '¿', '¿', '¿', '¿',
32
33       'À', 'Á', 'Â', 'Ã', 'Ä', 'Å', 'Æ', 'Ç',
34       'È', 'É', 'Ê', 'Ë', 'Ì', 'Í', 'Î', 'Ï',
35       'D', 'Ñ', 'Ò', 'Ó', 'Ô', 'Õ', 'Ö', 'x',
36       'Ø', 'Ù', 'Ú', 'Û', 'Ü', 'Y', 'P', 'ß',
37
38       'à', 'á', 'â', 'ã', 'ä', 'å', 'æ', 'ç',
39       'è', 'é', 'ê', 'ë', 'ì', 'í', 'î', 'ï',
40       'd', 'ñ', 'ò', 'ó', 'ô', 'õ', 'ö', '/',
41       'ø', 'ù', 'ú', 'û', 'ü', 'y', 'p', 'ÿ' );
42
43    type TIME is range -2147483647 to 2147483647
44       units
45          fs;
46          ps  = 1000 fs;
47          ns  = 1000 ps;
48          us  = 1000 ns;
49          ms  = 1000 us;
50          sec = 1000 ms;
51          min =   60 sec;
52          hr  =   60 min;
53       end units;
54    subtype DELAY_LENGTH is TIME range 0 fs to TIME'high;
55    impure function NOW return DELAY_LENGTH;
56    type FILE_OPEN_KIND is (READ_MODE, WRITE_MODE, APPEND_MODE);
57    type FILE_OPEN_STATUS is (OPEN_OK, STATUS_ERROR, NAME_ERROR,
58       MODE_ERROR);
59    attribute FOREIGN : string;
```

```
60    end package STANDARD;
```

A subtype DELAY_LENGTH is declared in line 54. The function NOW in line 55 has the reserved word **impure** to indicate the function is an impure function. Lines 56 and 57 declare types FILE_OPEN_KIND and FILE_OPEN_STATUS for the file I/O, which will be discussed in Section 14.8. Line 59 declares the new attribute FOREIGN, which can be used to interface with other simulators to provide simulation results of an entity or a function.

14.6 ATTRIBUTES

VHDL'87 attributes STRUCTURE (ENT'structure) and BEHAVIOR (ENT'behavior) are deleted in VHDL'93.

S'DRIVING returns a BOOLEAN value FALSE if the current value of the driver for signal S in the current process is determined by the *null transaction*, otherwise, it returns TRUE. If S is a composite signal, it returns TRUE when R'DRIVING is TRUE for every element R of S.

A new attribute DRIVING_VALUE associated with a signal is added. This attribute can also be attached to a port signal of mode **out**. In VHDL, a port signal of mode **out** cannot be read. A local variable or a temporary signal is usually created. The JKFF VHDL code in Chapter 4 exercises is an example. The following VHDL'93 code shows the JK flipflop model without using a variable or a temporary signal. Attribute DRIVING_VALUE is used in lines 17 and 18. Line 14 uses a subtype BITS2 to qualify the expression J & K,

```
1     entity JKFF93 is
2        port (
3           CLK, RSTn, J, K : in  bit;
4           Q               : out bit);
5     end entity JKFF93;
6     architecture RTL93 of JKFF93 is
7     begin
8        process (CLK, RSTn)
9           subtype BITS2 is bit_vector(1 downto 0);
10       begin
11          if (RSTn = '0') then
12             Q <= '0';
13          elsif (CLK'event and CLK = '1') then
14             case BITS2'(J & K) is
15                when "01" => Q <= '0';
16                when "10" => Q <= '1';
17                when "11" => Q <= not Q'driving_value;
18                when "00" => Q <= Q'driving_value;
19             end case;
20          end if;
21       end process;
22    end architecture RTL93;
```

T'ASCENTING returns a BOOLEAN value TRUE for an array or a subtype with an ascending range, otherwise, it returns FALSE.

T'IMAGE(X) returns a string image for the argument X of scalar or subtype T. For example, INTEGER'IMAGE(NUMBER) returns the string value of INTEGER NUMBER. This is useful in the assert and report statements that a string type expression is expected.

E'SIMPLE_NAME returns a string for the named entity E.

E'INSTANCE_NAME returns a string for the hierarchical path starting from the root of the design for the named entity E.

E'PATH_NAME returns a string for the hierarchical path to that named entity E.

Note that the named entity is an entity class as defined in Section 4. The preceding four attributes are useful for combining with *assert or report statements* to generate messages that show exactly where they are generated in the VHDL code.

14.7 DIRECT COMPONENT INSTANTIATION

VHDL'87 requires a component declaration either in the architecture declaration region or in a package which is referenced. The component declaration is rather lengthy and redundant. Also, a configuration of either hard or soft binding is required to bind the component. VHDL'93 introduces the direct instantiation which eliminates the requirement for a component declaration and combines the binding with the component instantiation statement. The following VHDL codes (PLUS2 and V93ENT) illustrate several VHDL'93 constructs that we have discussed.

VHDL code PLUS2 gets the input DIN of type integer and adds 2 to get output DOUT (line 8). Note that optional reserved word **inertial** is used. It has also the optional reserved word **postponed** to indicate a postponed concurrent signal assignment statement. A postponed process cannot introduce additional delta delay. Therefore, an after clause (after 2 ns) is required. Without the after clause, it would be an error. Lines 10 to 13 declare the configuration for PLUS2. It is an empty configuration since PLUS2 does not have any lower-level component (nothing between lines 11 and 12).

```
1     entity PLUS2 is
2        port (
3           DIN  : in  integer;
4           DOUT : out integer);
5     end entity PLUS2;
6     architecture V93ARCH of PLUS2 is
7     begin
8        label0 : postponed DOUT <= inertial DIN + 2 after 2 ns;
9     end architecture V93ARCH;
10    configuration CFG_PLUS2_V93ARCH of PLUS2 is
11       for V93ARCH
12       end for;
13    end configuration CFG_PLUS2_V93ARCH;
```

Referring to the following VHDL code (V93ENT), line 1 references the package V93PACK as discussed earlier. The ports are declared in lines 3 to 8. Line 11 declares a shared (global) variable. Note that a shared (global) variable cannot be declared in a region where a local variable can be declared (inside a process statement and a subprogram). A shared variable can be updated in more than one process, and the value will be updated immediately when the variable assignment statement is executed. Line 12 declares signals.

Lines 14 to 18 show a block statement with the VHDL'93 syntax (optional reserved word **is**). Lines 16 and 17 are for the direct component instantiation. The port map portion is the same. The direct instantiation in line 16 (underlined) replaces the component name with the binding use clause (library, entity, and architecture name in parenthesis). Note that there is no semicolon in line 16. Lines 19 to 23 illustrate a direct component instantiation statement inside a generate statement. VHDL'93 has an optional reserved word **begin** (line 20). Lines 21 and 22 are another example for direct instantiation with a reserved word **configuration** where line 16 uses the reserved word **entity**. Note that the component declaration for PLUS2 is not necessary if direct component instantiation is used. This reduces the amount of VHDL coding. The *direct component instantiation statement* is a *concurrent component instantiation statement*. It can exist without the enclosing *block statement* or the *generate statement*. The examples show that they are inside a *block statement* and a *generate statement* just to illustrate the VHDL'93 syntax for these two kinds (block and generate) of statements.

Line 24 depicts an example of a *postponed concurrent procedure call* TIME2 (in package V93PACK). Again, the *signal assignment statement* in the procedure requires an after clause to be legally postponed (line 20 of package body V93PACK).

```
1    use work.V93PACK.all;
2    entity V93ENT is
3       port (
4          EN     : in  bit;
5          SEL    : in  bit_vector(1 downto 0);
6          VALIN  : in  integer;
7          VALOUT : out integer;
8          NUMOUT : out integer);
9    end entity V93ENT;
10   architecture V93ARCH of V93ENT is
11      shared variable V93VAR  : integer;
12      signal SIG1, SIG2, SIG3 : integer;
13   begin
14      rqlabel0 : block is
15      begin
16         rqlabel1 : entity WORK.PLUS2(V93ARCH)
17            port map (DIN => VALIN, DOUT => SIG2);
18      end block rqlabel0;
19      rqlabel2 : if TRUE generate
20      begin
21         rqlabel3 : configuration WORK.CFG_PLUS2_V93ARCH
22            port map (DIN => SIG2, DOUT => SIG3);
23      end generate rqlabel2;
24      oplabela : postponed TIME2(SIG3, VALOUT);
25      oplabel1 : postponed process (VALIN, SEL) is
26      begin
```

```
27          oplabel0 : report "SIG1 Name = " & SIG1'simple_name;
28          oplabel2 : case SEL(0) is
29                        when '0' =>
30                           NUMBER := PLUS1(VALIN);
31                        when '1' =>
32                           NUMBER := MINUS1(VALIN);
33                     end case oplabel2;
34          oplabel3 : if SEL(1) = '0' then
35                        NUMBER := MINUS1(VALIN);
36                     else
37                        NUMBER := PLUS1(NUMBER);
38                     end if oplabel3;
39          oplabel4 : report "NUMBER = " & INTEGER'image(NUMBER);
40          oplabel5 : SIG1 <= inertial NUMBER after 5 ns;
41       end postponed process oplabel1;
42       oplabel6 : postponed process is
43       begin
44          wait for 10 ns;
45          oplabel7 : report "OPLABEL6 Path = " & OPLABEL6'path_name;
46          oplabela : report "SIG2 instance = " & SIG2'instance_name;
47          oplabel8 : NUMBER := NUMBER + 10;
48       end postponed process oplabel6;
49       oplabel9 : NUMOUT <= transport SIG1 + 2 when EN = '1'
50                            else unaffected;
51    end architecture V93ARCH;
```

Lines 25 to 41 show an example of a process statement with report, case, and if sequential statements inside. Line 27 is a report statement with an attribute SIMPLE_NAME associated with signal SIG1. Lines 28 to 33 show a VHDL'93 case statement syntax with an optional label. Lines 34 to 38 depict a VHDL'93 if statement syntax with an optional label. The shared variable NUMBER is updated in this process by calling various (pure and impure) functions. Line 39 is another report statement with an attribute IMAGE for NUMBER of type integer. Line 40 is a sequential signal assignment statement with optional reserved word **inertial** and an optional label.

Lines 42 to 48 are another postponed process statement example. Note that the shared variable NUMBER is updated in line 47 (and lines 30, 32, 35, 37) inside two different process statements. Lines 45 and 46 show examples of report statements with attributes PATH_NAME and INSTANCE_NAME associated with a label and a signal.

Lines 49 and 50 show an example of a concurrent signal assignment statement using the reserved word **unaffected**. Note that the reserved word **unaffected** can only be used in a concurrent signal assignment statement. It cannot be used in a sequential signal assignment statement.

The purpose of the preceding example is to show the VHDL'93 constructs which are different from VHDL'87 without considering much of the actual functionality. However, the waveform in Figure 14.2 shows the simulation results obtained by the following simulation commands:

force valin 10, force sel 00, force en 0, run 20

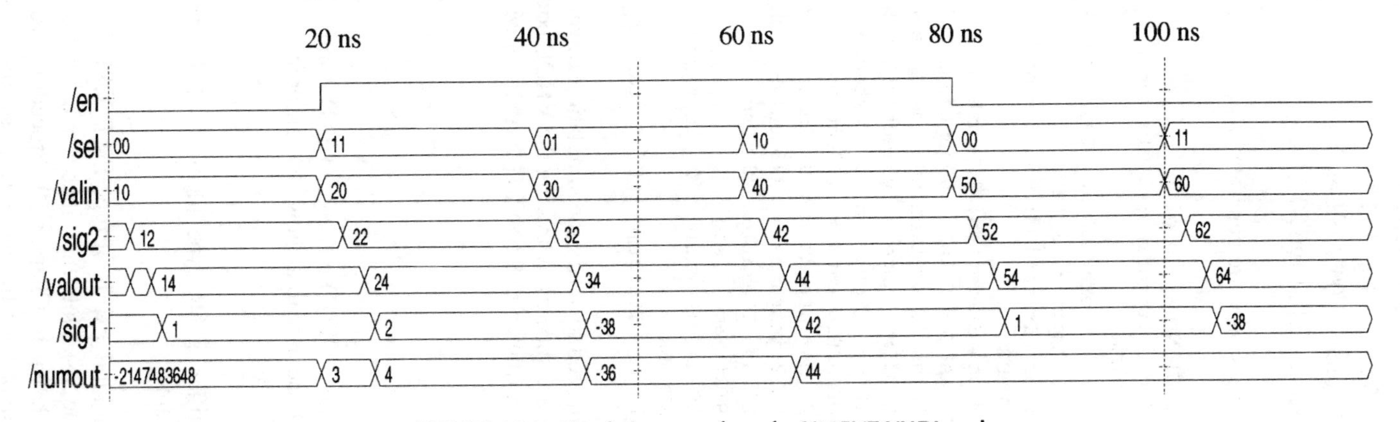

FIGURE 14.2 Simulation waveform for V93ENT VHDL code.

force valin 20, force sel 11, force en 1, run 20
force valin 30, force sel 01, run 20
force valin 40, force sel 10, run 20
force valin 50, force sel 00, force en 0, run 20
force valin 60 force sel 11 run 20

The following messages are generated from the various report statements:

```
1    # ** Note: SIG1 Name = sig1
2    #    Time: 0 ns  Iteration: 0  Instance:/
3    # ** Note: NUMBER = 1
4    #    Time: 0 ns  Iteration: 0  Instance:/
5    # ** Note: SIG1 Name = sig1
6    #    Time: 0 ns  Iteration: 1  Instance:/
7    # ** Note: NUMBER = 1
8    #    Time: 0 ns  Iteration: 1  Instance:/
9    # ** Note: OPLABEL6 Path = :v93ent:oplabel6:
10   #    Time: 10 ns  Iteration: 0  Instance:/
11   # ** Note: SIG2 instance = :v93ent(v93arch):sig2
12   #    Time: 10 ns  Iteration: 0  Instance:/
13   # ** Note: OPLABEL6 Path = :v93ent:oplabel6:
14   #    Time: 20 ns  Iteration: 0  Instance:/
15   # ** Note: SIG2 instance = :v93ent(v93arch):sig2
16   #    Time: 20 ns  Iteration: 0  Instance:/
17   # ** Note: SIG1 Name = sig1
18   #    Time: 20 ns  Iteration: 2  Instance:/
19   # ** Note: NUMBER = 2
20   #    Time: 20 ns  Iteration: 2  Instance:/
```

You are encouraged to verify the simulation waveform and the output messages (only the first 20 ns are shown) by tracing the VHDL code (this is left as an exercise). Note the format of attributes INSTANCE_NAME, SIMPLE_NAME, PATH_NAME, and IMAGE.

14.8 FILE AND TEXT I/O

VHDL'93 updated the TEXTIO package in STD library. The following VHDL code is part of the TEXTIO package that shows the differences. The file declaration has been changed as follows:

file file_identifier : file_type_identifier [[**open** file_open_kind_exp] **is** file_logical_name];

The file_open_kind_exp is of type FILE_OPEN_KIND declared in the STANDARD package (line 56). Note that lines 6 and 7 are different from lines 6 and 7 in TEXTIO package shown in Chapter 10. Procedures READLINE and WRITELINE (lines 8 and 14) cannot have the mode (in, out, inout) for the file parameter. Procedures READ and WRITE are the same as VHDL'87.

```
1     package TEXTIO is
2        type LINE is access string;
3        type TEXT is file of string;
4        type SIDE is (right, left);
5        subtype WIDTH is natural;
6        file input  : TEXT open READ_MODE  is "STD_INPUT";
7        file output : TEXT open WRITE_MODE is "STD_OUTPUT";
8        procedure READLINE(file f: TEXT; L: out LINE);
9        procedure READ(L:inout LINE; VALUE: out bit; GOOD:out BOOLEAN);
10       procedure READ(L:inout LINE; VALUE: out bit);
11       procedure READ(L:inout LINE; VALUE: out bit_vector;
12          GOOD : out BOOLEAN);
13       procedure READ(L:inout LINE; VALUE: out bit_vector);
14       procedure WRITELINE(file f : TEXT; L : inout LINE);
15       procedure WRITE(L : inout LINE; VALUE : in bit;
16          JUSTIFIED: in SIDE := right; FIELD: in WIDTH := 0);
17       procedure WRITE(L : inout LINE; VALUE : in bit_vector;
18          JUSTIFIED: in SIDE := right; FIELD: in WIDTH := 0);
19    end package TEXTIO;
```

The function ENDFILE is implicitly defined to return a BOOLEAN value TRUE when the end of file is reached. Procedures FILE_OPEN and FILE_CLOSE are also implicitly declared when a file is declared. Procedure FILE_CLOSE has a file name as a parameter of mode **in**. Procedure FILE_OPEN is overloaded as follows:

```
procedure FILE_OPEN(
   filename              : filetype;
   file_physical_name    : in STRING;
   open_kind             : in FILE_OPEN_KIND := READ_MODE);
procedure FILE_OPEN(
   status                : out FILE_OPEN_STATUS;
   filename              : filetype;
   file_physical_name    : in STRING;
   open_kind             : in FILE_OPEN_KIND := READ_MODE);
```

The following VHDL code shows examples of reading and writing files using procedures FILE_OPEN and FILE_CLOSE. Line 7 declares a file IFILE of type text without the open mode and file name of type string specified. Line 8 declares a file OFILE with everything specified. The VHDL code will try to open four data files (with file physical names "data0", "data1", "data2", "data3") in line 14 which is inside a while loop statement (lines 13 to 28). The while loop statement in lines 17 to 22 reads each line of the data file, multiplies the integer value by 2, and then writes out to the file "data.out". This is only executed when the data file is opened correctly (checked in line 15). Lines 23 and 24 call the procedure FILE_CLOSE to close the input data file and output data file.

```
1     use STD.TEXTIO.all;
2     entity FILES is
```

```
3     end entity FILES;
4     architecture BEH of FILES is
5     begin
6        parse : process is
7           file IFILE : text;
8           file OFILE : text open APPEND_MODE is "data.out";
9           variable LIN, LOUT : line;
10          variable D         : integer;
11          variable OK        : FILE_OPEN_STATUS;
12       begin
13          eachfile : for i in '0' to '3' loop
14             FILE_OPEN(OK, IFILE, "data" & i, READ_MODE);
15             if (OK = OPEN_OK) then
16                FILE_OPEN(OFILE, "data.out", APPEND_MODE);
17                eachdata : while not ENDFILE(IFILE) loop
18                   READLINE(IFILE, LIN);
19                   READ(LIN, D);
20                   WRITE(LOUT, D*2);
21                   WRITELINE(OFILE, LOUT);
22                end loop eachdata;
23                FILE_CLOSE(IFILE);
24                FILE_CLOSE(OFILE);
25             else
26                report "Cannot open file data" & i ;
27             end if;
28          end loop eachfile;
29          wait;
30       end process parse;
31    end architecture BEH;
```

The data files "data0" (12 and 0 in lines 1 and 2) and "data2" (23 and –2 in lines 1 and 2) exist and files "data1" and "data3" do not exist. After the simulation, file "data.out" has the content as follows. Note that the file "data.out" is opened with APPEND_MODE.

```
1     24
2     0
3     46
4     -4
```

The simulator generates the following messages. Lines 3 to 6 of the simulation messages are generated by the report statement in line 26 when the file is not opened successfully. Lines 1 and 2 of the simulation messages are caused by line 16 to open the file "data.out" at the first time. It is an error because file OFILE has been opened when it is declared in line 8 with everything specified. Note that the declaration of line 8 can be the same as in line 7 so that the error message in lines 1 and 2 will not be generated. This is done to show different ways of declaring a file and also to illustrate that the file is opened in line 8 but not in line 7. The error message in lines 1 and 2 can also be eliminated if line 16 and line 17 are deleted. The VHDL code is used to show that the file can be opened with the APPEND_MODE. When data file "data0" is processed, output file "data.out" has two lines with values of 24 and 0. When "data.out" is opened and written, the values 46 and –4 are appended.

```
1    # ** Error: Error opening file data.out, file already open
2    #    Time: 0 ns  Iteration: 0  Instance:/
3    # ** Note: Cannot open file data1
4    #    Time: 0 ns  Iteration: 0  Instance:/
5    # ** Note: Cannot open file data3
6    #    Time: 0 ns  Iteration: 0  Instance:/
```

14.9 EXTENDED IDENTIFIER

VHDL'93 extends the identifier definitions. An extended identifier can be specified with two enclosing back slashes '\'. This expands the flexibility of identifier names using special characters and reserved words. However, the letters in the extended identifiers are case sensitive. The following VHDL code shows examples of extended identifiers. Note that in line 3, \BEGIN\ and \begin\ are different identifiers. Special characters, spaces, and more than one continuous underscore can be used in an extended identifier. The underscore can also be the last character.

```
1    entity EXTEND is
2       port (
3          \BEGIN\, \begin\, \END \ : in  bit;
4          \BEGIN% nand $END___\    : out bit);
5    end entity EXTEND;
6    architecture \be___gin\ of EXTEND is
7    begin
8       \BEGIN% nand $END___\ <= \begin\ nor (\BEGIN\ nand \END \);
9    end architecture \be___gin\;
```

14.10 EXERCISES

14.1 What are the new reserved words in VHDL'93?

14.2 What is the difference between a local variable and a shared variable?

14.3 Where can a shared variable be declared? What are the differences between a local variable and a shared variable? What applications can best utilize the concept of shared variables?

14.4 What is changed in VHDL'93 for sequential statements?

14.5 What is a postponed process? What are the advantages of using a postponed process compared to a normal process? What is the restriction for a postponed process? In addition to a process statement, what other statements can be postponed?

14.6 What attributes are deleted and added in VHDL'93?

14.7 What is the direct component instantiation? What are the advantages of a direct component instantiation?

14.8 Read the VHDL code V93ENT carefully. Verify the relationships among the VHDL code, the simulation commands, waveform, and messages generated from report statements.

14.9 How is the reserved word **unaffected** used? What is the advantage of using it?

14.10 What is a rejection pulse width? How is that specified? Give an example. Can it be in a sequential signal assignment statement or a concurrent signal assignment statement or both?

14.11 What functions and procedures are implicitly declared for the file I/O?

Appendix A

VHDL'87 Quick Reference

```
1     library IEEE;                                  -- library reference
2     use IEEE.std_logic_1164.all;                   -- package reference
3     use STD.TEXTIO.all;
4        ---------------------------- package declaration
5     package PACK_NAME is
6           -- declaration of subprogram, type, subtype, constant,
7              -- signal, alias, file, component, attribute,
8              -- specification of attribute, disconnetion, use clause are OK.
9         -- variable, subprogram body, configuration specification NOT ALLOWED
10
11       ---------------------- type and subtype declaration
12       type     LOGIC4    is ('X', '0', '1', 'Z');  -- enumerated type
13       type     STATE     is (IDLE, ADDR, DATA);
14       type     ADDRESS   is range 0 to 16#FFFF#;
15       subtype RAM_ADDR  is ADDRESS range 16#F000# to 16#FFFF#;
16       subtype BIT_INDEX is integer range 7 downto 0;
17       subtype BYTE      is BIT_VECTOR(BIT_INDEX);
18       type     B_BYTE    is array (7 downto 0) of std_logic;
19       subtype DBUS      is std_logic_vector(0 to 7);
20       type    TWO_DIM  is array (NATURAL range <>, NATURAL range <>) of bit;
21       subtype WINDOW    is TWO_DIM (0 to 80, 0 to 100);
22       type CELL;                     -- forward reference for access type
23       type POINTER is access CELL; -- access type
24       type CELL is record            -- record type
25          ID   : INTEGER;
26          PTR  : POINTER;
27       end record;
28
29       ---------------------- function and procedure declaration
30       function REDUCE_AND (DIN: in std_logic_vector) return std_logic;
31       procedure DECODER24 (
32          signal DIN  : in  std_logic_vector(1 downto 0);
```

```
33          signal DOUT : out std_logic_vector(3 downto 0)) ;
34
35       --------------------- objects declarations
36       constant DEFER  : integer; -- deferred constant
37       constant PERIOD : time := 30 ns; -- a space between 30 and ns
38       signal   ONEBIT : std_logic := '0'; -- single quote '
39      signal  GLOBAL : std_logic_vector(7 downto 0) := "00000000"; -- use "
40       alias    OPCODE : std_logic_vector(2 downto 0) is GLOBAL(7 downto 5);
41       file     FILEP  : text is in  "rom.in";
42       file     FILEQ  : text is out "rom.out";
43
44       --------------------- attribute declaration
45       attribute NET_WT : integer;
46       --------------------- attribute specification
47       attribute NET_WT of GLOBAL : signal is 23;
48       attribute NET_WT of others : signal is  1;
49    end PACK_NAME;
50
51       ------------------- package body declaration
52    package body PACK_NAME is
53       constant DEFER  : integer := 32; -- deferred constant
54          -- declaration of subprogram, subprogram body, type, subtype,
55             -- constant, alias, file, and use clause are OK.
56       -- variable, signal, attribute, attribute specification NOT ALLOWED
57      -- component, disconnetion and configuration specification NOT ALLOWED
58       --------------------- function body
59       function REDUCE_AND (DIN: in std_logic_vector) return std_logic is
60          variable result: std_logic;
61          -- declaration of subprogram, subprogram body, type, subtype,
62          --    constant, variable, alias, file, attribute,
63          -- specification of attribute, and use clause.
64          -- signal, component, disconnection specification NOT ALLOWED
65          -- configuration specification                    NOT ALLOWED
66       begin
67             -- only sequential statements allowed
68          result := '1';
69          oplabel : for i in DIN'range loop -- optional label
70             result := result and DIN(i);
71          end loop oplabel;                 -- optional label
72          return result;
73       end REDUCE_AND;
74       --------------------- procedure body
75       procedure DECODER24 (
76          signal DIN  : in  std_logic_vector(1 downto 0);
77          signal DOUT : out std_logic_vector(3 downto 0)) is
78          -- declaration part : same as in function body
79       begin
80             -- only sequential statements allowed in procedure body
81          DOUT <= "0000";
82          case DIN is
83             when "00"   => DOUT(0) <= '1';
```

```
84            when "01"    => DOUT(1) <= '1';
85            when "10"    => DOUT(2) <= '1';
86            when "11"    => DOUT(3) <= '1';
87            when others => null; -- null statement
88          end case;
89       end DECODER24;
90    end PACK_NAME;
91
92    --------------------------------------------- entity
93    library IEEE;                  -- library and use clause does not go beyond
94    use IEEE.std_logic_1164.all; -- primary design unit
95    entity NAND_GATE is
96       generic ( DELAY : time := 2 ns ); -- semicolon required ***
97       port (
98          A, B : in  std_logic;
99          Z    : out std_logic);
100         -- declaration of subprogram, subprogram body, type, subtype,
101            -- constant, signal, alias, file, attribute,
102           -- specification of attribute, disconnetion, and use clause are
OK.
103         -- variable, component, configuration specification NOT ALLOWED
104      begin
105         assert (A = B) report "A = B" severity NOTE;
106   end NAND_GATE;
107   --------------------------------------------- architecture
108   architecture ARCH_NAME of NAND_GATE is
109         -- declaration of subprogram, subprogram body, type, subtype,
110            -- constant, signal, alias, file, attribute, component,
111            -- specification of attribute and disconnetion, use clause, and
112            -- configuration specification are OK.
113         -- variable NOT ALLOWED
114   begin
115      Z <= transport A nand B after DELAY; -- optional keyword transport
116   end ARCH_NAME;
117   --------------------------------------------- entity
118   library IEEE;                  -- library and use clause does not go beyond
119   use IEEE.std_logic_1164.all; -- primary design unit
120   use work.PACK_NAME.all;     -- reference package PACK_NAME in library work
121   use STD.TEXTIO.all;
122   use IEEE.std_logic_textio.all;
123   entity ENT_NAME is
124      port (
125         RSTn, CLK, A, B, C, D, E, F : in  std_logic;
126         DBUS             : in  std_logic_vector(7 downto 0);
127         SEL              : in  std_logic_vector(1 downto 0);
128         signal W, X, Y, N, K : out std_logic; -- optional keyword signal
129         signal Z0        : in  std_logic_vector(31 downto 0);
130         signal Z3        : out std_logic_vector(31 downto 0));
131   end ENT_NAME;
132   --------------------------------------------- architecture
133   architecture ARCH_NAME of ENT_NAME is
```

```
134      signal EN, LATCH, TRISTATE1, TRISTATE2 : std_logic;
135      signal Y2 : std_logic_vector(1 downto 0) := "00";
136      signal Y4 : std_logic_vector(3 downto 0);
137      signal Z1, Z2, SZ1, SZ2 : std_logic_vector(31 downto 0);
138
139      --------------------- guarded signal declaration
140      signal GS1 : std_logic bus := '0';         -- guarded signal
141      signal GS2 : std_logic register;           -- guarded signal
142     disconnect GS1    : std_logic after 2 ns; -- disconnect specification
143     disconnect others : std_logic after 1 ns; -- disconnect specification
144
145      --------------------- component declaration
146      component NAND_GATE
147      generic ( DELAY : time := 2 ns ); -- semicolon required ***
148      port (
149         A, B : in  std_logic;
150         Z    : out std_logic);
151      end component;
152
153      --------------------- hard binding configuration specification
154      for nand0 : NAND_GATE
155         use entity work.NAND_GATE(ARCH_NAME);
156
157   begin
158         -- only concurrent statements are allowed here
159      --------------------- concurrent signal assignment statements
160      X  <= A or B;
161      Y  <= '0', '1' after 20 ns, A nor B after 30 ns;
162      Z1 <= (31 downto 4 => '0', others => '1');
163      Z2 <= (Z1'range => '1');
164
165      W  <= A xor REDUCE_AND(Z0)  when C = '1' else -- function call
166            B   after 3 ns        when D = '0' else
167            '0' ;
168
169      oplabel : with SEL select                -- optional label
170         Y2 <= DBUS(1 downto 0) when "00",     -- use comma, not semicolon
171               DBUS(5 downto 4) when "01",     -- slice of an array
172               A & DBUS(7)      when others;  -- end with semicolon
173
174      --------------------- concurrent procedure call, optional label
175      conp : DECODER24(Y2, Y4);
176      --------------------- concurrent assert statement,
177                         -- optional label, optional severity
178      conasrt : assert C = D report "Just checking" severity WARNING;
179
180      --------------------- component instantiation statement
181      nand0 : NAND_GATE -- label required
182         generic map (DELAY => 0.8 ns) -- NO semicolon ***
183         port map (A => E, B => F, Z => K); -- name mapping
184
185      --------------------- if generate statement
```

```
186      ifgen : if TRUE generate   -- label required
187          -- only concurrent statement can be inside generate statement
188        nand0 : NAND_GATE
189           generic map (DELAY => 0.5 ns) -- NO semicolon ***
190           port map (A => C, B => D, Z => N); -- name mapping
191      end generate ifgen;        -- optional label at the end
192
193      --------------------- for generate statement
194      forgen : for i in Z3'range generate -- label required
195          -- only concurrent statement can be inside generate statement
196        nand1 : NAND_GATE
197           generic map (DELAY => 1.2 ns) -- NO semicolon ***
198           port map (Z1(i), Z2(i), Z3(i)); -- positional mapping
199      end generate;          -- optional label at the end
200
201      -------------------------------- infer flip flop, latch, tristate
202      --------------------- process statement with sensitivity list
203                           -- wait statement is NOT ALLOWED
204      oplabel1 : process (RSTn, CLK) -- optional label
205             -- declaration part same as in subprogram declaration part
206      begin
207            -- only sequential statement is allowed inside the process body
208           -------------- if statments
209         if (RSTn = '0') then
210            ONEBIT <= '0';
211         elsif (CLK'event and CLK = '1') then -- infer flip flop
212            if (EN = '1') then
213               ONEBIT <= Y2(0);
214            end if;
215         end if;
216      end process oplabel1; -- optional label
217
218      oplabel2 : process (D, EN) -- optional label
219             -- declaration part same as in subprogram declaration part
220      begin
221            -- only sequential statement is allowed inside the process body
222         if (EN = '1') then
223            LATCH     <= D;   -- infer a latch, not all choices specified
224            TRISTATE2 <= D;
225         else
226           TRISTATE2 <= 'Z'; -- infer a three state buffer with 'Z' assigned
227         end if;
228      end process oplabel2; -- optional label
229
230      TRISTATE1 <= D when EN = '1' else 'Z'; -- infer three state buffer
231
232      --------------------- process statement with no sensitivity list
233                           -- required a wait statement
234      oplabel4 : process  -- optional label
235             -- declaration part same as in subprogram declaration part
236         variable BUFF : line;
237         variable VAR  : std_logic;
```

```
238         variable DONE : boolean;
239      begin
240            -- only sequential statement is allowed inside the process body
241
242         ------------------------- sequential assert statement
243         assert C = D report "Just checking" severity ERROR; -- no label
244
245         ------------------------- sequential procedure call statement
246         DECODER24(GLOBAL(5 downto 4), GLOBAL(3 downto 0)); -- no label
247
248         ------------------------- variable assignment statement
249         VAR := A and B;           -- use :=, not <=
250
251         ------------------------- signal assignment statement
252         X   <= A or B;
253         GS1 <= '0', '1' after 20 ns, null after 30 ns; -- null waveform
254         SZ1 <= (31 downto 4 => '0', others => '1');
255         SZ2 <= (Z1'range => '1');
256
257         ------------------------- if statements, see above
258
259         ------------------------- case statements
260         case SEL is
261            when "00" | "10" => GLOBAL(7 downto 6) <= "01";
262            when "01"        => GLOBAL(7 downto 6) <= "11";
263            when others      => GLOBAL(7 downto 6) <= "00";
264         end case;
265
266         ------------------------- loop statements
267            -- index variable i is not visible outside of the loop
268            -- it should not be declared as a variable or signal
269         oplabel5 : for i in GLOBAL'range loop -- optional label
270               -- sequential statements
271            VAR := VAR xor GLOBAL(i);
272            next when SEL(0) = '0';    -- next statement
273            exit when SEL(1) = '1';    -- exit statement
274            DONE := TRUE;
275         end loop oplabel5; -- optional label
276
277         oplabel6 : while not DONE loop -- optional label
278               -- sequential statements
279            readline(FILEP, BUFF);
280            read(BUFF, VAR);
281            write(BUFF, VAR);
282            writeline(FILEQ, BUFF);
283         end loop oplabel6; -- optional label
284
285         ------------------------- various wait statements
286         wait on EN                                   ; -- a. signal'event
287         wait            until CLK'event              ; -- b. until
288         wait                              for 20 ns; -- c. period of time
289         wait on EN                        for 30 ns; -- a and c
```

```
290         wait on EN      until CLK = '1'          ; -- a and b
291         wait            until CLK'event for 50 ns; -- b and c
292         wait on EN, SEL until CLK = '1' for 50 ns; -- a, b, c
293         wait;                                      -- wait forever
294      end process oplabel4; -- optional label
295
296      --------------------- block statement with no guarded expression
297      blabel : block  -- label required
298         -- declaration part the same as in architecture declaration part
299      begin
300         -- can only have concurrent statement
301      end block blabel; -- optinal label
302
303      --------------------- block statement with guarded expression
304      clabel : block (EN = '1') -- label required
305         -- declaration part the same as in architecture declaration part
306      begin
307         -- can only have concurrent statement
308         GS2 <= guarded A xor B; -- guarded signal
309      end block clabel; -- optinal label
310
311   end ARCH_NAME;
312
313   configuration CFG_NAME of ENT_NAME is
314      for ARCH_NAME
315         for ifgen
316            for all : NAND_GATE
317               use entity work.NAND_GATE(ARCH_NAME);
318            end for;
319         end for;
320         for forgen (10 downto 0)
321            for nand1 : NAND_GATE
322               use entity work.NAND_GATE(ARCH_NAME);
323            end for;
324         end for;
325         for forgen (31 downto 12)
326            for nand1 : NAND_GATE
327               use entity work.NAND_GATE(ARCH_NAME);
328            end for;
329         end for;
330         for forgen (11)
331            for all : NAND_GATE
332               use entity work.NAND_GATE(ARCH_NAME);
333            end for;
334         end for;
335      end for;
336   end CFG_NAME;
```

VHDL'87 RESERVED WORDS

abs	access	after	alias	all
and	architecture	array	assert	attribute
begin	block	body	buffer	bus
case	component	configuration	constant	disconnect
downto	else	elsif	end	entity
exit	file	for	function	generate
generic	guarded	if	in	inout
is	label	library	linkage	loop
map	mod	nand	new	next
nor	not	null	of	on
open	or	others	out	package
port	procedure	process	range	record
register	rem	report	return	select
severity	signal	subtype	then	to
transport	type	units	until	use
variable	wait	when	while	with
xor				

VHDL'87 OPERATORS WITH PRIORITY IN ASCENDING ORDER

1	Logical	and	or	nand	nor	xor	
2	Relational	=	/=	<	<=	>	>=
3	Adding and Concatenation	+	-	&			
4	Sign (unary)	+	-				
5	Multiplying	*	/	mod	rem		
6	Exponent, absolute, complement	**	abs	not			

Appendix B

Declaration Part Table

	Entity declarative part	Block and architecture declarative part	Process and subprogram declarative part	Package declarative part	Package body declarative part	Configura-tion declarative part
Subprogram declaration	OK	OK	OK	OK	OK	
Subprogram body	OK	OK	OK		OK	
Type and subtype	OK	OK	OK	OK	OK	
Constant	OK	OK	OK	OK	OK	
Variable			OK			
Signal	OK	OK		OK		
Alias	OK	OK	OK	OK	OK	
File	OK	OK	OK	OK	OK	
Attribute declaration	OK	OK	OK	OK		
Attribute specification	OK	OK	OK	OK		OK
Disconnection specification	OK	OK		OK		
Component declaration		OK		OK		
Use clause	OK	OK	OK	OK	OK	OK
Configuration specification		OK				

Appendix C

VHDL'93 Grammar and Syntax Reference

The grammar and syntax are reprinted here with the permission from the IEEE. All keywords are in **bold** face. The additions and updates from VHDL'87 are underlined. Refer to Appendix A for lists of VHDL'87 and VHDL'93 keywords.

abstract_literal ::= decimal_literal | based_literal

access_type_definition ::= **access** subtype_indication

actual_designator ::= expression | signal_name | variable_name | file_name | **open**

actual_parameter_part ::= parameter_association_list

actual_part ::= actual_designator | function_name (actual_designator)
| type_mark (actual_designator)

adding_operator ::= + | - | &

aggregate ::= (element_association { , element_association })

alias_declaration ::= **alias** alias_designator [: subtype_indication] **is** name [signature] ;

alias_designator ::= identifier | character_literal | operator_symbol

allocator ::= **new** subtype_indication | **new** qualified_expression

architecture_body ::=
architecture identifier **of** entity_name **is**
architecture_declarative_part
begin
architecture_statement_part
end [**architecture**] [architecture_simple_name] ;

architecture_declarative_part ::= { block_declarative_item }

architecture_statement_part ::= { concurrent_statement }

array_type_definition ::= unconstrained_array_definition | constrained_array_definition

assertion ::= **assert** condition [**report** expression] [**severity** expression]

assertion_statement ::= [label :] assertion ;

association_element ::= [formal_part =>] actual_part

association_list ::= association_element { , association_element }

attribute_declaration ::= **attribute** identifier : type_mark ;

attribute_designator ::= attribute_simple_name

attribute_name ::= prefix [signature] ' attribute_designator [(expression)]

attribute_specification ::=
    attribute attribute_designator **of** entity_specification **is** expression ;

base ::= integer

base_specifier ::= B | O | X

base_unit_declaration ::= identifier ;

based_integer ::= extended_digit { [underline] extended_digit }

based_literal ::= base # based_integer [. based_integer] # [exponent]

basic_character ::= basic_graphic_character | format_effector

basic_graphic_character ::= upper_case_letter | digit | special_character | space_character

basic_identifier ::= letter { [underline] letter_or_digit }

binding_indication ::= [**use** entity_aspect] [generic_map_aspect] [port_map_aspect]

bit_string_literal ::= base_specifier " bit_value "

bit_value ::= extended_digit { [underline] extended_digit }

block_configuration ::=
    for block_specification
        { use_clause }

{ configuration_item }
end for ;

block_declarative_item ::= subprogram_declaration | subprogram_body | type_declaration
| subtype_declaration | constant_declaration | signal_declaration
| shared_variable_declaration | file_declaration | alias_declaration
| component_declaration | attribute_declaration | attribute_specification
| configuration_specification | disconnection_specification | use_clause
| group_template_declaration | group_declaration

block_declarative_part ::= { block_declarative_item }

block_header ::=
[generic_clause [generic_map_aspect ;]] [port_clause [port_map_aspect ;]]

block_specification ::= architecture_name | block_statement_label
| generate_statement_label [(index_specification)]

block_statement ::=
block_label : **block** [(guard_expression)] [**is**]
block_header
block_declarative_part
begin
block_statement_part
end block [block_label] ;

block_statement_part ::= { concurrent_statement }

case_statement ::=
[case_label :] **case** expression **is**
case_statement_alternative
{ case_statement_alternative }
end case [case_label] ;

case_statement_alternative ::= **when** choices => sequence_of_statements

character_literal ::= ' graphic_character '

choice ::= simple_expression | discrete_range | element_simple_name | **others**

choices ::= choice { | choice }

component_configuration ::=
for component_specification
[binding_indication ;]
[block_configuration]
end for ;

component_declaration ::=
    component identifier [**is**]
        [local_generic_clause]
        [local_port_clause]
    end component [component_simple_name] ;

component_instantiation_statement ::=
    instantiation_label : instantiated_unit [generic_map_aspect] [port_map_aspect] ;

component_specification ::= instantiation_list : component_name

composite_type_definition ::= array_type_definition | record_type_definition

concurrent_assertion_statement ::= [label :] [**postponed**] assertion ;

concurrent_procedure_call_statement ::= [label :] [**postponed**] procedure_call ;

concurrent_signal_assignment_statement ::=
    [label :] [**postponed**] conditional_signal_assignment |
    [label :] [**postponed**] selected_signal_assignment

concurrent_statement ::= block_statement | process_statement
    | concurrent_procedure_call_statement
    | concurrent_assertion_statement | concurrent_signal_assignment_statement
    | component_instantiation_statement | generate_statement

condition ::= boolean_expression

condition_clause ::= **until** condition

conditional_signal_assignment ::= target <= options conditional_waveforms ;

conditional_waveforms ::= { waveform **when** condition **else** } waveform [**when** condition]

configuration_declaration ::=
    configuration identifier **of** entity_name **is**
        configuration_declarative_part
        block_configuration
    end [**configuration**] [configuration_simple_name] ;

configuration_declarative_item ::= use_clause | attribute_specification | group_declaration

configuration_declarative_part ::= { configuration_declarative_item }

configuration_item ::= block_configuration | component_configuration

configuration_specification ::= **for** component_specification binding_indication ;

constant_declaration ::= **constant** identifier_list : subtype_indication [:= expression] ;

constrained_array_definition ::= **array** index_constraint **of** element_subtype_indication

constraint ::= range_constraint | index_constraint

context_clause ::= { context_item }

context_item ::= library_clause | use_clause

decimal_literal ::= integer [. integer] [exponent]

declaration ::= type_declaration | subtype_declaration | object_declaration
| interface_declaration | alias_declaration | attribute_declaration
| component_declaration | group_template_declaration
| group_declaration | entity_declaration | configuration_declaration
| subprogram_declaration | package_declaration

delay_mechanism ::= **transport** | [**reject** time_expression] **inertial**

design_file ::= design_unit { design_unit }

design_unit ::= context_clause library_unit

designator ::= identifier | operator_symbol

direction ::= **to** | **downto**

disconnection_specification ::=
disconnect guarded_signal_specification **after** time_expression ;

discrete_range ::= discrete_subtype_indication | range

element_association ::= [choices =>] expression

element_declaration ::= identifier_list : element_subtype_definition ;

element_subtype_definition ::= subtype_indication

entity_aspect ::= **entity** entity_name [(architecture_identifier)]
| **configuration** configuration_name | **open**

entity_class ::= **entity** | **architecture** | **configuration** | **procedure** | **function** | **package**
| **type** | **subtype** | **constant** | **signal** | **variable** | **component** | **label** | **literal**
| **units** | **group** | **file**

entity_class_entry ::= entity_class [<>]

entity_class_entry_list ::= entity_class_entry { , entity_class_entry }

entity_declaration ::=
    entity identifier **is**
        entity_header
        entity_declarative_part
    [**begin**
        entity_statement_part]
    end [**entity**] [entity_simple_name] ;

entity_declarative_item ::= subprogram_declaration | subprogram_body | type_declaration
    | subtype_declaration | constant_declaration | signal_declaration
    | shared_variable_declaration | file_declaration | alias_declaration
    | attribute_declaration | attribute_specification | disconnection_specification
    | use_clause | group_template_declaration | group_declaration

entity_declarative_part ::= { entity_declarative_item }

entity_designator ::= entity_tag [signature]

entity_header ::= [formal_generic_clause] [formal_port_clause]

entity_name_list ::= entity_designator { , entity_designator } | **others** | **all**

entity_specification ::= entity_name_list : entity_class

entity_statement ::= concurrent_assertion_statement
    | passive_concurrent_procedure_call_statement | passive_process_statement

entity_statement_part ::= { entity_statement }

entity_tag ::= simple_name | character_literal | operator_symbol

enumeration_literal ::= identifier | character_literal

enumeration_type_definition ::= (enumeration_literal { , enumeration_literal })

exit_statement ::= [label :] **exit** [loop_label] [**when** condition] ;

exponent ::= E [+] integer | E - integer

expression ::= relation { **and** relation } | relation { **or** relation } | relation { **xor** relation }
    | relation [**nand** relation] | relation [**nor** relation] | relation { **xnor** relation }

extended_digit ::= digit | letter

extended_identifier ::= \ graphic_character { graphic_character } \

factor ::= primary [** primary] | **abs** primary | **not** primary

file_declaration ::= **file** identifier_list : subtype_indication file_open_information] ;

file_logical_name ::= string_expression

file_open_information ::= [**open** file_open_kind_expression] **is** file_logical_name

file_type_definition ::= **file of** type_mark

floating_type_definition := range_constraint

formal_designator ::= generic_name | port_name | parameter_name

formal_parameter_list ::= parameter_interface_list

formal_part ::= formal_designator | function_name (formal_designator)
| type_mark (formal_designator)

full_type_declaration ::= **type** identifier **is** type_definition ;

function_call ::= function_name [(actual_parameter_part)]

generate_statement ::=
generate_label : generation_scheme **generate**
[{ block_declarative_item }
begin]
{ concurrent_statement }
end generate [generate_label] ;

generation_scheme ::= **for** generate_parameter_specification | **if** condition

generic_clause ::= **generic** (generic_list) ;

generic_list ::= generic_interface_list

generic_map_aspect ::= **generic map** (generic_association_list)

graphic_character ::= basic_graphic_character | lower_case_letter | other_special_character

group_constituent ::= name | character_literal

group_constituent_list ::= group_constituent { , group_constituent }

group_template_declaration ::= **group** identifier is (entity_class_entry_list) ;

group_declaration ::= **group** identifier : group_template_name (group_constituent_list) ;

guarded_signal_specification ::= guarded_signal_list : type_mark

identifier ::= basic_identifier | extended_identifier

identifier_list ::= identifier { , identifier }

if_statement ::=
    [if_label :] **if** condition **then**
        sequence_of_statements
    { **elsif** condition **then**
        sequence_of_statements }
    [**else**
        sequence_of_statements]
    end if [if_label] ;

incomplete_type_declaration ::= **type** identifier ;

index_constraint ::= (discrete_range { , discrete_range })

index_specification ::= discrete_range | static_expression

index_subtype_definition ::= type_mark **range** <>

indexed_name ::= prefix (expression { , expression })

instantiated_unit ::= [**component**] component_name
    | **entity** entity_name [(architecture_identifier)] | **configuration** configuration_name

instantiation_list ::= instantiation_label { , instantiation_label } | **others** | **all**

integer ::= digit { [underline] digit }

integer_type_definition ::= range_constraint

interface_constant_declaration ::=
    [**constant**] identifier_list : [**in**] subtype_indication [:= static_expression]

interface_declaration ::= interface_constant_declaration | interface_signal_declaration
    | interface_variable_declaration | interface_file_declaration

interface_element ::= interface_declaration

interface_file_declaration ::= **file** identifier_list : subtype_indication

interface_list ::= interface_element { ; interface_element }

interface_signal_declaration ::= [**signal**] identifier_list : [mode] subtype_indication [**bus**]
    [:= static_expression]

interface_variable_declaration ::=
    [**variable**] identifier_list : [mode] subtype_indication [:= static_expression]

iteration_scheme ::= **while** condition | **for** loop_parameter_specification

label ::= identifier

letter ::= upper_case_letter | lower_case_letter

letter_or_digit ::= letter | digit

library_clause ::= **library** logical_name_list ;

library_unit ::= primary_unit | secondary_unit

literal ::= numeric_literal | enumeration_literal | string_literal | bit_string_literal | **null**

logical_name ::= identifier

logical_name_list ::= logical_name { , logical_name }

logical_operator ::= **and** | **or** | **nand** | **nor** | **xor** | **xnor**

loop_statement ::=
    [loop_label :] [iteration_scheme] **loop**
        sequence_of_statements
    end loop [loop_label] ;

miscellaneous_operator ::= ** | **abs** | **not**

mode ::= **in** | **out** | **inout** | **buffer** | **linkage**

multiplying_operator ::= * | / | **mod** | **rem**

name ::= simple_name | operator_symbol | selected_name | indexed_name | slice_name
    | attribute_name

next_statement ::= [label :] **next** [loop_label] [**when** condition] ;

null_statement ::= [label :] **null** ;

numeric_literal ::= abstract_literal | physical_literal

object_declaration ::= constant_declaration | signal_declaration | variable_declaration
    | file_declaration

operator_symbol ::= string_literal

options ::= [**guarded**] [delay_mechanism]

package_body ::=
package body package_simple_name **is**
package_body_declarative_part
end [**package body**] [package_simple_name] ;

package_body_declarative_item ::= subprogram_declaration | subprogram_body
| type_declaration | subtype_declaration | constant_declaration
| shared_variable_declaration | file_declaration
| alias_declaration | use_clause | group_template_declaration | group_declaration

package_body_declarative_part ::= { package_body_declarative_item }

package_declaration ::=
package identifier **is**
package_declarative_part
end [**package**] [package_simple_name] ;

package_declarative_item ::= subprogram_declaration | type_declaration
| subtype_declaration | constant_declaration | signal_declaration
| shared_variable_declaration | file_declaration | alias_declaration
| component_declaration | attribute_declaration | attribute_specification
| disconnection_specification | use_clause | group_template_declaration
| group_declaration

package_declarative_part ::= { package_declarative_item }

parameter_specification ::= identifier **in** discrete_range

physical_literal ::= [abstract_literal] unit_name

physical_type_definition ::=
range_constraint
units
base_unit_declaration
{ secondary_unit_declaration }
end units [physical_type_simple_name]

port_clause ::= **port** (port_list) ;

port_list ::= port_interface_list

port_map_aspect ::= **port map** (port_association_list)

prefix ::= name | function_call

primary ::= name | literal | aggregate | function_call | qualified_expression | type_conversion
| allocator | (expression)

primary_unit ::= entity_declaration | configuration_declaration | package_declaration

procedure_call ::= procedure_name [(actual_parameter_part)]

procedure_call_statement ::= [label :] procedure_call ;

process_declarative_item ::= subprogram_declaration | subprogram_body | type_declaration
  | subtype_declaration | constant_declaration | variable_declaration | file_declaration
  | alias_declaration | attribute_declaration | attribute_specification | use_clause
  | group_template_declaration | group_declaration

process_declarative_part ::= { process_declarative_item }

process_statement ::=
  [process_label :] [**postponed**] **process** [(sensitivity_list)] [**is**]
    process_declarative_part
  begin
    process_statement_part
  end [**postponed**] **process** [process_label] ;

process_statement_part ::= { sequential_statement }

qualified_expression ::= type_mark ' (expression) | type_mark ' aggregate

range ::= range_attribute_name | simple_expression direction simple_expression

range_constraint ::= **range** range

record_type_definition ::=
  record element_declaration
    { element_declaration }
  end record [record_type_simple_name]

relation ::= shift_expression [relational_operator shift_expression]

relational_operator ::= = | /= | < | <= | > | >=

report_statement ::= [label :] **report** expression [**severity** expression] ;

return_statement ::= [label :] **return** [expression] ;

scalar_type_definition ::= enumeration_type_definition | integer_type_definition
  | floating_type_definition | physical_type_definition

secondary_unit ::= architecture_body | package_body

secondary_unit_declaration ::= identifier = physical_literal ;

selected_name ::= prefix . suffix

selected_signal_assignment ::=
    with expression **select**
        target <= options selected_waveforms ;

selected_waveforms ::= { waveform **when** choices , } waveform **when** choices

sensitivity_clause ::= **on** sensitivity_list

sensitivity_list ::= signal_name { , signal_name }

sequence_of_statements ::= { sequential_statement }

sequential_statement ::= wait_statement | assertion_statement | report_statement
    | signal_assignment_statement | variable_assignment_statement
    | procedure_call_statement | if_statement | case_statement | loop_statement
    | next_statement | exit_statement | return_statement | null_statement

shift_expression ::= simple_expression [shift_operator simple_expression]

shift_operator ::= **sll** | **srl** | **sla** | **sra** | **rol** | **ror**

sign ::= + | -

signal_assignment_statement ::= [label :] target <= [delay_mechanism] waveform ;

signal_declaration ::=
    signal identifier_list : subtype_indication [signal_kind] [:= expression] ;

signal_kind ::= **register** | **bus**

signal_list ::= signal_name { , signal_name } | **others** | **all**

signature ::= [[type_mark { , type_mark }] [**return** type_mark]]

simple_expression ::= [sign] term { adding_operator term }

simple_name ::= identifier

slice_name ::= prefix (discrete_range)

string_literal ::= " { graphic_character } "

subprogram_body ::=
    subprogram_specification **is**
        subprogram_declarative_part

begin
subprogram_statement_part
end [subprogram_kind] [designator] ;

subprogram_declaration ::= subprogram_specification ;

subprogram_declarative_item ::= subprogram_declaration | subprogram_body
| type_declaration | subtype_declaration | constant_declaration | variable_declaration
| file_declaration | alias_declaration | attribute_declaration | attribute_specification
| use_clause | group_template_declaration | group_declaration

subprogram_declarative_part ::= { subprogram_declarative_item }

subprogram_kind ::= **procedure** | **function**

subprogram_specification ::=
procedure designator [(formal_parameter_list)]
| [**pure** | **impure**] **function** designator [(formal_parameter_list)] **return** type_mark

subprogram_statement_part ::= { sequential_statement }

subtype_declaration ::= **subtype** identifier **is** subtype_indication ;

subtype_indication ::= [resolution_function_name] type_mark [constraint]

suffix ::= simple_name | character_literal | operator_symbol | **all**

target ::= name | aggregate

term ::= factor { multiplying_operator factor }

timeout_clause ::= **for** time_expression

type_conversion ::= type_mark (expression)

type_declaration ::= full_type_declaration | incomplete_type_declaration

type_definition ::= scalar_type_definition | composite_type_definition
| access_type_definition | file_type_definition

type_mark ::= type_name | subtype_name

unconstrained_array_definition ::=
array (index_subtype_definition { , index_subtype_definition })
of element_subtype_indication

use_clause ::= **use** selected_name { , selected_name } ;

variable_assignment_statement ::= [label :] target := expression ;

variable_declaration ::=
[**shared**] **variable** identifier_list : subtype_indication [:= expression] ;

wait_statement ::=
[label :] **wait** [sensitivity_clause] [condition_clause] [timeout_clause] ;

waveform ::= waveform_element { , waveform_element } | **unaffected**

waveform_element ::= value_expression [**after** time_expression]
| **null** [**after** time_expression]

Index

About the Author

K. C. CHANG, PhD

K. C. Chang received his B.S. degree in electrical engineering from the National Taiwan University in 1979. His MS and PhD degrees in computer science were earned at the University of Minnesota in 1984 and 1986. In 1986 he joined The Boeing Company where he teaches VHDL and synthesis courses in addition to his regular responsibilities of computer-aided design algorithm development, VHDL, synthesis, and ASIC design. He has designed several ASICs with VHDL and synthesis and gives conference tutorials on these same subjects. Chang has served as a reviewer of technical papers for the ACM/IEEE Design Automation Conference since 1987. He holds three US patents and has published 12 technical papers, including three papers in *IEEE Transactions on Computers* and *Transactions on Computer-Aided Design.*

IEEE Computer Society Press Editorial Board

Practices for Computer Science and Engineering

The IEEE Computer Society now publishes a broad range of practical and applied technology books and executive briefings in addition to proceedings, periodicals, and journals. These titles provide practitioners and academicians alike the tools necessary to become "instant experts" on current and emerging topics in computer science.

3/4/96

http://computer.org

12/12/96